Precipitation History of the Rocky Mountain States

Raymond S. Bradley

PRECIPITATION HISTORY OF THE ROCKY MOUNTAIN STATES

WESTVIEW SPECIAL STUDIES: INSTITUTE FOR ARCTIC AND ALPINE RESEARCH—STUDIES IN HIGH ALTITUDE GEO-ECOLOGY

To my parents

Published 1976 in the United States of America by

Westview Press, Inc.
1898 Flatiron Court
Boulder, Colorado 80301
Frederick A. Praeger, Publisher and Editorial Director

Published in cooperation with the

Institute of Arctic and Alpine Research
University of Colorado

Library of Congress Cataloging in Publication Data

Bradley, Raymond S 1948-
Precipitation history of the Rocky Mountain States.

Bibliography: p.
1. Precipitation (Meteorology)--Rocky Mountain region. I. Title.
QC925.1.U8R622 1976 551.5'772'78 75-45279
ISBN 0-89158-026-3

Printed in the United States of America.

CONTENTS

FIGURES

Figure		Page

Figure		Page

Figure		Page

TABLES

Table		Page

Table | Page

SERIES EDITOR'S PREFACE

The current volume by Dr. Raymond S. Bradley represents the first of a new series venture based upon the collaboration of Westview Press and the Institute of Arctic and Alpine Research (INSTAAR), University of Colorado. INSTAAR is a research and graduate teaching unit of the University of Colorado Graduate School that concentrates its research and publication energies on the atmospheric, earth, and life science aspects of cold-stressed environments. In effect, this research embraces the present polar and alpine landscapes and the environmental changes that have occurred over the last few million years of earth history.

The research ranges from individual doctoral dissertation topics to large-scale interdisciplinary and multidisciplinary endeavors including cooperation with other research institutions. The subject matter includes pure as well as applied research. INSTAAR faculty are very much involved with the applications of their expertise and research findings to the decision-making process in the development of rational land-management policies and this brings Man and his appurtenances into our purview. This area of application is receiving growing emphasis as a result of our affiliation with the Unesco Man and the Biosphere Program (MAB), Project 6: study of the impact of human activities on mountain and tundra ecosystems.

It is hoped that the current volume will be followed during 1976 by a parallel study of temperature fluctuations in the United States Southwest by the same author, a monograph on geophysical hazards (snow avalanches, landslides, mudflows, etc.) in the Colorado Rocky Mountains and their relationship to land-management problems, and a collection of integrated research papers on the ecology of the Indian Peaks section of the Colorado Front Range. Other titles will be announced in due course.

I would like to take this opportunity to thank Frederick A. Praeger, President and Publisher of Westview Press, for proposing the series venture and for making it feasible.

Jack D. Ives
Series Editor
Director, INSTAAR, University of Colorado
December 1975

FOREWORD

During the past decade interest in climatic change and in the reconstruction of past climates has grown considerably. In part this is due to realization that shortage of food and fuel on a worldwide scale could be intensified by drier or cooler weather in the years ahead. This in turn has led to serious discussion on the possibilities of long-range weather forecasting--even the forecasting of climate--and it is generally conceded that an understanding of changes in the past will be an essential ingredient. At the Institute of Arctic and Alpine Research, University of Colorado (INSTAAR), interdisciplinary studies of cold-stressed environments, in general, and of the last ice age, in particular, have invariably contained a paleoclimatic or climatic element. More specific to the present, study, analysis of the ecological impact of winter cloud-seeding in the San Juan Mountains, Colorado, led to an attempt to determine climatic trends in the western United States over the past 150 to 200 years.

This monograph is an analysis of the earliest meteorological observations for the Rocky Mountain states. It is the first comprehensive attempt to document the quite extensive information on changing climatic conditions from the second half of the nineteenth century. Work which is continuing on the temperature records from the same area as well as the related synoptic pressure patterns since 1899 will be published at a later date. The documentation of the range of climatic variability over the span of more than a century serves to sharpen our appreciation of the extent to which past decades have differed from recent ones and is essential for assessing the probabilities of various departures from the average within a given time interval. Moreover, the results provide a direct means of checking and also improving the reconstruction of past climatic conditions based on dendroclimatic and other proxy data sources.

R. G. Barry
INSTAAR
University of Colorado
December 1975

ACKNOWLEDGMENTS

The study was initially supported by Bureau of Reclamation contract 14-06-D-7052 to Colorado State University and University of Colorado (San Juan Ecology Project). Subsequently, it was supported by a grant-in-aid from the Society of the Sigma Xi and a National Science Foundation grant for dissertation research in meteorology (GA-33177). Most of the study was supported under National Science Foundation grant GA-40256 to R. G. Barry, INSTAAR, for research on "The Secular Climatic History of the Rocky Mountain Area." Early data collection was greatly facilitated by William Hodge, Clyde Collier, and Vincent Haggarty, Washington, D.C., and Richart Lytle, Smithsonian Institution, Washington, D.C. Other assistance, provided by INSTAAR, the Department of Geography, and the Computing Center at the University of Colorado, and the Department of Geology-Geography and the Computing Center at the University of Massachusetts is also gratefully acknowledged. Programming and technical help, without which the study would not have been possible, was provided by Margaret Eccles, INSTAAR, and Gregory Tsoucalas and Robert Gonter, University of Massachusetts. Jane Bradley drew many of the illustrations and typed the original dissertation. Marie Litterer, Department of Geology-Geography, University of Massachusetts, and Marilyn Joel, INSTAAR, painstakingly drew most of the figures in the Appendixes. My work was also assisted by the very useful discussions with Dr. H. C. Fritts and Dr. T. J. Blasing, Laboratory for Tree Ring Research, University of Arizona. I am extremely grateful to all these people for their invaluable assistance. I also acknowledge the following for permission to include previously published maps and diagrams: the Swiss Federal Observatory, Zurich, for Figure 36 from _Sunspot Activity in the Years 1610-1960_ by M. Waldmeier; the Editor, _Tellus_, for Figure 108 from D. L. Dzerdzeevski, _Tellus_ 18(4) 1966; and the Editor, _Weatherwise_, for Appendix C from R. D. Elliott, _Weatherwise_ 2, 1949.

Raymond S. Bradley
Department of Geology and Geography
University of Massachusetts
Amherst
December 1975

ABSTRACT

Using early historical records for states west of 105°W (excluding California) precipitation variations during the latter half of the nineteenth century (1851-1890) are documented. Long-term records (>50 years) at over 160 Rocky Mountain stations were used to analyze secular changes of precipitation in that area since records began around 1850. Nineteenth-century data were placed "in context" by comparing records with 1951-60 averages at the nearest comparable stations. Winter and spring precipitation was above 1951-60 averages over much of the western United States from at least 1865-1890. Summers 1865-90 were drier than in the 1950s. Falls were very dry at times over large areas in the latter part of the nineteenth century.

A distinct four- to six-year interval between winter precipitation maxima in the Southwest and corresponding minima in the Northwest was characteristic of the period 1851-1990, with peaks (or troughs) in 1851?, 1855, 1859, 1868, 1873, 1883 or 1884, and 1888 or 1889. Such a periodicity is not apparent in other seasons or in Rocky Mountain data during the twentieth century.

Careful screening and testing for homogeneity of long-term precipitation records in the Rockies was carried out. Maps of average seasonal precipitation and coefficients of variations 1941-70 are presented. Those series considered homogeneous were used to document seasonal decadal changes in precipitation in relation to 1941-70 "normals." Generally precipitation was high in the 1890s, 1910s and/or 1920s, 1940s, and 1960s. Other decades were relatively dry, particularly the 1930s. Springs 1941-70 were wetter than the previous 30 years but the period 1881-1910 was wetter than the 1941-70 average. Considering summer precipitation, 1941-70 was probably the most anomalously wet period for at least 110 years and perhaps much longer. Over much of the Rockies the 1960s were extraordinarily wet. Fall precipitation was exceptionally low in the 1950s but the 1941-60 average was still relatively high, though generally drier than the 1890s, 1910s, and 1920s. Winter precipitation in recent decades has been considerably less than was characteristic of the late nineteenth and early twentieth centuries, except in Idaho. Recurrent anomaly patterns, centered over southern Idaho are characteristic of the winter record.

When the entire secular period is considered the recent "normal" is anomalous; in particular summer precipitation has been unusually high and winter precipitation relatively low. One hundred years ago the climate of the Rockies was characterized by wetter winters and springs but much drier summers (and in some areas) drier falls compared to the post-1940 climate of the region. Temperatures were generally warmer. The recent anomalous normal is related to a change towards more meridional circulation types since c. 1941. A catalog of synoptic types was used to demonstrate this change which was hemispheric in extent. Some evidence that the change may be related to solar activity (sunspot area) is presented. Changes in precipitation from the 1930s to the 1940s and 1950s are related to changing frequencies of certain synoptic types.

Power spectrum analyses of over 200 homogeneous seasonal precipitation records (>65 years in length) from throughout the Rockies are discussed. The main feature is a marked 2- to 2.2-year periodicity in summer and late summer precipitation records from Utah and western Colorado only. Few other statistically significant features were revealed. A simplistic comparison of nineteenth-century instrumental precipitation data and dendroclimatic reconstructions for the area tends to support the reconstructions.

Despite problems of recording precipitation in mountainous areas and great topoclimatic variations in the Rockies the study suggests that topography plays a secondary role to regional synoptic controls which determine the broad-scale secular changes observed.

CHAPTER 1

OUTLINE OF THE STUDY

Introduction and Objectives

In recent years a number of studies of the climatic history of different parts of the United States have been published, utilizing to varying extents early historical data (Mitchell, 1961; Roden, 1966; Wahl, 1968; Rosendal, 1970; Wahl and Lawson, 1970). At the present time, a gap in our knowledge of climatic fluctuations in the United States exists for the Rocky Mountain and Great Basin area (here defined as the states of Montana, Idaho, Wyoming, Utah, Nevada, Colorado, northern New Mexico, and northern Arizona) particularly the central and northern part of the region. Most of the information on climatic fluctuations in these states, particularly variations in precipitation, has been inferred from dendroclimatic reconstructions (Fritts, 1965; Fritts *et al*., 1971; LaMarche and Fritts, 1971); little attention has been paid to the instrumental data which have been collected throughout the area over the past 100 years or more.

The author's work on the San Juan Ecology project[1] indicated that a considerable amount of information about the climate of the mid- to late nineteenth century could be obtained from early instrumental records (Bradley and Barry, 1973a). It was also clear that very little was known about the climatic history of the Rocky Mountain and Great Basin region, yet there appeared to be a large number of stations where observations have been kept for over half a century. Furthermore, there are few, if any comparable areas of the world where such a large number of climatological stations have operated at relatively high elevations for such long

[1]The San Juan Ecology Project is a Bureau of Reclamation-sponsored multidisciplinary attempt to examine the ecological impact of cloud seeding in the San Juan Mountains of southwestern Colorado. The author examined the secular climatic history of the area to provide a perspective on the natural climatic variability of the area.

periods of time. Hence this study arose out of a desire to document climatic fluctuations in a mountainous area of the United States, the climatic history of which was hitherto unknown.

It is also worth noting in this regard that the climatic characteristics of the area even today are not well known (Bryson and Hare, 1974) though recent studies (e.g., Mitchell, 1969) have attempted to simplify the climate by removing the effects of topography. There has also been much recent work connected with short-term forecasting and this has indirectly shed light on the synoptic characteristics associated with particular precipitation and temperature patterns (see, for example, Klein, 1965; Sullivan and Severson, 1966a, 1966b; Jorgensen, et al., 1967; Augulis, 1969, 1970a, 1970b; Korte, et al., 1969, 1972). Nevertheless, compared with the rest of the United States the area has been relatively neglected in terms of climatological studies, particularly those concerned with climatic variations in the secular period. The work presented here has been directed towards this gap in our knowledge.

In the case of nineteenth century climate, this study has examined both temperature and precipitation data for all the western states (excluding California, in order to reduce the number of stations). For continuous data (see Chapter 3) the study has concentrated on a large number of precipitation stations throughout the Rocky Mountain area (Montana, Idaho, Wyoming, western Colorado, and Utah). Precipitation was studied in detail because of its significance in terms of potential cloud seeding operations throughout the Rockies in the next decade and because of its importance to tree growth in the area (e.g., Fritts, 1966) and hence to dendroclimatic reconstructions. Because of the diverse topography of the area the general philosophy of the study was to examine as many stations as possible in order to distinguish between regional and local climatic effects.

Literature Review

As mentioned above, many studies of the climatic history of the western United States have been based on statistical inference from dendroclimatic reconstructions. These are discussed in Chapter 5; studies using mainly instrumentally recorded and historical data are the focus of this section.

As instrumental observations for the western United States are only available for the last 100-120 years it is first necessary to look elsewhere, to examine evidence of hemispheric climatic fluctuations over the last 200-300 years and hence "set the scene" within which changes in the climate of the western United States have occurred. Evidence for a cool period affecting the entire northern hemisphere from about 1550 to the mid- or late nineteenth century (the Little Ice Age or Neoboreal episode) has been presented on numerous occasions (e.g., Lamb and Johnson, 1959; Lamb, 1963, 1966, 1972; Barreis and Bryson, 1965; Le Roy Ladurie, 1971). In the western United States, this relatively cool period was associated with the growth of glaciers in high mountain regions as evidenced by moraines variously named Gannett Peak (Front Range, Benedict, 1968; Wind Rivers, Moss, 1951), Garda (Mt. Rainier, Crandell, 1969, Sigafoos and Hendricks, 1973), Matthes (Sierra Nevada, Porter and Denton, 1967), and Eagle Cap (Wallowa Mountains, Williams, 1974).

Some instrumental evidence from the eastern United States is also available for the early nineteenth century indicating the main characteristics of climate which may have predominated over the previous 200-300 years. Wahl (1968) has compared seasonal temperature and precipitation conditions in the 1830s and 1840s in the United States east of 95°W with the 1931-60 averages at analogous stations. In all seasons, over all areas except Gulf and Atlantic coasts, temperatures were markedly cooler in the 1830s and 1940s with a maximum difference (>4°F) in "early fall" (September/October). In all seasons the maximum difference was centered on the southern Great Lakes area. Wahl suggests this is related to more frequent and more persistent influxes of polar air into the Midwest in this early period. Increases in precipitation close to the core areas of temperature change were related to a more southern mean storm track which concurs with Lamb's (1963) suggestion that July storm tracks over the United States in the early nineteenth century were 4°-5° farther south than in recent decades. Brunk (1968) supports Wahl's conclusions of a cooler and wetter climate in the eastern United States in the 1830s and 1840s with records of extremely high levels of the Great Lakes for a number of years in the 1830s. Similarly, Wing (1943) has shown that lakes and rivers in Wisconsin, Minnesota and Michigan were frozen for much longer periods in the early to mid-nineteenth century than in the mid-1930s, and Baker (1960) examined the St. Paul, Minnesota, temperature record from 1819 to the 1950s and found period minima in the 1850s (winter) and 1860s (summer). All these studies tend to support Wahl's conclusions that "the climate that our forefathers lived under was somewhat more severe than that which was normal in the first half of this century."

Further studies by Wahl and Lawson (1970) have suggested that climatic conditions in the eastern United States in the 1850s and 1860s were similar to the 1830s and 1840s (e.g., cooler and wetter than the period 1931-60). However, conditions in the western United States appear to have differed from those to the east. In fact, Wahl and Lawson's maps indicate a marked positive temperature anomaly over the mountain states associated with a similar precipitation anomaly. These patterns were "quite consistent throughout the whole year" and led them to conclude that the climate of the western United States in the 1850s and 1860s was "distinctly warmer and decisively wetter" (except for Pacific coastal regions) compared to conditions in 1931-60. However, generalizations made by Wahl and Lawson for the western United States are based on a few widely scattered observations, most of which were kept for less than five years within the 20-year period. In fact, many states had almost no records for this period (see Chapter 2, Table 1 and Figures 2 to 17). Furthermore, some data included in Schott (1881) (the source of Wahl and Lawson's precipitation data) were found to be erroneous when original (microfilmed) records were consulted. This led Bradley and Barry (1973a) to suggest that (at least in southwestern Colorado and northwestern New Mexico) the climate of the 1860s was at least as dry as 1931-60 normals.

It is interesting to compare all these results, as Wahl and Lawson did, with the work of Lamb and Johnson (1959) who studied historical pressure data for the Northern Hemisphere. The latter show that the longitude of the western Atlantic surface trough at 45°N moved steadily eastward from the early 1800s to c. 1845-84, from 56°W to between 49 and 45°W. The trough in the early to mid-1800s was thus 3-5° <u>nearer</u> the east coast of the United States than in the period 1910-49 which would indicate increased flow of polar air into the eastern United States in the earlier period. However, pressure data from the mid-continent has been nearly constant since the 1850s with a maximum at 100-105°W (Lamb and Johnson, 1959). The more westerly position of the western Atlantic trough in the early to mid-nineteenth century might thus be interpreted as a period of shorter wavelength of the stationary long-wave pattern, with a stronger trough over eastern North America (i.e., more meridional conditions) with reduced zonal flow in the upper westerlies. This suggestion is supported by two other lines of reasoning. Wahl and Lawson (1970) note that temperature deviations over the United States for 1959-68 from 1931-60 averages had a very similar pattern to the anomalies of the 1850s and 1860s. Sanchez and Kutzbach (1974) and Namias (1969, 1970) have also examined temperature anomalies in the 1960s from 1931-60 means, for the United States and for tropical and subtropical North and South America, respectively. They also report on circulation changes associated with these anomalies noting that all seasons were characterized by a strengthened trough over eastern North America. This is precisely what was inferred for the early nineteenth century and lends some support to the suggestions of Lamb and Johnson that the circulation of the early to mid-nineteenth century was more meridional than in the 1931-60 period. However, this 1931-60 pattern appears to have ended in the recent decade(s) (see Chapter 6) and may indeed indicate a reestablishment of climatic conditions "which governed our climate for the last three to four centuries and which [were] . . . interrupted only briefly by a minor climatic fluctuation" (Wahl and Lawson, 1970).

Historical Evidence for Climatic Change in the Western United States

Although instrumental data for the western United States cover a relatively brief period, some inferences about the climate of the area over the last few hundred years have been obtained from early diaries and explorers' accounts (Ives, 1954) and from a variety of biological and geological evidence (Hubbs, 1957). Ives has described historical records dating from as long ago as 1540-41 which indicate that 200-400 years ago the climate of the Southwest was considerably wetter than in recent years. In fact, he describes travel accounts "replete with details of difficult river crossings, descriptions of safe fords and records of troubles encountered while swimming horses across the San Pedro River" in areas now cut only by dry river valleys. In one interesting case a boat was being built (around 1695) to sail down the Rio Concepcion to the Gulf of California; today the entire journey can be made along the dry river bed by jeep (Ives, 1954).

Hubbs (1957) has placed the observations of Ives in a longer time-frame noting that a trend towards aridity in the Southwest has been characteristic of the entire postglacial period. He notes "evidence of desiccation" over large areas "scores of great lakes dried to remnants or to dust; former streams now traceable by dry physiographic features, or by chains of related fishes living in isolated springs along the old course; many sites of an abundant aboriginal population in areas now devoid of surface water . . .," etc. Hubbs acknowledges that many of the features he describes probably date from the Pluvial period but he maintains that there is much evidence from more recent times to suggest that the droughts of the twentieth century (post-1930) "have been the most severe of any since the Pluvial period." Hubbs also compares fish fauna caught at San Diego by naturalists of the Pacific Railroad Survey (1853-7) with present-day species. The warm-water, nonmigratory species taken at that time "virtually demands warm coastal sea temperature--and by correlation warm air temperature--over a considerable time preceding the middle of the last century." Analysis of temperature data for the area of southern and south-central California from 1850 indicates a downward trend, particularly in late spring and summer until at least 1910 after which urbanization affected the data being used (Hubbs, 1957). This suggestion of a warmer mid-nineteenth century in the western United States concurs with the suggestions of both Bradley and Barry (1973a) and Wahl and Lawson (1970). Bradley and Barry also note a cooling trend in seasonal mean temperatures for southwestern Colorado similar to that of southern and south-central California, with cooling from c. 1867 to about 1930 (particularly in spring, fall, and winter seasons) and warming thereafter. This is supported by data on the length of the growing season and degree day totals for the area (Bradley and Barry, 1973b).

In Montana, studies by Dightman (1956) and Dightman and Beatty (1952) also point to cooling from the early 1900s to c. 1920 followed by marked warming (+2°F for "weighted areal average for Montana") to c. 1940 and even more marked cooling thereafter (-3°F). These changes were reflected in significant glacial mass loss in all 60-80 glaciers of Glacier National Park, Montana, during the 1930s and 1940s (Dyson, 1948) and in a renewal of glacial activity in the 1950s (Dightman, 1956). For example, the Sperry Glacier was 840 acres in extent in 1901 and only 330 acres by 1946; the Grinnell Glacier decreased in volume by 50% between 1850 and 1937 and by a further 25-30% between 1937-46 (Dyson, 1948). Ives (1954) also notes marked changes in the extent of Front Range ice bodies in the early twentieth century. He notes that the Arapahoe Glacier lost 50% of its volume between 1938 and the early 1950s and describes other ice bodies (in "Ice Lake Valley, north of Arapahoe Glacier" and "at the head of Hauge Creek in Rocky Mountain National Park") which melted out in the 1940s or early 1950s. This period of ablation in the Front Range continued through to the early 1950s after which a change to more positive net balances has occurred.

Finally, Landsberg (1960) has compared the two 25-year periods (1905-30 and 1931-55 for changes in seasonal mean temperatures and precipitation totals over the United States, and used 10 stations west of 105°W. Nearly all station season records showed increases between the two periods and in some cases these changes were larger than the standard deviation for 1905-30. Changes in summer temperatures were particularly marked. Precipitation generally increased in Oregon/Washington/western Idaho/Nevada and California and decreased in the Rocky Mountain states, Arizona and north-central New Mexico. However, almost none of these changes was statistically significant. Only in northern Arizona was a significant decrease in precipitation apparent.

Changes in precipitation in Arizona and New Mexico have been examined by Macdonald (1956), Sellers (1960), and Von Eschen (1958). Macdonald notes that Arizona has experienced two major dry periods and one major maximum of winter precipitation. The minima were centered around the turn of the century (cf. Bradley and Barry, 1973a) and from 1942 to (at least) 1956; the maxima were centered around 1916-25. The most recent dry period he suggests is one of the worst droughts in the past 350 years, perhaps being exceeded in severity only by the Great Drought of the late thirteenth century. Sellers (1960) has examined precipitation from a number of stations in Arizona and western New Mexico for the period 1898-1959 and notes a definite downward trend in annual precipitation beginning in approximately 1905 and continuing to the late 1950s, averaging 1 inch/30 years and due almost entirely to a decrease in winter precipitation. Von Eschen (1958) came to a similar conclusion for the New Mexico area showing that total annual precipitation in the area has fallen steadily since 1915 accompanied by rising temperatures. Further evidence pointing to the significance of this recent trend towards less precipitation comes from river outflow data for the Upper Colorado River Basin (Thomas, 1959) which drains much of Utah, western Colorado and part of southwestern Wyoming. Thomas estimates annual virgin outflow from the basin from 1897-1929 at 17 million acre-feet whereas in 1930-56 a similar estimate of virgin flow is only 14 million acre-feet, a decrease of 18% between the two periods.

One of the few studies pertaining to climatic fluctuations in the Great Basin is that of Hardman and Venstrom (1941) who reconstruct changes in the levels and volumes of Pyramid and Winnemucca lakes from the early nineteenth century to 1940. They indicate drought conditions prevailed in the Truckee River watershed "for many years" prior to about 1840 and the period 1840-1860 was also relatively dry. Beginning in the early 1860s, however, a period of greatly increased precipitation began with the levels of Pyramid and Winnemucca lakes rising 9 and 22 feet, respectively, between 1867 and 1871 according to expedition accounts. Lake levels fell slightly from 1882-89 and then the wet period resumed, lasting until c. 1917. Considering outside consumptive use, maximum levels would have been reached in 1911, after which precipitation declined so that in the early 1930s virgin levels were probably slightly less than in 1840. This record is of interest in that it points to a generally wet period around the latter part of the 1860s, which is seen in the instrumental records for other areas also, particularly in winter months (see Figure 17). This is significant because many of the records used by Wahl and Lawson (1970) which were used to form area averages representative of the 1850s and 1960s were from this latter half of the

1860s. Hence the wetter period of the 1850 and 1860s may be unduly weighted toward this abnormal and relatively short period.

Summary

The sparse historical data for the western United States point to three main conclusions: 1) temperatures in the mid- to late nineteenth century were warmer than in the period 1931-60. Temperatures slowly fell during the late nineteenth century to some time in the period 1910-1930 and rose markedly over the next two or three decades. In some areas lower temperatures were recorded in the 1950s but the 1960s were generally warmer than the previous 30 years particularly in winter months (Namias, 1970). 2) Secular temperature variations in the western United States have been opposite to those over the midwestern and eastern United States and, in fact, the generally recognized hemispheric warming from 1880 to c. 1940 and cooling thereafter (Veryard, 1963; Mitchell, 1963) does not appear to be a feature of the climatic record in the western United States. 3) Precipitation studies are generally based on data from the southwestern states rather than the western states as a whole. The northern and central states can be considered as virtually unknown in this regard. In the Southwest, precipitation has been much heavier in the recent past (early seventeenth to at least the late eighteenth centuries) though the early nineteenth century (prior to 1840) may have been quite dry. The instrumental period has been characterized by periods of prolonged drought (up to 25 years in duration) the most recent example of which (1930-55) may have been the most severe for several centuries (if not longer) particularly in the Southwest.

Problems of Historical Data Sources

Since instrumental meteorological observations were first kept on a regular basis in the western states, many thousands of observing stations have been in operation. One of the principal problems faced by this study was to survey these masses of data and select those stations which have operated in more or less the same location for as long a period as possible. Because methods of tabulating data have changed many times over the past 100 years this procedure is not as easy as it might appear. The major problems are

1. Station histories of most stations are incomplete and sometimes inaccurate. The substation histories of each state often do not cover the entire period of record (generally ending in the early 1950s or late 1940s) and what information they do provide is frequently difficult

to interpret. "Distance and direction to Post Office," for example, is not a great deal of help if the Post Office has been moved!

2. Prior to the introduction of index numbers, stations were identified by name only. A station relocation was occasionally accompanied by a change in the station name but a change in station name did not necessarily indicate a change in location.

3. The precise geographical location, and more particularly the elevation, of some of the early stations was frequently not accurately known. Thus when a subsequent check of station locations and elevations "up-dates" the original station data, it is often not clear whether the "new" station data are simply corrections or if they indeed indicate a real change in station location.

4. Monthly precipitation totals are readily available in a number of publications. However, monthly mean temperature data are much harder to obtain. Apart from data for first order stations (which are generally located in cities and therefore subject to the urban "heat-island" influence) (Parry, 1966) monthly mean temperatures are not tabulated prior to 1951 (Bulletin W, Supplement 2) other than in the annual summaries for each state. Surveying long-term temperature stations is thus considerably more difficult than is the case for precipitation.

Data sources themselves are many and varied, ranging from the individual daily weather records (e.g., those listed in Darter's 1942 inventory at the National Archives) to the monthly, seasonal or annual means for periods of different length (e.g., Schott, 1873, 1876, 1881) averaged without regard to months or years when data are missing. This study has attempted to use the best data sources which provide abstracted data summaries, i.e., monthly average temperatures or precipitation totals tabulated month by month, year by year. An annotated list of the principal sources of data and data inventories is given in Appendix A. Other studies (e.g., Wahl and Lawson, 1970) have used only average values for periods of different length adjusted to a standard period.

For twentieth century data, monthly precipitation totals are available in two major sources: a) U.S.W.B., Bulletin W, "From the establishment of stations to 1930," and b) on magnetic tapes, Card Index Format 912/932, available for each state from the National Climatic Center (NCC). The latter is the result of cooperative agreements between various state agencies and the NCC to inventory monthly climatic data resources in each state. In most cases data are not available prior to 1930; however, some longer-term stations are recorded. First order stations are not included on the tape but these data (for 1931-70) are available in the publication "Local Climatological Data."

Although Bulletin W claims to include data "from the establishment of stations . . ." this is not always the case. The best nineteenth century data source was found to be Greely (1891). Many other sources were consulted (see Appendix A) but frequently, data were not tabulated in monthly form, only in terms of average values for widely varying periods. Records were extended back in time as far as possible using all available data sources. However, no attempt was made to combine early records in one location with later records in the same area unless there was evidence that the sites were reasonably close throughout the recording period. Two sets of data were thus obtained: data from stations operating for long periods in approximately the same location and still operating in the late 1960s (hereafter referred to as <u>continuous data</u>) and data from stations which operated for a season or more in the mid- to late nineteenth century but which were discontinued (hereafter referred to as <u>nineteenth century data</u>). The two sets of data begin to overlap in the 1870s to 1890s enabling the pattern of climatic fluctuations over the last 120 years (1851-1970) to be established. In the next three chapters the two sets of data are considered separately. An overview of the entire period is given in Chapter 7.

CHAPTER 2

NINETEENTH CENTURY DATA

Summary of Early Meteorological Observations in the United States

Instrumental meteorological records in the United States date from the year 1715 when the first measurements of precipitation were made at Cambridge, Massachusetts. Over the next century a network of observing stations gradually became established east of the Mississippi River but records generally cover only short periods. In fact by 1818 there were only 67 temperature and 23 precipitation records of more than one year in length (Havens, 1958).

The first steps towards an organized meteorological network were taken in 1817 by the Commissioner General of the Land Office who established a system of thrice-daily observations at the various Land Offices. In 1819 the Surgeon General of the Army also instituted a system of meteorological observations. These national networks were supplemented by the Patent Office in 1841 and the Smithsonian Institution in 1974 (Smithsonian Institution, 1893; Weber, 1922). Under these authorities observing procedures and instruments were standardized, resulting in comparable observations at widely scattered parts of the country.

Earliest instrumental meteorological observations west of the Mississippi date from 1821 when all daily temperatures were recorded at Fort George.[1] However, no observations in the mountain and desert states (Idaho, Montana, Wyoming, Utah, Nevada, Colorado, Arizona and New Mexico) were kept until the mid-nineteenth century, reflecting the late settlement of these areas. Observers were either surgeons at various military posts or voluntary observers reporting

---

[1]According to Scouler (1827), Fort George (46°18'N, 123°00'W), operated in what is today the state of Washington. However, Roden (1966) claims that the site of Fort George is in what is today Astoria, Oregon (46°11'W, 123°48'W). Either the original location of the station is incorrect or Roden (1966) is in error.

their observations to the Smithsonian Institution. This observing network continued until 1870 when Congress made the Chief Signal Officer responsible for observing and reporting storms. At this point the Smithsonian handed over its coordination duties to the Signal Service who continued to maintain the observing network until July 1, 1891, when the Weather Bureau was established in the Department of Agriculture. Virtually all observations since then have been under the auspices of the Weather Bureau.

In the area of concern to this study the earliest measurements were made at Fort Marcy (Santa Fe), New Mexico, in January of 1849 (Table 1) followed later that year by other stations in the same area (Albuquerque, Socorro, Ceboletta, Laguna) and at Cantonment Loring (Fort Hall), Idaho (at that time, part of Oregon territory). Blodget (1857, p. 50) notes mean temperature data from "Lapwai (Kooskooskia) Oregon [now Idaho], 46°27', 117°00', elevation 1,000 ft.?, 1837; 40-June 41" [sic] however, these data have not been located elsewhere. Blodget cites "Spalding, Wilkes' Exp. Expedition" as the source of the data but does not indicate if the observations were made at regular observation times which would make the data comparable with later observations. If these observations should prove to be carefully maintained according to the military observation procedures of the day then they may indeed represent the first official instrumental meteorological observations in the Rocky Mountain area.

Precipitation in the Latter Half of the Nineteenth Century

Clearly the climate of the first half of the nineteenth century in the western United States can only be inferred from secondary sources rather than direct instrumental observations. Such a study would involve historical records, diaries, expedition accounts, etc., and perhaps dendroclimatological techniques. In this section the focus of study is on the latter half of the nineteenth century when a network of recording stations gradually became established over much of the mountain and desert West. However, even during this period the observing network was extremely sparse and concentrated in certain favored regions. During the 1850s and 1960s, for example, nearly all observations of five years or more were made in New Mexico or the Pacific coastal states. Generalizations about the climate of the western United States at this time (e.g., Wahl and Lawson, 1970) must therefore be treated with caution.

Methodology

For the entire nineteenth century, records of more than twenty years' duration are extremely rare and the vast majority were kept for less than five years. Furthermore the distribution of stations reflects the settlement pattern of the area and is thus weighted towards certain regions. This study has attempted to maximize the information available by examining all available records which span a season or more at all stations in the study area. This is an attempt to avoid "over-weighting" the importance of the few long-term climatic stations by placing the information from these in the context of other station records. However, it is important that any generalizations made about the late-nineteenth century climate of an area be viewed with the limited distribution of early recording stations in mind (Figure 1). For example, in most of Utah and Nevada, there are virtually no records at all prior to the 1890s or 1900s.

Several phases of data reduction were undertaken. First, for those stations which had operated for five years or more in the same location, five-year averages were computed, beginning in 1851. When stations had records of ten years or more, adjacent five-year periods could thus be compared. In most states, the averages so derived were only available from the late 1860s. To supplement these data, individual seasonal values were computed for all stations for the early decades, irrespective of the length of record available. Thus, even if a station operated for two summers, some comparative information was available (i.e., summer 1 wetter or drier than summer 2). In this way it was hoped that some general statements about different time periods could be made, supplementing information from the long-term records with that from shorter term records. However, this amalgamation of individual season records with pentad averages proved to be difficult to use in practice as there was generally no means of fixing the precipitation amounts in one period at one station to precipitation amounts in another (later) period at a different station. Continuity of records through the 1850s and early 1960s to a period when more long-term records were beginning was generally not available. In addition, the number of stations operating in the early period in most states was frequently too small to make meaningful generalizations and in the later decades the availability of pentad averages made the technique superfluous. Nevertheless, the data from this type of analysis are presented whenever some useful information was obtained.

A second major part of the study was to try and place the climate of this period in the context of the twentieth century climate about which we know much more. This involved the technique of Wahl (1968) and Wahl and Lawson (1970) who located stations which have operated during recent decades in locations comparable to the nineteenth century stations. This "analog" method is not easy for a number of reasons. Early station locations were often not accurately known (particularly the longitudinal positions) and elevations appear to have been frequently estimated to the nearest 500 feet. In mountainous regions, even a distance of a mile between two stations may mean a considerable difference in the climate of each location, and elevation differences will almost certainly have important consequences. Twentieth-century urbanization of some areas is also a problem and the influence of cities on downwind precipitation patterns has been noted in other areas (Huff and Changnon, 1973). Furthermore, it is often difficult to find comparable stations in the twentieth century which have operated for a long enough period of time to establish a climatic "normal" according to WMO standards (WMO, 1966). Because of this and because of the over-riding importance of comparing observations at sites as close together as possible, a shorter period (1951-60) was chosen as the comparable twentieth-century period. It was felt that placing this decade in the context of the twentieth-century normals is relatively simple compared to the problems of finding long-term twentieth-century stations close to nineteenth-century equivalents. Tables 33 to 43 (Appendix B) list the nineteenth-century stations used and their twentieth-century equivalents and Figure 1 shows the locations of the nineteenth-century stations. It should be noted that while care was taken to find the closest twentieth-century station at an elevation approximately similar to the nineteenth-century stations, quite often no equivalent stations could be found. These cases are designated in Tables 33 to 42 by a question mark. In other cases, equivalent stations have been chosen which may not be perfect analogs but they have nevertheless been used with caution. Clearly meaningful generalizations about comparisons with the 1950s averages can only be made when several stations show similar characteristics. It should be noted that by the use of this analog method early records of only one season or individual pentad averages can be placed in the context of each other by use of the 1951-60 average as a "fix" on the absolute magnitude of the precipitation amounts. This sometimes allowed the use of early individual seasonal values and single pentads which could not otherwise have been placed in context.

Precipitation Fluctuations 1851-1890 by State

Using the methods outlined above, precipitation variations in the western states are discussed in this section. The years 1851 to 1890 were chosen as the period of study for several reasons. First, as outlined above, observations prior to 1851 are virtually nonexistent. After 1890 (with the establishment of the Weather Bureau in 1891) the observing station network became much more dense and by 1895 almost 800 stations were in operation in the western United States; however, over 40% of these were in California. Also, in the early 1890s regional offices of the Weather Bureau were established in most states and by 1893 all the western states were publishing regular state Weather-Crop Bulletins (Table 2). Hence, it is the period before 1891 about which very little is known. After 1890, the observing network becomes so large that a great deal of selectivity is needed to reduce the data to manageable proportions.

Each state is discussed in turn in Appendix B though often, as noted above, large areas of the states may be unrepresented. Furthermore, state boundaries may not reflect distinct climatological regions. However, it was felt that discussion by state would be the most suitable method of presenting the data. All conclusions and generalizations are based on the data presented in the histograms accompanying each state discussion, supplemented by early season by season comparisons and graphs of individual long-term station records. Whenever possible, examples of typical seasonal records for the state are presented to illustrate the year to year variability and overall characteristics of the seasonal record in that state. The terms "wetter" and "drier" are used in a relative not an absolute sense, and the unqualified use of the term "normal" or "average" refers to the 1951-60 average at the equivalent analog station. Winter seasons extend from November to March and are identified by the year in which November occurs. Winter of 1881 thus extends from November 1881 to March 1882.

Regional Overview of Seasonal Precipitation in the Period 1850-1890

In order to bring together in a manageable manner the large amounts of data discussed in Appendix B and to clarify the regional patterns of precipitation without regard to state boundaries, maps of seasonal precipitation as a percentage of 1951-60 averages at analogous stations were prepared. Only those stations which were considered to have a meaningful equivalent during the period 1951-60 were chosen (see Tables 33 to 42) and inevitably some of these are probably not good analogs due to local site factors with which the author is not familiar. Because of these problems the patterns discussed below must be considered in broad-scale terms. Each pentad will be discussed initially, by season, and where feasible a general summary of the entire period will follow.

1851-55 (Figures 2 to 5)

The only area with sufficient stations to allow meaningful generalizations is New Mexico. Only one other station (in Washington) is available for the pentad. Precipitation was above average in New Mexico in all seasons averaging only 105% of the 1951-60 mean in spring but 147% in summer, 174% in fall and 134% in winter. Overall, precipitation was about 140% of 1951-60 averages.

1851-60 (Figures 6 to 9)

Only data for the New Mexico area and the Columbia River area of Washington and Oregon are available.

<u>Spring</u>. Well below average in the New Mexico area (overall, less than 50% of 1951-60 averages), but not clear in the Northwest, where half the stations are below and half above average and to about the same degree.

<u>Summer</u>. Above average in the New Mexico area at levels which were overall very close to those of the previous pentad. In the Washington/Oregon area stations within 100 miles of the coast were below average (averaging 85% of the 1951-60 mean) but the eastern part of the states may have been above average.

<u>Fall</u>. Markedly above average over most of the New Mexico area and also the Washington/Oregon area excluding two stations in northwestern Oregon.

<u>Winter</u>. Below average in north-central and southern New Mexico/Arizona and above average elsewhere. Maximum values are found over south-central New Mexico where precipitation is twice 1951-60 averages.

<u>Pentad and Decade Summary</u>. Apart from spring, most of New Mexico was again above average in this pentad (the only exception is Fort Fillmore); the 1850s as a whole were thus relatively wet in this area. The stations in Oregon and Washington indicate below average precipitation along the coast but above average inland. This pattern of "coastal" precipitation being the opposite of that further inland is a characteristic feature of this area.

1861-65 (Figures 10 to 13)

Apart from the Columbia River Valley, few stations are available.

<u>Spring</u>. Coastal Oregon and Washington were below average with higher amounts to the east. There are insufficient data to generalize for other areas.

Summer. The inverse of the spring pattern characterizes precipitation in Oregon and Washington.

Fall. Markedly below average at all stations in the Northwest (75% of 1951-60 averages).

Winter. Below average at stations within 100 miles of the coast but well above average to the east, and south into Nevada (?).

1866-70 (Figures 14 to 17)

Spring. Arizona, New Mexico, and southern Colorado were below average (~70%) except for southeastern New Mexico. The northern states were generally above average (~120%).

Summer. Almost all stations were above average with maximum precipitation anomalies over northern New Mexico and southern Colorado.

Fall. Generally below average although the few above average stations in northern Nevada, southern Colorado, and northwestern New Mexico may reflect a large anomolously wet area across the center of the region.

Winter. Considerably above average at nearly all stations (~130% of 1951-60 mean in the northern half of the area and ~125% in the southern half).

Pentad Summary. Summer and winter averages in the latter half of the 1860s were greater than in the 1950s. Fall precipitation was generally below average in the southern and northern parts of the region. In the spring, precipitation was above average except in the area of northwestern New Mexico/central Arizona.

1871-75 (Figures 18 to 21)

Spring. Generally above average in New Mexico, Arizona, Colorado, Utah, southern Wyoming, and northeastern Nevada (overall average of~120%), and in the Oregon/Washington, northern Idaho, and northwestern Montana area (average of ~110%). In the intervening area precipitation averaged only 60% of 1951-60 means.

Summer. Precipitation was relatively heavy along the eastern margins of the region averaging 125% of 1951-60 averages; southwestern Arizona was also above average. The rest of the region was markedly below average, however, with amounts less than 50% of average in Nevada and 80% of average to the north and northwest.

Fall. A pattern similar to that of summer with precipitation ~125% of average in the east and ~70% in the north and west.

Winter. Precipitation was markedly above average in all areas except central Montana and eastern Colorado. In the northern half of the area precipitation averaged 143% of 1951-60 means and in the southern half ~150%.

Pentad Summary. Precipitation was generally above average in all seasons in the southern and eastern parts of the region and in the northwest in spring and winter. The western half of the area was well below average in summer and fall.

1876-80 (Figures 22 to 25)

Spring. Much of the area was above average with the exception of two main zones--southern Arizona/most of New Mexico/ eastern Colorado (67%) and northern Nevada/northern Utah and southern Idaho (70%). Precipitation was considerably above average in the northernmost five states (particularly in northern Idaho and Montana) (~150%) and less anomalous to the south (~ 130% of 1951-60 means).

Summer. Well below average over most of the area (~ 70%) but considerably above average in the southeast (~127%) and slightly above average in the Northwest and central Montana.

Fall. Above average in the north (130%) and in southern Arizona (136%) and below average in the remainder of the area (75% of 1951-60 means), particularly in the Great Basin area of Nevada (~60% of average).

Winter. Above average in most areas, the main exception being a zone extending from central Arizona to northeastern Nevada and northern Utah (~70% in the northern part of this zone and ~90% in the south). The area with highest precipitation anomalies is centered in Idaho.

Pentad Summary. Precipitation was above 1951-60 averages in the five northern states. In Colorado, New Mexico, and southeastern Arizona amounts were generally above average in summer and winter. Other areas received amounts close to or below the overall average for 1951-60.

Decade Summary. The 1870s were generally above average in spring (except for an area around northern Nevada/southern Idaho/western Oregon and southern Arizona/south and central New Mexico) below average in summer and fall for all areas except northern New Mexico, southeastern Arizona, and Colorado, and generally above average in all areas in winter.

1881-85 (Figures 26 to 29)

Spring. Three main zones of below average precipitation (each averaging about 75% of 1951-60 means) are apparent: southern Arizona and New Mexico extending northward into Colorado, a zone across north-central Nevada, and northern Idaho/northern and eastern Montana. Elsewhere precipitation is markedly above average--generally about 140% of 1951-60 means in the north (Washington/Idaho) and averaging about 130% overall.

Summer. Above average in eastern Arizona/northern New Mexico/Colorado, parts of northern Nevada and southern Oregon (?) and most of Montana/northern Idaho (overall average of 120-125% in each area). Elsewhere precipitation was below average being ~70% of 1951-60 means in western and southern Arizona/southern New Mexico and ~80% over the rest of the region.

Fall. Well above average precipitation in the northern half of the region particularly northern Nevada and Utah/southern Idaho where precipitation averaged ~185% of 1951-60 means. To the north and west averages were still high, generally 140% of amounts in the 1950s. Parts of eastern Arizona and northern New Mexico were also relatively wet (125% of average). Elsewhere precipitation was only about 70% of 1950s averages.

Winter. Above average (125-135%) in all areas except the eastern margins (~85%) and the northwestern coast (~72%).

Pentad Summary. Precipitation was generally above 1950s levels in most areas except parts of the Great Basin (particularly western Nevada), a zone along the Oregon/Washington coast, and parts of eastern Montana.

1886-70 (Figures 30 to 33)

Spring. Predominantly below average, particularly in the extreme south where precipitation was only ~55% of 1951-60 means. In most other areas precipitation was ~80% of average.

Summer. Again, most areas were below average, the main exceptions being southern Arizona/New Mexico, eastern Colorado, Washington and southern Oregon. However, most of these areas were only slightly above average (except the Oregon and Washington zones which were ~120% of the average). Precipitation was particularly low in the Great Basin (~ 60% of 1950s averages).

Fall. There are three main zones of above average precipitation--Arizona/New Mexico, Utah, and parts of eastern Nevada and east-central Montana. Elsewhere precipitation is predominantly below average.

Winter. Unlike the other three seasons, winter precipitation is predominantly above average in all areas except Colorado and eastern Wyoming/southeastern Montana (?), parts of Oregon and Washington, and south-central Arizona. The only significant zone, however, is in Colorado (~80% of 1951-60 averages). Precipitation was ~120% of average in the Great Basin area, slightly less to the north and ~130% in Arizona and New Mexico.

Pentad Summary. Arizona and New Mexico received above average precipitation throughout the year but most other areas were below average in all but winter months.

Decade Summary. The 1880s were characterized by relatively wet winters in the study area in all regions except the extreme northwest coast and the eastern margins. In the other seasons the two pentads were generally quite different.

The above pentad discussions are briefly summarized in Table 3 though it must be noted that any generalizations for such a large area will inevitably omit much detail which may be significant in individual areas, and/or particular time periods. The table should be viewed in light of the discussions above.

Recurrent Peaks and Troughs in the Winter Precipitation Record 1850-89

A feature of considerable interest in the winter precipitation records of some states is a tendency for peaks to recur with a periodicity of from four to six years. The feature is most noticeable in Arizona and New Mexico records and is particularly well illustrated at Fort Lowell, Arizona, (Figure 34) with peaks in 1868(?), 1873, 1878, and 1883 (see also Fort McDowell, Figure 127). At Fort Bowie to the east, peaks occurred in 1867, 1873, 1878, 1883, and 1888, and at Fort Verde to the north (Figure 126) in 1873, 1879, 1883, and 1888 or 1889. This apparent four- to six-year periodicity can also be seen at other stations throughout Arizona, even at the extreme western stations of Yuma and Fort Mojave. Furthermore, extrapolating back in time one would expect to find peaks in the early and late 1850s if this periodicity had been operating in earlier decades. It is therefore interesting that at the only two records for the state, winter precipitation was exceptionally heavy at Fort Defiance in the winter of 1854 and 1855 (the wettest and third wettest winters in the period) and at Fort Buchanan in the winter of 1850 (the wettest winter in the record). Winter precipitation at Fort Defiance in 1854 and 1855 was 180% and 126% of the mean (1852-1859), respectively, and at Fort Buchanan the 1859 winter precipitation was 166% of the 1857-60 average. It seems very probable then that a five- or six-year periodicity was apparent in Arizona winter precipitation for most of the latter half of the nineteenth century. However, no similar periodicity is apparent in the other seasonal records for Arizona.

In New Mexico, synchronous precipitation maxima were also recorded. At Fort Fillmore, for example, peaks in the record (1851-60) occurred in 1851, 1855, and 1859. High precipitation totals were also recorded in 1855 at Santa Fe, Fort Thorn, and Fort Craig (the maximum for the period at each station) and at Fort Union (exceeded only by the high precipitation total for 1887) (Figure 170). The 1859 peak is illustrated by the records for Fort Stanton, Fort Fillmore and Fort Craig. In 1868 heavy winter precipitation was recorded at stations as far apart as Fort Seldon, southern New Mexico, and Fort Garland, Colorado, where precipitation was next to the maximum for the periods 1866-76 and 1858-83, respectively (Figure 35). Other stations with relatively high winter precipitation in this year include Fort McRae, Fort Wingate, and Santa Fe (Figure 170).

Several stations show relatively high winter precipitation totals in 1873 or 1874 (Santa Fe, Fort Union, Fort Garland [Colorado], Fort Craig) and in 1878 (La Mesilla, Silver City, Fort Union, Fort Wingate, Santa Fe). In fact, the only record for 1878 which does not show a significant peak in that year is from Fort Craig.

Although there are very few data for the state in 1883 or 1884 it does not appear that these winters were characterized by heavy precipitation, as was found in Arizona. 1888 was relatively wet at some stations (e.g., Fort Selden, Lordsburgh, Lava) and at Lava, the winter precipitation for 1888 was more than half of the total winter precipitation from 1885-1889. Similarly, at Lordsburgh it was almost one third of the total for 1881-1889. However, at several stations (Fort Union, Fort Stanton, Fort Wingate?) the winter of 1887 was wetter than the following winter. Hence, like Arizona, the late nineteenth-century winter precipitation record of New Mexico is characterized by recurrent peaks at approximately five-year intervals.

When the winter records for Arizona and New Mexico are compared with the corresponding data for Oregon and Washington, it is clear that quite the opposite situation prevailed there; where peaks in the Arizona and New Mexico records are found, Oregon and Washington records show marked troughs.

At Astoria (Figure 184), for example, exceptionally low precipitation amounts were recorded in the years 1855, 1859, 1868, 1873, 1883,[1] 1884, and 1888. Precipitation in the years 1859, 1883(?), 1884, and 1888 was more than two standard deviations below the 1856-75 average at Astoria. Other stations also show minima in these years. At Fort Yamhill and Fort Hoskins, for example, winter precipitation in 1859 was the minimum for the entire record at each station (1856-65 and 1856-63, respectively). At Fort Yamhill the 1859 value was 1.44 standard deviations below the mean and at Fort Hoskins it was 1.66 standard deviations below the period average. There are virtually no data for Oregon to investigate the suggestion of a peak in 1868 but several stations indicate troughs in 1873, 1883 or 1884, and 1888. Table 4 gives some examples.

Clearly the trough in 1888 is most noticeable with precipitation amounts generally only 50% of the average for the 1870s and 1880s. Thus, recurrent troughs in the Oregon record occur every four to six years (excluding the early 1860s and the late 1870s) and these correspond closely to precipitation maxima in the Southwest.

As might be expected, recurrent troughs in the Oregon winter precipitation record are also found in the data for Washington. At Fort Vancouver, low winter precipitation amounts were recorded in 1850 and 1855 (cf. Astoria, Oregon) and even lower amounts in 1859 (1.3 standard deviations below the 1852-64 average). The winter of 1859 was dry over much of the Northwest, with low precipitation totals recorded at Fort Steilacoom, Fort Walla Walla, and Fort Cascade.

Some support for the Astoria trough in 1868 comes from the records of San Juan Island (where winter precipitation for that year was the period minimum, 1.2 standard deviations below the mean for 1861-73) and Fort Colville (where the 1868 winter total was the second lowest in the period 1860-78). However, 1868 was relatively wet at Fort Canby (Cape Disappointment), so the areal extent of this dry winter is unclear.

The winters of 1873, 1883 and/or 1884, and 1888 were also exceptionally dry in Washington, as Table 5 indicates. It

---

[1]Although there are no data for this station from September 1876 to December 1883, the total for the winter of 1883, excluding November, is only 22.22 inches, less than half of the average winter precipitation for 1856-1875. As November precipitation for the same period averaged 10.80 inches, it seems very likely that 1883 was also a year of exceptionally low winter precipitation.

thus appears that recurrent dry winters occurred in Washington and Oregon in the years 1850 or 1851(?), 1855, 1859, 1868, 1873, 1883 or 1884, and 1888. Clearly with the exception of 1863 or 1864 and 1878 or 1879 there is strong evidence for a periodicity of low winter precipitation in these northwestern states approximately every four to six years.

The fact that precipitation maxima in the Southwest correspond to minima in the Northwest is not totally unexpected. Sellers (1968) in an analysis of the dominant precipitation anomaly patterns for the western United States, notes "in each month there is one eigenvector with precipitation anomalies of opposite signs in the Pacific Northwest and the Arizona-New Mexico-Texas area. This opposition is probably associated with shifts in the location of the storm track." LaMarche and Fritts (1971) have also noted such an anomaly pattern in tree ring data for the same period (1931-62), and also in the period 1700-1930 (see their Figure 1; eigenvector 2, for November). This would suggest that such a relationship has been characteristic of the climate in the western United States for at least 200-300 years. Klein (1965) using data for 1949-68 examined five-day precipitation amounts in different areas of the United States in terms of the associated five-day mean 700-mb height anomaly patterns. Heavy winter precipitation in southern New Mexico is associated with anomalously low 700-mb heights centered over northern Mexico, with the region dominated by air from the Gulf of Mexico. At the same time the Northwest is associated with positive height anomalies (Klein, 1965, Figure 5, map 29). Maps centered on the Four Corners area and on eastern Arizona and western New Mexico show that heavy precipitation in these areas is associated with negative 700-mb height anomalies centered to the west of the region, with dominant air flow from the southwest. At the same time the Northwest is under the influence of northeasterly air flow (Klein, 1965, Figure 5, maps 33 and 34). Finally, maps centered on the Oregon and Washington coast (Klein, 1965, Figure 5, map 38) show that very low winter precipitation is associated with very strong positive 700-mb height anomalies over the area, again with dominantly northeasterly airflow, whereas the Southwest is under negative 700-mb height anomalies and is influenced by air from the West. Klein (1965) also gives surface pressure anomaly maps for five-day precipitation totals and these have been compared with reconstructed sea-level pressure anomaly maps for January 1883, 1884 and 1888 prepared by Roden (1966 from data in Bigelow, 1902). Little correspondence is found between the anomaly maps of Roden for 1884 and 1888 and the expected anomalies for periods of low precipitation as shown by Klein (Klein, 1965, Figure 7, map 38). However, the map for 1883 shows much more correspondence, with positive anomalies over the Northwest, centered off the coast of northern California/southern Oregon. It should be noted, however, that the Klein maps are for winter months (mid-December to mid-March) whereas Roden's maps are for January only.

Clearly then, an inverse relationship between precipitation in Arizona/New Mexico and Oregon/Washington is to be expected but it seems probable, in view of the very strong inverse relationships shown between 1850 and 1890, that the particular anomaly patterns associated with this type of precipitation distribution were extremely common at approximately five-year intervals in the latter half of the nineteenth century.[1] What is much more difficult to explain is the regularity of occurrence of the peaks and troughs. It is interesting that Sellers (1968) also notes "when the track is north of its normal position as it was during much of 1943, 1948, 1953 and 1956, heavy rain falls in the Northwest while drought occurs in the Southwest." Hence, a five-year pattern in the (annual) precipitation record of the area has also been noted for at least part of the twentieth century.

A periodicity of four to six years might be considered to have a subharmonic relationship with the "eleven-year cycle" of solar activity (Waldmeier, 1961). Figure 36 shows this cycle as described by Wolf Relative Sunspot Numbers. This is similar to other indices of solar activity including that of faculae and umbral areas. Clearly there is no simple relationship between the winters of high (low) precipitation and solar activity. The winter of 1888 was characterized by very low solar activity whereas in 1883 and 1884 activity was relatively high. The winter of 1868 was during the rising phase of solar activity while the winter of 1873 occurred during the falling phase. Hence solar activity, at least as expressed in simple indices, does not suggest an explanation for the pattern of recurrent precipitation peaks and troughs.

An alternative possibility is that the four- to six-year periodicity is in some way a function of the well known quasi-biennial oscillation (Berlage, 1957) though the

---

[1]It should be noted that by analogy, using the maps of Klein (1965) one could expect to "predict" what the circulation was like over the rest of the U.S. in these particular years, and hence what precipitation conditions were like on other areas. In a sense this would be essentially a forecasting technique with the objective being retrospect rather than prospect.

relationship of this phenomenon to solar activity is still not fully clear (Shapiro and Ward, 1962; Berson and Kulkarni, 1968). At present there appears to be no simple explanation for this regularity and the problem seems worthy of more detailed research, particularly into the synoptic conditions characteristic of the seasons in question.

The periodicity is discussed further in the section on spectral analysis of precipitation data for the Rocky Mountain states (Chapter 4).

Suggestions for Further Work

In the preceding sections, all available seasonal pentad precipitation data from 1851-1890 have been discussed, and over 150 station records from the ten states have been presented. Clearly, in spite of the sparse data coverage in some areas and at different time periods, a great deal of useful information can be obtained from these early climatological records. Furthermore, in the course of this study the author has become aware of the extensive sources of other data which, as yet, have hardly been touched. In particular, observations of pressure, wind direction and speed, and cloud cover reported several times a day at all army posts could provide fairly detailed maps of circulation patterns over the area (cf. Roden, 1966) for a period as yet almost unknown. Temperature data are also available and observations appear to have been made even more conscientously than were observations of precipitation; however, differences in recording times make comparisons with contemporary observations difficult. In spite of this, Roden (1966) has shown that detailed studies of historical temperature data can provide valuable information on the historical climatology of the United States and on the stability or lack of stability of "present" climatic conditions; his contribution in this respect is unique. Other useful studies using temperature data would be to examine the dates of first and last frosts and length of the growing season (cf. Bradley and Barry, 1973b).

Precipitation data could also be studied in more detail, particularly with regard to precipitation size-frequency classes. Leopold (1951) has shown that significant changes in the frequency of certain precipitation size classes occurred in the New Mexico area. This has important implications for erosion studies and in semiarid areas can be of critical importance to the growth of vegetation. Indeed, the particular changes noted by Leopold (1951) have been attributed to equally marked changes in the characteristics of tree growth increments in trees of the area (W. Glock, personal communication). Dates of the first and last snows would be a valuable index of changes in climate, as would studies of maximum snow depth during winter months. This would be particularly valuable to ecological studies related to winter precipitation augmentation in many parts of the western United States today.

Finally, many of the early records on microfilm record other discrete phenomena--occurrences of thunder, hail, fogs, dust devils, auroras, smoke from fires, and even occasional phenological reports. In short, sources of early historical data, not only for the western United States but also for other areas,[1] can provide quite detailed information on many aspects of the climate of the late nineteenth century or earlier without resorting to inferences based on dendroclimatic or other reconstructions. Studies of these data should be prerequisites to attempts at reconstructing past climatic variability from secondary sources. However, the potential researcher should be aware that studies using historical sources of climatic data are extremely time-consuming and results will only be achieved slowly and painstakingly.

---

[1]Areas particularly rich in historical climatic data are the Gulf of Mexico coast states, Florida and "Indian Territory" (Texas). No studies of these data have been attempted to date.

Author's Note: For complete maps, including California, see Bradley, R.S., 1976, Seasonal precipitation fluctuations in the western United States during the late 19th century. Monthly Weather Review (in press).

CHAPTER 3

CONTINUOUS DATA: I

Initial Data Survey

In the initial data survey (using a criterion of at least 50 years of record to 1970) approximately 260 precipitation records were noted in the study area. These stations were then carefully screened on two counts:

1. to assess the amount of missing data,

2. to assess the number of times the recording station was moved during the period of record.

In the former case, monthly values were divided into seasonal units and the number of missing months in each season for the entire record length was assessed. Any station/season record with more than 5% of the monthly values absent was rejected. Any interpolated values already in the record as published were assumed to be correct and were not recomputed or considered in calculating the 5% missing data limit. This method thus took into account the different number of months in the seasons used and allowed a "good" seasonal record to be used even though the other seasons might have much missing data.

On the question of station relocations, considerably more difficulty was encountered; station histories are poor in most cases. At nearly all stations, the original sites of observations are poorly known or not known at all, and station movements were seldom carefully noted. The publication "Substation History" for each state was an attempt to remedy this situation but unfortunately many of the entries therein are ambiguous or uninformative. For example, the section "distance and direction from previous site" is absent on many of the earlier forms and a change in the elevation listed may be the only indication of an apparent move. Also, in some cases a "new location" entry is made each time there is a new observer, but the information given is frequently insufficient to tell whether this event was also associated with a new location of the observing station. Furthermore, the Substation Histories have not been updated since their publication in the early 1950s. In view of these problems, attempts to screen out those stations which may have experienced frequent moves were inevitably inadequate. Nevertheless, a procedure was followed which did result in the elimination of some stations whose station histories indicated large and/or frequent moves which were likely to result in inhomogeneities in the record. The criteria for rejection were arbitrarily chosen; any move of greater than one mile was considered too large and more than an average of one move per fifteen years of record was considered too many. In either case the station was rejected. When doubt existed as to the character or authenticity of a station move, the record was assumed to be acceptable pending subsequent statistical investigations of homogeneity (see below).

As a result of these somewhat arbitrary data quality standards, the initial selection of 260 precipitation station records was reduced by about one third giving an average of approximately 30 stations per state. These stations are shown by state in Figures 37 to 40.

Missing Data

Because only a few stations in the study area have no missing data, some monthly values had to be interpolated at most stations. As noted above, this ranged from one or two months to 5% of a seasonal record.

Interpolation is normally carried out using two records, one from the station (Q) which has missing data, the other from an adjacent station (A). In the case of precipitation, the ratio of the means for corresponding periods ($\bar{Q}/\bar{A}$) is first calculated. Then the missing monthly value at station Q is estimated to be the product of the corresponding monthly value at station A and the ratio of the means (Conrad and Pollack, 1950).

In this study a slightly different procedure was followed which was found to give estimates on average as good as, or better than, the above method. Suppose there are two series:

$$Q_1, Q_2, Q_3, Q_4, \ldots, Q_k, \ldots, Q_n$$

$$A_1, A_2, A_3, A_4, \ldots, A_k, \ldots, A_n$$

a new series of ratios is computed such that

$$Z_1 = \frac{Q_1}{A_1}, \; Z_2 = \frac{Q_2}{A_2}, \; \ldots, \; A_k = \frac{Q_k}{A_k}, \; \ldots, \; Z_n = \frac{Q_n}{A_n}$$

In the case of the Q and A series being precipitation totals, the distribution of the Z series is highly positively skewed.

Hence, the median ratio is taken and used to compute an estimate of the missing value in the same way as the ratio of the means is used above ($Q_k = A_k \times Z$ [median]). In choosing "adjacent stations," selection was limited to those long-term stations already isolated in the earlier data survey. Shaded topographic relief maps were then used to examine the location of stations nearest the key station. By assessing the three factors, topographic situation (with respect to mountain ranges and divides in particular), elevation and distance, the most suitable record to be used for interpolation was chosen. However, this may not be the optimum method; in most cases it would have been possible, and may have been more satisfactory, to select a station closer than the nearest long-term station. However, this would have involved considerably more data handling than was feasible. By using data files already constructed there were also computational advantages. In general the particular long-term reference station selected was chosen to be as close as possible to the station in question and to be at a similar elevation. In computing the ratio series, Z, the longest possible parallel records were used.

It was felt that the interpolation method used was adequate for this study because:

1) The amount of missing data had been previously restricted to a maximum of 5% of any one season,

2) Monthly values were being interpolated whereas in the statistical analyses which followed, seasonal totals were examined. The importance of any one monthly value was thereby reduced, except in cases where all months of a season were missing. However, even in these cases if errors in estimation are assumed to be random the overall error of the seasonal estimate is likely to be reduced by summation of the individual monthly estimates.

3) All seasonal series were to be tested for homogeneity subsequent to the interpolation of missing values (see next section).

The Homogeneity of Records

A fundamental prerequisite for studies of secular climatic change is that the station records being examined are homogeneous. In general terms this implies that a given record is equally representative of climatic conditions in the vicinity of the station throughout the entire period of its operation. In practice, of course, a number of factors mitigate against such a condition. Stations are often moved, some observers may be less diligent than others, and exposures may not be ideal. Indeed, even if a station had an ideal exposure in 1900 and was never moved, its exposure may have deteriorated over time due to the construction of buildings nearby or simply the natural growth of trees (e.g., Lawrence, 1970). Furthermore, many of the longest records are associated with the growth of an urban center where the local climate has gradually been modified over time (Landsberg, 1970). Similarly, modification of the natural environment in other ways (the construction of artificial lakes and reservoirs, agricultural development, deforestation, etc.) may also introduce inhomogeneities into a record. However, identifying inhomogeneities in a climatological series is an extremely difficult task.

Conrad and Pollack (1950) distinguish two types of inhomogeneity--absolute and relative--and the terms are discussed in great detail by Mitchell (1961). A climatological series is said to be relatively homogeneous with respect to a synchronous series at another place "if the . . . ratios of pairs of homologous averages constitute a series of random numbers that satisfies the law of errors" (Conrad and Pollack, 1950, p. 226). However, as pointed out by Mitchell, et al. (WMO, 1966), the evaluation of relative homogeneity is only a step towards ascertaining if a record is absolutely homogeneous, i.e., if, by itself, the record is representative of purely natural variations in the climate of the area immediately surrounding the recording site.

Unfortunately, the evaluation of absolute homogeneity is very difficult, if not impossible, in practice. Tests of absolute homogeneity usually require the construction of a "regional climatological series" (comprising the average of a number of adjacent records) with which the key station in question is tested. This generally involves computing a series of ratios (for precipitation) or differences (for temperature) between the regional and key station records and testing the resulting series for nonrandomness at a suitable significance level. However, in constructing a "regional climatological series" one is faced, inevitably, with assuming that some, if not all of the station records to be used in constructing the regional series are homogeneous and of course this may not be so. There is, then, a danger of circular reasoning in the concept of absolutely homogeneous records, and in practice the best that can be achieved is a somewhat stronger evaluation of relative homogeneity than a simple test of relative homogeneity between one or more individual records.

In this study, further problems arise in using the classical tests of both relative and absolute homogeneity. It was pointed out above that evaluation of homogeneity rests on the idea that the station record being evaluated is representative of its surrounding area. In an area of little relief and uniform topography such a concept can be applied with a certain amount of success. However, in a mountainous region the problems of determining how large an area a particular station represents are formidable. Furthermore, when selecting the longest records in an area it is generally impossible to find stations in the immediate vicinity which have records as long as the record in question. Finally, in dealing with a large number of records, as this study does, the problems of acquiring long-period data from a dense network of stations close to each key station in question makes detailed studies of absolute homogeneity at each station impossible.

Because of all these problems the procedure used here to evaluate homogeneity was a compromise between computational costs, practical considerations of data availability and the potential meaningfulness of exhaustive tests, given the extremely diverse topography of the study area. Nevertheless, it is felt that the careful (perhaps even overcautious) application of the criteria outlined below were sufficiently stringent to preclude the incorporation of inhomogeneous records into subsequent statistical analyses. As pointed out earlier, the general philosophy of this study was to use as many station records as possible in the initial analyses and discard questionable records whenever necessary. Consequently, no attempts were made to adjust records which were thought to be inhomogeneous.

Testing for Homogeneity

Tests of homogeneity generally involve the null hypothesis that series of ratios between stations exhibit characteristics of a random series. This may be assessed by graphical techniques (Kohler, 1949; Conrad and Pollack, 1950, p. 231; WMO, 1966) or by statistical tests for alternatives to randomness. In this study, the latter were preferred as graphical analyses are often difficult to evaluate objectively. The following procedure was used: for each station in question (hereafter referred to as the key station), the nearest long-term stations which had records as long as, or longer than, the key station were identified. Five seasons were then chosen[1] ("Spring," April and May; "Summer," June,

---

[1]Seasons were selected to be comparable with other studies in the area. "Late summer" was added after an examination of intermonthly correlations showed a distinct break at some stations between the months of June and July. This reflects the July singularity noted previously by Bryson and Lowry (1955) and Bryson and Lahey (1958).

July and August; "Late Summer," July, August and September; "Fall," September and October; "Winter," November, December [Year 1], January, February and March [Year 2]) and seasonal totals of precipitation were computed.

In the case of precipitation, seasonal ratio series were constructed by dividing one of the series term by term, into the other. This new series was tested for nonrandomness using the Mann-Kendall nonparametric (rank) statistic (Mann, 1945; Kendall, 1948). This is a general test of randomness against the alternative of trend (linear or otherwise), as trend is the form of nonrandomness most likely to appear in the ratio series. The statistic is computed, thus. The first term of the ratio series (X_1) is compared with all subsequent terms, and the number of terms greater than the first is noted (n_1). This procedure is repeated with the second term (X_2) and so on to X_{N-1}. The statistic, τ is then:

$$\tau = \frac{4P}{N(N-1)} - 1$$

$$\text{where} \quad P = \sum_{i=1}^{N-1} n_1$$

τ has an approximately Gaussian normal distribution with an expected value of 0 and a variance of 4N + 10/9N(N-1) (WMO, 1966). The computed values of τ were thus tested for significance by comparison with the values:

$$(\tau)_t = 0 \pm t_g \sqrt{\frac{4N + 10}{9N(N-1)}}$$

where t_g is the (two-tailed) 95% probability point of the Gaussian normal distribution (τ is positive for an upward trend and negative for a downward trend, hence a two-tailed test was used). A computed τ value exceeding the 95% probability point was taken to indicate that one of the two stations had an inhomogeneity in the seasonal record being examined. The problem was, which one? By comparing each seasonal record with two or more other seasonal records it was possible in many cases to isolate the inhomogeneous record in the following way. If a seasonal record was found to be inhomogeneous with respect to another record of comparable length <u>and</u> a second record of comparable length or shorter, the record was rejected. These criteria were generally adequate to isolate the most questionable records, and by removing these from the matrix of significant τ values generally only a few questionable pairs remained. These remaining pairs of records were noted as "Suspect" and were rejected from many of the subsequent statistical analyses. Such precautions eliminated some records which might have been in reality acceptable but it was felt that stringency was necessary in view of the problems of evaluating homogeneity in the area. To illustrate the procedure used, the following example for Montana Spring season precipitation records is given.

The Mann-Kendall rank statistic was computed for all stations as outlined above. Table 6A shows those station-pairs which had a τ value significantly greater than the 95% significance level. The statistics resulting from all other station-pairs were not significantly different from zero. Stations listed in column 1 were the key stations in each calculation, and thus the entire period of record at those stations was used in the calculation. Stations in row one may have had only part of their record involved in the calculation. Examination of this table shows that spring records for stations 1202, 2409, and 3489 appear to be inhomogeneous with respect to two or more adjacent records and should therefore be rejected. Removal of these stations results in the matrix shown in Table 6B.

Further analysis would probably lead to a decision as to the homogeneity or otherwise of each of these records. However, in general the added computation was not considered to be profitable and the eight stations were designated "Suspect."

It should be noted that in Table 6, the station 6157 could also have been rejected. However, in view of the fact that one of the apparent relative inhomogeneities resulted from an analysis with station 2409, which is clearly unacceptable (two tests using the complete record suggest inhomogeneities and four cases when part of the record was used to point to a similar conclusion), station 6157 (spring) was not rejected but still remained "Suspect."

Tables 7 to 11 list those seasonal records which were rejected or designated Suspect in this manner. These tables include only those stations which survived the initial screening for station movement and missing data, as outlined above. The stations are arranged by record length and it should be noted that the longest station record in each state was inevitably only partially tested. This method of evaluating homogeneity would not be suitable for studies concerned with a few specific climatic stations. However, in this study, where the number of stations being examined was large, the methodology was found to be quite satisfactory in most cases.

In all seasons there are no clear regional patterns of inhomogeneity. Table 12 summarizes the results of the homogeneity tests. It is interesting to note that the number of station records rejected or suspected is considerably higher in winter than in summer months (an average of 44% of winter records compared with 18% of summer or late summer records). As most of the precipitation in the area falls as snow during winter months, it is likely that problems of accurately recording snowfall (Leaf, 1962) are often responsible for inhomogeneities in the records. Snowfall recording problems may also be reflected in the number of records rejected or suspected for spring and fall seasons when precipitation falls in the form of both rain and snow. Considering all the seasonal records examined, 26% of the records were rejected or suspected of inhomogeneities.

Average station elevation is 4,866 feet, ranging from 1,260 feet at Kooskia, Idaho, to 9,322 feet at Silverton, Colorado (Table 13). Clearly the stations chosen do not reflect in detail the diverse topography of the area; however, they can be considered to represent a topographic surface across the region as shown by the contours of station elevation in Figure 41. Analyses of the data must be interpreted in terms of this "station-elevation" surface within the context of the "real" topography of the area.

It is of interest to see how representative of different elevation categories the stations are within each state. Table 14 shows a hypsometric division of the study area into area-elevation categories. As might be expected, the zone 2,000-5,000 feet is over-represented by the long-term station network, whereas the higher zones (>5,000 feet) are under-represented. In particular, that part of the study area above 10,000 feet is not represented at all as there are no long-term stations at such elevation. In fact, there are very few high elevation stations with even 30 years of record (1971-70), as shown in Table 15. Only one station >10,000 feet in the entire study area has operated over a 30-year period and there are only eight such records for elevations >9,000 feet (seven in Colorado, one in Wyoming). In the statistical analyses that follow, this problem of sampling the "real" topographic and climatological diversity of the region must be considered; conclusions reached may not represent the very high elevation parts of the region, for example. Nevertheless, the author is confident that the stations chosen represent the optimum long-term station network for the area both geographically and with elevation.

Continuous Precipitation Data Descriptive Statistics

Using the precipitation data bank described above, statistical analyses of the records were carried out with four main aims in view: 1) to document the regional patterns of seasonal precipitation totals and variability in the Rocky Mountain area; 2) to examine changes in seasonal precipitation amounts and to characterize their spatial dimensions; 3) to examine the data for evidence of trends, quasi periodicities and cycles and to see if such features are characteristic of particular regions; 4) to see if there is a relationship between changing synoptic activity and seasonal precipitation amounts in the Rocky Mountain area over the secular period. Aims 1 and 2 are dealt with in this chapter, while 3 and 4 are treated in Chapters 4 and 6, respectively.

Seasonal Precipitation Totals 1941-70

In Figures 42 to 46, isopleths of seasonal precipitation totals are drawn. Detailed maps for each state are available (generally for 1931-60) (see "Climates of the States" publications, U.S. Weather Bureau) and Mitchell (1969) has given maps of average January, July and annual precipitation with the topographic factors standardized to a common level of 5,000 feet. The maps presented here are meant to represent only the broad patterns of precipitation distribution and not detailed values for particular areas. They serve as brief introduction to the other studies which follow.

Spring (April, May): Figure 42

Precipitation is lowest in (a) the Upper Colorado River Basin, (b) the Great Salt Lake Desert and Snake River Plains and (c) the northern plains of Montana, along the Missouri River Valley. Precipitation is relatively heavy in (a) the Idaho panhandle (Columbia plateau), (b) parts of Montana just east of the Continental Divide (particularly the upper Beaverhead River area), (c) along the Wasatch Front in Utah and (d) in eastern Wyoming. The precipitation maxima reflect the importance of moisture from the Pacific (particularly the Idaho and Montana maxima) and from the Gulf of Mexico (particularly the eastern Wyoming maxima).

Summer (June, July, and August): Figure 43

The distribution is characterized by two broad zones of precipitation--one extending from western Montana to the Wasatch mountains of Utah, the other marking the eastern margins of the region. In these areas precipitation is greater than five inches.

Precipitation amounts of less than 3 inches are recorded over a broad area of the Snake River Plains, Utah (excluding the Wasatch Front stations), the Upper Colorado River Basin in Colorado and an extension of this valley system into Wyoming (the Green River and Bighorn River basins).

Late Summer (July, August, and September): Figure 44

In all areas except the San Juans of southwestern Colorado precipitation is less than in "summer" months. The difference varies from one to two inches over most of the area, but the overall pattern is very similar, with lowest amounts in the central Colorado River Basin, the eastern Snake River Valley and the Bighorn Basin of Wyoming.

Fall (September, October): Figure 45

Over much of the area this is the season of lowest precipitation amounts with most stations recording less than 2.5 inches. Only two areas have relatively high precipitation--the northern panhandle of Idaho and southwestern Colorado. Over the rest of the area precipitation totals vary very little.

Winter (November, December, January, February, and March): Figure 46

Precipitation in winter months over the area is mostly in the form of snow and is expressed on the maps as water equivalent. The most striking feature of the maps is the extremely heavy precipitation totals in the relatively low mountains of northern Idaho. Precipitation amounts exceed ten inches over a wide area reflecting the dominant influence of storms from the Pacific in these winter months. Precipitation amounts decrease very rapidly eastwards, however, and most of Montana east of the Continental Divide receives less than three inches of precipitation. This is also true for most of Wyoming. Precipitation is relatively heavy along a belt from the western mountains of Wyoming and southeastern Idaho southwards along the Wasatch mountain front and its extension in southern Utah. Finally, the high mountains of Colorado also stand out as a region of heavy precipitation, though precipitation amounts are far less than in the much lower mountains of northern and western Idaho. Winter precipitation totals thus reflect the major topographic divisions of the area and the influence of Pacific storms crossing the northwestern part of the region. It should be noted that most of Wyoming and Montana receive less precipitation in the five winter months than in the two spring months which follow.

Precipitation Variability

A useful index of precipitation variability is the coefficient of variation,

$$V = \frac{s}{\bar{x}} \times 100\%$$

where $\bar{x}$ is the mean and s is the standard deviation for the period in question (in this case, 1941-70). Coefficients of variation were calculated for each season at all stations. Figures 47 to 51 show the resulting isopleths of variability.

Spring: Figure 47

Variability generally reflects the pattern of precipitation with high coefficients in those areas which receive low precipitation amounts (Colorado River Basin, Snake River Basin, etc.). Hence, as might be expected, precipitation in these regions at this time of year is not only light but extremely variable from year to year.

Summer: Figure 48

Again variability of precipitation reflects summer totals (Figure 43) with the four-inch isohyet approximately equivalent to the 45% isopleth of variability. The 60% isopleth approximates the 2 - 2.5-inch isohyets.

Late Summer: Figure 49

This is the period of maximum variability in the Snake River Valley and western Utah. Apart from this area, however, values are very similar to those in the "summer" period in spite of the lower precipitation amounts in most areas at this time.

Fall: Figure 50

Precipitation variability in this season is at a maximum over most of the area (excluding those regions mentioned above--see Late Summer). Over almost all of the area coefficients are greater than 55% reflecting the low and uncertain precipitation amounts at this time of year (Figure 45).

Winter: Figure 51

Variability is least in this season; coefficients as low as 18% are found in northwest Idaho and few stations have coefficients of more than 45%. Even in the Snake River Valley, variability averages only 30%, though in the Upper Colorado River Basin coefficients average about 42%.

In short, variability of precipitation seems to vary inversely with precipitation totals over most of the area, with maximum variability in late summer or fall and minimum variability in winter (i.e., $V = f[1/x]$). Such an inverse relationship does not appear to be simply linear where regions of very high precipitation occur (e.g., the Idaho panhandle in winter). Variability does not seem to be very dependent on station elevation except in as far as precipitation totals reflect topgraphic influences.

Decadal Seasonal Precipitation Maps (1891-1970) and Seasonal Precipitation Trends (1920-1970)

In order to investigate the spatial distribution of precipitation change over time in the study area, maps of decadal seasonal precipitation from 1891-70 to 1961-70 were prepared. The seasonal average precipitation at each station for each decade was compared with the average for 1941-70 and expressed as a deviation from the 1941-70 mean, in terms of the standard deviation for 1941-70 at each station. The value plotted at each station ($\hat{x}$) is thus a standardized or z-score thus,

$$\hat{x}_1 = \frac{\bar{x}_1 - \bar{x}}{s}$$

where $\bar{x}_1$ is the decadal seasonal average precipitation for decade 1, $\bar{x}$ is the seasonal average for 1941-70 and s is the standard deviation for 1941-70.

This procedure was carried out in order to take into account differences in variability from one part of the area to another (see above). The period 1941-70 was chosen as this is the most recent "normal" period (according to WMO definition; WMO, 1966). Isopleths separating areas of above and below normal precipitation have been drawn without oversimplifying the patterns. Care was taken to take all data points into account; however, generally at least two adjacent stations were needed before "anomalous" points would be identified by isopleths. All rejected stations (see above) were omitted from the analysis; suspect stations were used to supplement the other data points but were not used to construct isopleths unless other "acceptable" stations in the area appeared to support the suspect stations. Statistical significance of the decadal averages from 1941-70 "normals" was assessed by means of "Student's" t statistic,

$$t = \frac{\bar{x}_1 - \bar{x}}{\frac{s_1}{n_1} + \frac{s}{n}}$$

where $\bar{x}_1$ and $\bar{x}$ are the averages for the decade in question and normal period, respectively, s_1 and s are the standard deviations of the decadal average and normal period, respectively, and n_1 and n are the number of observations in the decade and normal period, respectively. Those decadal averages which were significantly different from 1941-70 averages at ≤5% level are identified on the maps, indicating those areas and periods in which the most important moisture surpluses and deficits have occurred.

In addition, seasonal precipitation records were examined for trend using the Mann-Kendall rank statistic (described above). Although the data bank contained records of varying length, all records span the 51-year period 1920-70 and so trends during this period were examined. Only those stations which were not rejected or "suspect" were chosen for this part of the study. Those records which gave statistically significant τ values at the 10% level or above (two-tailed test) are listed in Table 16. All τ values are mapped for each season in Figures 61, 71, 72, 82, and 91 and these are commented on after discussion of the decadal average maps, below.

Spring: Figures 52, 56, 60, 62, 66. 77, 83, and 87

In most areas precipitation in the period 1941-70 was relatively heavy compared to the previous thirty-year period (1911-40). Increases were most marked in Idaho, Montana and northern Wyoming whereas in Utah and Colorado about half of the stations showed increases and half decreases. Overall, more than 70% of stations with 60 years or more of record showed increases 1911-40 to 1941-70.

The relatively high averages for 1941-70 are mainly the result of heavy spring precipitation in the 1940s and 1960s (Figures 77 and 87). In the 1940s (Figure 60) precipitation was the maximum for sixty years (1911-1970) over most of Colorado, southern Wyoming, southeastern Idaho and the Idaho panhandle. In the 1960s (Figure 60) maximum precipitation was recorded at stations in central Utah, southern Montana, northern Wyoming and the upper Snake River Valley of Idaho. In most areas, therefore, the period 1941-70 was anomalous in terms of spring precipitation amounts over the last sixty years. However, precipitation in the 1890s and 1900s was relatively high at most stations with records for this period (Figures 52 and 56), the main exceptions being Idaho stations in the 1900s. As a result the few records for the thirty-year period (1881-1910) indicate relatively high spring precipitation amounts compared to the subsequent sixty years.

Precipitation was well below average over much of the area in the 1930s and over Montana, Idaho, and Utah no other decade in the entire period 1891-1970 was as dry. In fact, ∿70% of stations in these states recorded lowest spring precipitation in the 1930s. The most significant "anomalies" were in northern Idaho and western Montana (Figure 73). In

Wyoming and Colorado, however, no single decade stands out as a clear minimum.

When precipitation trends for 1920-70 are examined (Figure 61) there is a general pattern of negative values (i.e., an area of decreasing spring precipitation) south of a line across southern Idaho and the northern Wyoming border (excluding the Yellowstone Park area), and positive values (increasing spring precipitation) to the north. Isolated departures from this pattern do occur, particularly in the south, the largest anomaly occurring in part of the Colorado River Basin in Utah. Although the distribution of positive and negative values suggests a latitudinal influence it should be noted that the area of positive values corresponds quite well to that part of the study where station elevations are generally low (less than 4,000 feet--see Figure 41). The main exceptions to this are the stations in southwestern Montana and adjacent parts of Idaho and Wyoming. The average station elevation north of the main zero line (across south Idaho and north Wyoming) is 3,800 feet and south of the line is 5,400 feet. However, this is a problem of interpretation in all cases where differences occur between "north" and "south"; all the very high elevation stations in the study area are in the southeast (Figure 41) and so it is frequently difficult to assign a given pattern to factors of geographical location or influences of elevation. For example, stations with precipitation maxima in the 1940s (apart from the Idaho panhandle area) are generally those above 5,500 feet, but whether this is a major factor in causing this period to be a maximum in the area (e.g., due to a particularly strong orographic effect in those years) is debatable. However, in the case of spring precipitation trends, it seems likely that the latitudinal factor is predominant and that positive tau values to the north represent a slightly increased frequency of storms crossing the area from the Pacific in the latter part of the period. This is hardly a major change though, considering the very few significant tau values (Table 16).

Summer 1891-1970: Figures 53, 57, 63, 67, 74, 78, 84, and 88

The summer precipitation record is quite different from other seasons being characterized by three periods of high precipitation which affected different parts of the study area to different extents. These three periods are, in order of increasing importance, the 1920s, 1940s and 1960s (Figures 67, 78 and 88). Considering the entire period 1891 to 1970, precipitation was at a maximum in the 1960s over a wide area including the Snake River Basin, northwestern Utah and the Colorado River Basin, northwestern Wyoming and southern Montana (Figure 70). In northern Idaho and most of Montana, precipitation was heaviest in the 1940s whereas maximum precipitation was reached in the 1920s in the Colorado mountains and in eastern Wyoming. It should be noted, however, that precipitation was well below "normal" at many stations in the northwestern half of the region in the 1920s, particularly western Idaho and western Montana (Figure 67). These patterns may reflect the dominant influence of storms from the Pacific across northern parts of the region in the 1940s whereas in the 1960s storms may have entered the region more from the Great Basin area and the southwest, to be channeled along the Snake River and Colorado River basins (see Chapter 6). The 1920s maxima over eastern parts of the region are probably related to the predominant influence of storms from the Gulf of Mexico during this period.

Because of the relatively high precipitation totals over much of the area in the 1940s and 1960s, other decadal maps show a pattern of predominantly below 1941-70 average values. All stations in the 1890s and over 75% of stations in the 1900s were below "normal" (Figures 53 and 57). In the 1910s, precipitation was also generally below average except for the Colorado River Basin in Utah and parts of the southwestern Colorado mountains (Figure 63). In the 1930s and 1950s precipitation was well below average over virtually the entire area. In particular, the 1930s were extremely anomalous in the northwestern part of the region; 70% of stations in Montana and almost all stations in the Idaho panhandle showed highly statistically significant differences from 1941-70 averages (Figure 74). This is similar to the spring pattern of the 1930s, but even more anomalous.

In view of these facts, it is clear that the 30-year "normal" for 1941-70 is probably the most anomalously wet period in this area for at least the past 80 years. If the historical data for 1866-1890 are considered (see Chapter 2, Table 3) it is clear that most stations were below or only slightly above 1951-60 averages during this period (virtually no Rocky Mountain data are available for the period 1851-65). As the 1950s were markedly below the 1941-70 averages in all areas except the extreme northern parts of Idaho and northwestern Montana it is very probable that summers in the period 1941-70 were, on average, wetter than for any comparable period in at least the last 110 years.

Because of the anomalous character of the 1941-70 period, summer tau values (Figure 71) are positive over almost the entire region (85% of stations), indicating a tendency for increasing precipitation amounts from 1920-1970. Furthermore, eighteen stations showed statistically significant increases, particularly in Idaho and the bordering areas of adjacent states. One third of Idaho stations showed a statistically significant increase in summer precipitation. Only two stations, one in northern Montana and one in western

Colorado showed a significant decrease in summer precipitation.

Late Summer

No decadal maps were examined for this season but tau values were computed to examine trends over the period 1920-70 (Figure 72). Negative tau values are found in nearly all parts of Utah, Wyoming and Colorado (80% of stations in these states) and eight stations show statistically significant decreases in precipitation. A second major area of negative values is found in Montana just east of the Continental Divide and along the upper Missouri River, though only one station in this area shows a statistically significant decrease in precipitation. Most of Idaho excluding the southeastern corner, Montana went of the Divide and southern Montana show positive tau values, though only one station shows a statistically significant increase in precipitation.

Late summer trends are thus quite different from those of the "summer" period in most areas except Idaho and southern Montana. The late summer map is similar to that of spring (with the addition of more negative values over central and northern Montana), fall (obviously reflecting the duplication of September values) and rather surprisingly, winter months. In short, all seasons except summer proper show similar features with decreases over most of the area except Idaho and parts of Montana. This is discussed further below.

Fall: Figures 54, 58, 64, 68, 75, 79, 85, and 89

As in other seasons already discussed, fall precipitation was relatively heavy in the 1910s and 1920s (in all areas except northern and western Idaho) and in the 1940s (except in northwestern Colorado) (Figures 64, 68, and 79). The 1960s were also relatively wet over much of the region, the principal exceptions being northern Idaho, northern Montana, and central Wyoming (Figure 89). The 1950s on the other hand were extremely low in fall precipitation in all areas except the Idaho panhandle (where the 1920s were lowest) and part of the Colorado River Basin in Utah (where the 1930s were lowest) (Figure 81). Excluding these two areas over 85% of stations recorded in the 1950s their lowest decadal totals for at least 60 years (the 1900s were even drier in some cases). Data for the 1890s are scarce, particularly in the eastern half of the region, but it appears that Idaho, Utah, and western Montana, at least, were above average. In short, the 1890s(?), 1910s, 1920s, 1940s and in some areas the 1960s were relatively wet (particularly the 1910s), and the 1900s, 1930s and 1950s were mostly dry (particularly the 1950s).

The distribution of tau values in the fall (Figure 82) is very similar to that in late summer though the area of positive tau values has decreased. Zones of positive values along the central mountains of Utah and central Colorado are also apparent. However, the only state with statistically significant trends is Wyoming where four stations show marked decreases in fall precipitation.

Winter 1891-1969:[1] Figures 55, 59, 65, 69, 76, 80, 86, and 90

Over much of the western region--Idaho, western Montana, western and central Utah and western Wyoming--precipitation in the period 1941-69 was considerably above the average for the previous 30 years. This is mainly due to relatively high precipitation receipts over Utah, western Montana and western Wyoming in the 1940s and 1950s, and over Idaho in the 1940s and 1960s. The pattern of anomalously high precipitation in the 1950s is almost the inverse of that in the 1960s when precipitation was below average in almost all areas except eastern Idaho, central and southern Montana and southern Utah. However, it is of interest that about 24 stations received their highest decadal average of winter precipitation for 50 years in the 1960s. Further examination of these stations indicates they are all at relatively high elevations in each state. The four stations in Wyoming and Colorado, for example, are amongst the eight highest elevation stations in these states. A similar pattern though with more low elevation stations included is also seen in Idaho and Utah and to some extent in Montana. Further study of high elevation precipitation receipts in the 1960s would be of interest, particularly in view of suggestions that the climate of the 1960s in the northern hemisphere may have resembled that of the Little Ice Age (Lamb, 1969; Sanchez and Kutzbach, 1974).

For the period 1891-1910, although data are scarce, it appears that precipitation was relatively heavy (above 1941-70 averages) in both decades for most of the area. At a number of stations, in fact, maximum decadal precipitation for the 70 (or 80) year period was recorded in these decades. This

---

[1]Although "decadal" averages are referred to below, winter seasons were examined in nine-year intervals such that data for the "1890s" extends from November 1891-March 1892 to November 1889-March 1890, etc. It is felt that this is representative of each decade and so the term decadal has been retained for convenience. The normal period for comparison extended from November 1941-March 1942 to November 1969-March 1970 (i.e., 29 years).

is discussed further in the section on winter precipitation trends (see below).

Over the last 70 years, one pattern of precipitation deficit has frequently recurred. This is the pattern seen in the 1910s, 1920s, 1930s, and 1950s with low precipitation amounts centered over southern Idaho and extending to different extents northward over northern Idaho and western Montana, eastward over western Wyoming and southward over northern and central Utah. It is interesting that a pattern similar to this is shown by Sellers (1968) as the dominant December eigenvector of precipitation anomaly over the western United States (actually the inverse of Sellers' example). A similar pattern is shown by the second eigenvectors for November and March, and certain features of the pattern can be seen in eigenvectors for other winter months. As Sellers' analysis covered only the period 1931-1966, it appears that this anomaly pattern has been common for at least the past 70 years and perhaps longer (cf. LaMarche and Fritts, 1971). It is also of interest that the low precipitation receipts centered in the area of Idaho were lowest in the 1920s and 1930s and, although the pattern has recurred, the magnitude of the deficit has been less in recent decades. Over 70% of Idaho stations not suspect or rejected, for example, recorded lowest winter precipitation (1921-1969) in the 1920s (55%) or 1930s (18%). As a result, winter precipitation in this area has shown a general upward trend over the past fifty years, a pattern which is quite different to other parts of the region (Figure 91). Three Idaho stations, in fact, show significant increases in winter precipitation as shown by tau values for the season (Table 16). In general, however, values are negative over most of the region and the overall distribution resembles that of fall and late summer, although the zone of positive values across southern Montana is absent. Negative values predominate in Montana, Wyoming, western Colorado and Utah (70% of stations in these states). Twenty-five percent of stations south and east of the main zero line (circumscribing Idaho and parts of western Montana) show statistically significant decreases in precipitation (Table 16).

Although tau values discussed here refer to the 1920-70 period, heavy precipitation over most of the area in the 1890s and 1900s (and over much of the area in the 1910s) suggests that a general decrease may have affected the region for the last 80 years. Furthermore, consideration of the nineteenth century data, Chapter 2, Table 3) indicates that precipitation was generally above 1951-1960 averages for the 1870s and 1880s at least and it therefore seems likely that winter precipitation over much of the region (except Idaho) has been considerably less in recent decades than was characteristic of the late nineteenth and early twentieth centuries.

Summary

The main points outlined in the above discussions have been incorporated in summary form in Table 17. However, by definition this summary omits much detail and the serious reader should consult the relevant sections above rather than rely implicitly on this overview.

CHAPTER 4

CONTINUOUS DATA II: POWER SPECTRUM ANALYSIS

Introduction

Most statistical tests used to analyze data assume that successive observations in a series are independent of one another (i.e., the series is assumed to be random or represent a random process). However, quite often in time series analysis (observations arranged sequentially with respect to time) such an assumption is invalid and successive observations are not statistically independent due to the presence of trends, quasi periodicities, cycles or simple persistence. In fact, it is precisely these nonrandom factors which are of interest in studies of climatic change. Various tests are available to examine specific types of nonrandomness (e.g., the Mann-Kendall rank statistic for tests of randomness versus trend alternatives or the autocorrelation coefficient for tests of persistence in a series) but no one test is available which can fully describe or identify all forms of nonrandomness in a climatological series. However, the approach which comes closest to such an ideal is that of power spectrum analysis with which one can identify a number of different types of nonrandomness in a series.

Power spectrum analysis "is based on the premise that time series are not necessarily composed of a finite number of oscillations, each with a descrete wavelength (as one tacitly assumes when one applies classical harmonic analysis) but rather that they consist of virtually infinite numbers of small oscillations spanning a continuous distribution of wavelengths" (WMO, 1966). The resulting spectrum thus represents the variance associated with different wavelengths, ranging from infinite wavelength (linear trend) to the shortest resolvable wavelength (twice the interval between successive observations in a series).

Computational Procedure

Briefly, the following procedure was followed in obtaining the power spectral estimates. Computations were carried out by means of a modified BMD program (Dixon, 1973) coupled to a plot routine.

All series were initially standardized by subtracting the mean and dividing each term by the standard deviation. Autocovariances were computed for each series according to the noncircular definition,

$$C_\tau = \frac{1}{n-\tau} \sum_{i-1}^{n-\tau} x_i \, x_{i+\tau}, \quad \tau = 0,1,2, \ldots m$$

where C_τ is the autocovariance at lag τ and m is the maximum number of lags.

Raw spectral estimates ($\hat{s}_k$) were then obtained from the autocovariance as follows,

$$\hat{s}_k = \frac{2\Delta t}{\pi} \sum_{\tau=0}^{m} \varepsilon_\tau \, C \, \frac{\cos k\tau\pi}{m}$$

where

$$\varepsilon_\tau \begin{cases} 1 & 0 < \tau < m \\ 0.5 & \tau = 0,m \end{cases}$$

Δt is the time interval and k is the harmonic number (0,1, 2,...,m). Smoothing of the spectral estimates by "hamming" was carried out thus,

$$s_o - 0.54\hat{s}_o + 0.46\hat{s}_1$$

$$s_k = 0.23\hat{s}_{k-1} + 0.54\hat{s}_k + 0.23\hat{s}_{k+1}, \quad 0<k<m$$

$$s_m = 0.54\hat{s}_m + 0.46\hat{s}_{m-1}$$

Finally, the smoothed spectral estimates were plotted against frequency. Periods associated with each estimate are simply the reciprocal of frequency, hence the shortest resolvable period (p) corresponds to the spectral estimate at a frequency of 0.5 (a periodicity of 2 years) which is twice the interval between successive observations. Conversely $p \to \infty$ as $f \to 0$.

Procedure for Evaluating Nonrandomness in a Series

In this section, the steps followed in evaluating and defining the nature of nonrandomness in the precipitation time series are outlined.

Initially autocorrelation coefficients were computed for different lags by standardizing the autocovariances with respect to the autocovariance at lag 0 (i.e., with respect to the variance of the series). The autocorrelation coefficient at lag 1 (r_1) can be used to evaluate persistence in a series which is represented by relatively high positive autocorrelation at small lags. Anderson (1941) has shown (for a circularly defined autocorrelation coefficient) that in a random series with a normal (Gaussian) distribution r_1 is approximately normally distributed with a mean of $(-1)/(n-1)$ and variance $n-2/(n-1)^2$. Confidence limits can thus be defined for r_1 based on probability points of the Gaussian distribution curve such that,

$$(r_1)_\alpha = \frac{-1 \pm Z_\alpha \sqrt{n-2}}{n-1}$$

where Z is the standard normal variate corresponding to significance level α (Joseph, 1973). Generally in a climatological series most alternatives to randomness would tend to increase the value of r_1 (WMO, 1966) and hence a one-tailed significance test is used with the plus sign chosen in the numerator. However, if r_1 is negative, the series contains a high frequency oscillation and a one-tailed test with a negative sign in the numerator would be chosen.

If r_1 does not differ significantly from zero, then the series can be considered to exhibit no persistence, i.e., successive observations are statistically independent of one another. In such cases the appropriate null continuum with which to test the power spectrum for statistical significance is that of "white noise" or a horizontal straight line equal to the average value of all m+1 "raw" (unsmoothed) spectral estimates. Confidence limits for such a continuum are discussed below.

If r_1 is $>(r_1)_\alpha$, the series contains persistence ("red noise," Gilman *et al.*, 1963) such that the value of one point in the series is dependent on the preceding value, etc. If the coefficient r_τ approximates to $(r_1)^\tau$, that is,

$$r_2 \sim r_1^{\,2}, \quad r_3 \sim r_1^{\,3}, \quad \text{etc. . . .}$$

the persistence is first-order linear Markovian and this must be taken into account in defining the null continuum with which to test the power spectrum (i.e., a "red noise" null continuum is the appropriate hypothesis). If $r_1 > (r_1)_\alpha$, but $r_k \neq r_1^{\,k}$ then the persistence is more complex than simple first-order linear Markovian. In such cases the red noise null continuum is still used for the first half of the spectrum but the white noise continuum is used for the other spectral estimates (Jagannathan and Parthasarathy, 1973). The "red noise" continuum can be estimated by evaluating the following equation for various choices of harmonic number k (where k = 0,1,2, ..., m).

$$\mathcal{S}_k = \bar{s} \left(\frac{1-r_1^{\,2}}{1+r_1^{\,2} - 2r_1 \cos \frac{\pi k}{m}} \right)$$

where $\bar{s}$ is the average of all k "raw" spectral estimates, (s_k) (i.e., the "white noise continuum), r_1 is the autocorrelation coefficient at lag 1 and m is the number of lags. The resulting values are plotted on the graph of spectral estimates against frequency and a smooth curve is drawn. Various examples of this continuum for r_1 from 0.1 to 0.9 are given in Gilman et al. (1963) and WMO (1966).

Confidence limits for the spectrum of a series are obtained by multiplying values of the null continuum by $\chi^2_{1-\alpha}/\upsilon$ and χ^2_{α}/υ (for upper and lower confidence limits respectively), where υ is the appropriate number of degrees of freedom and α is the chosen probability level. In a series of N values with maximum lag m,

$$\upsilon = \frac{2N - \frac{m}{2}}{m}$$

As the confidence limits are a fixed multiple of each value in the null continuum (>1 for upper limits, <1 for lower limits), it is a relatively simple matter to compute the appropriate confidence limits for any given white or red noise continuum. These are plotted on the graph and can be compared with each spectral estimate to identify those significantly different from the null continuum. Further analysis depends on the features shown by the spectrum but might involve low-pass filtering for analysis of trends or band-pass filtering for analysis of periodicities and cycles (see below). If features are noted in the spectrum which have not been identified hitherto and which are not immediately explicable in physical terms, more stringent levels of statistical significance need to be applied (as discussed in WMO, 1966, p. 41). Generally, however, it was felt that while such precautions would be called for in the evaluation of an apparently significant feature in the spectrum of one or two series, if the feature was identified in a large number of records in a geographical region then this would be of equal importance in evaluating its significance, whether or not the feature had been noted hitherto (cf. Landsberg et al., 1963). This is the philosophy also noted by Blackman and Tukey (1958) and followed by Bharghava and Bansal (1969).

Analysis of Seasonal Precipitation Data

Although all the seasonal records being considered in the study area were at least fifty years in length (i.e., each comprised a series of at least 50 points), this is barely adequate for meaningful time series analysis. Spectral resolution is directly dependent on the number of lags, m, such that the larger the number of lags selected, the greater the number of spectral estimates (from a frequency in the range 0.5 to∞). It is obvious then that the higher the value of m which can be used, the more meaningful will be the spectral estimates. This is particularly true in terms of the lower frequency estimates where resolution is much coarser than in the case of higher frequency estimates (because frequency is the reciprocal of the period). However, an equally important consideration concerns the tests of significance of the spectral estimates and the degrees of freedom available for the test. As explained above, the 5% and 95% significance levels are defined as the white or theoretical red noise spectrum multiplied term by term by the χ^2/υ values corresponding to the desired confidence limit. As the number of degrees of freedom

$$\upsilon = \frac{2N - \frac{m}{2}}{m}$$

it is obvious that when m is close to N/2 (i.e., the uppermost limit of m), υ is correspondingly small. Conversely when m is only a small fraction of N, υ is large and the χ^2/υ estimates are small resulting in significance levels relatively close to the spectral estimates. When υ is small, χ^2/υ is large and hence confidence limits for the spectral estimates are correspondingly wide.

Thus a balance has to be found between spectral resolution and degrees of freedom. Generally a value of one-third the number of points in a series is chosen as the maximum number of lags (WMO, 1966) though a smaller fraction is desirable. In the first analysis (I) conducted, lags up to a maximum of ~N/7 were chosen. Subsequently (analysis II) lags up to a maximum of ~N/4 were selected and the results were compared. Furthermore, only those records greater than 65 years in length were analyzed in order to have as good a spectral resolution as possible.

Analysis I

In the first analysis all those seasonal records which were not suspect or rejected were selected. These are listed in Table 18 and summarized in Table 19. Two hundred and ten seasonal time series, with an average length of 73 years, and from 63 stations were thus chosen.

Results

Initially all series were examined for normality by computing the coefficient of skewness (α_3) for each series

$$\alpha_3 = \frac{\mu_3}{\sigma^3}$$

where σ is the standard deviation and μ_3 is the third moment about the mean. In nearly all cases significant skewness was not noted and in no cases was it deemed necessary to adjust the series to a more normal distribution. All series were then examined for lag one autocorrelation coefficients significantly different from zero. As explained above, such a case would indicate persistence and require a different procedure for evaluating the significance of the spectral estimates. Of the 210 series examined, 22 (10%) showed statistically significant positive lag one autocorrelation coefficients (Table 20). However, there was a strong tendency for winter and fall seasons to show high $+r_1$ values compared to other seasons. Thirty-three percent of winter and 14% of fall records examined showed persistence compared to 8% and 4% for summer and late summer records respectively. No spring records showed statistically significant positive r_1 values.

Perhaps more interesting was the fact that over half of the records examined showed <u>negative</u> lag one autocorrelation coefficients and nine of these were statistically significant. This indicates that many stations exhibit quite a regular high frequency oscillation (in this case a quasi-biennial oscillation) such that relatively high precipitation one year tends to be followed by relatively low precipitation the following year and <u>vice versa</u>. Unlike the high <u>positive</u> lag one autocorrelation values, the significant negative values were associated mainly with summer and late summer months (Table 20). It is also of interest that this feature shows a distinct geographical distribution with summer records from the western slope of Colorado and the Upper Colorado River Basin of Utah tending to exhibit the oscillation. This is discussed further below. Maps of all the lag one autocorrelation coefficients computed for different seasons generally do not show any distinct geographical distributions. However, as mentioned above there is a tendency for statistically significant values of r_1 to cluster in different areas at different seasons of the year. In winter, for example, positive values of r_1 are mainly in northern parts of the region, particularly in Montana where five of the eight stations examined showed statistically significant (positive) r_1 values.

For the majority of cases where r_1 values significantly different from zero were not found, spectral estimates were tested against the white noise null continuum and appropriate confidence limits were computed.

The results of all these tests of the significance of spectral estimates are given in Tables 21 to 25. Overall only about 30% of records showed a statistically significant (>95% level) spectral peak at any frequency although this varies by season, ranging from 42% of summer records to only 19% of winter records. All principal spectral peaks are summarized in Table 26 and from this three main periods can be identified. These are 2-2.2 years, 2.2-4 years and >20 years (long period trend), together accounting for 67% of all statistically significant spectral peaks observed. Other periods cluster around 2.7-3.00 years and 5.0-6.0 years, accounting for a further 16% of the main spectral peaks. Some examples of records exhibiting these periodicities and their corresponding power spectra with confidence limits are given in Figures 92 to 94.

Discussion

A periodicity of approximately 2 to 2.2 years has been noted in a variety of atmospheric phenomena (Berlage, 1957; Landsberg, 1962) most notably in the wind and temperature field of the equatorial stratosphere (Reed *et al*., 1961; Angell and Korshover, 1962) but also in mean monthly temperatures at widely separated parts of the earth's surface (Landsberg *et al*., 1963). An approximately two-year cycle in Indian precipitation has also been noted by Bhargava and Bansal (1969) and other studies are reviewed by Landsberg (1962). It must be realized, however, that the periodicity noted here is based on seasonal, not annual contiguous standardized monthly precipitation data (WMO, 1966).

The most interesting feature of the periodicity in this study is the marked tendency for it to occur mainly (a) in summer and/or late summer months and (b) in a restricted geographical area corresponding roughly to the inland drainage basins of Utah and the Upper Colorado River Basin (eastern Utah/western Colorado). Engelen (1972a, 1972b) has also noted a two-year cycle in the "hydrological subsystems" of the Upper Colorado River Basin, most notably in net soil moisture recharge during *winter* months, maximum *winter* snowpack water equivalent and *total* discharge of the Colorado River at Lee's Ferry during the snow melt period (March-August). Engelen also noted a similar cycle in mean annual air temperature and relative humidity at Grand Junction but did not examine summer precipitation data for any stations. It is thus very interesting that the analysis of summer precipitation data presented here also points to an approximately two-year periodicity which may help to explain the high frequency variation in Colorado River flow that Engelen notes and which is in phase with summer precipitation in the area (Figure 95).

To investigate whether this period is widespread in the region, all other summer and late summer stations records, regardless of record length and including some that were suspect, were analyzed. Table 27 presents the results of this further study and it is clear that the periodicity is noted at a large number of stations in the region (Figure 96) and spanning a wide elevational range (4,342 feet at Utah, 5,610 to 9,322 feet at Colorado 7656). The reasons for this are not obvious; Engelen suggests that "regional atmospheric subsystems are important factors in the moisture exchange balance of the area and his thesis can be summarized as follows. "Moisture which evapotranspirates into the overlying atmosphere is not completely carried away during the summer and early fall but . . . stays partly in place . . . this is highly improbable in wide open areas with large movements of air masses, where the moisture may be carried far away within hours or days. However, in rugged mountain areas like the Colorado and New Mexico Rockies there is a much greater chance that rotating cells of moist air remain at the spot in the sheltered valleys between the high mountain ridges if continent-wide movements of air masses move over the mountain ranges." On the basis of this hypothesis he goes on to construct a model of moisture exchange in the area. However, if such an effect <u>were</u> important in controlling summer precipitation amounts <u>then</u> it would be expected that similar periodicities would be found in other "rugged mountain areas" and this is clearly not the case. Evidently there is some factor unique to this area which has not at present been isolated. Further analysis to the south in Arizona and western New Mexico may help to delimit the areal extent of this phenomena and hence point to the causative factors involved. At present, however, an adequate hypothesis is lacking.

The less important group of spectral peaks at 3.3-4.0 years show no coherent geographical distribution but most are found in summer and/or late summer records. It is possible that this is a higher harmonic of the 2.0-2.2 year feature already discussed.

In an earlier section on nineteenth century precipitation data (Chapter 2) it was noted that winter records in the Northwest and Southwest showed a periodicity of from 4 to 6 years. No comparable period is found in the winter precipitation records examined for the Rocky Mountain region. A few records do exhibit a 5 to 6 year periodicity but these are all in the summer to fall seasons. However, as the apparent periodicity in the nineteenth century records was from a different, albeit adjacent, geographical area, the absence of the feature in Rocky Mountain records may not be significant. Further analysis of records from Oregon and Washington, and Arizona and New Mexico are needed before the reality of this feature over longer time periods can be confirmed or denied.

A number of records showed a low frequency periodicity indicating a marked long period oscillation or trend in the record. These records are Idaho 6542 (summer and winter); Montana 4055 (spring), and 5015 (summer and late summer); Utah 5826 (spring and summer), 5402 late summer, 2101 and 6357 (fall and winter); and Wyoming 2995 (late summer). Comparison of these records with the analysis of randomness against trend alternatives outlined above, using the Mann-Kendall statistic (Chapter 3), indicates that only the summer and winter records at Idaho 6542 also had statistically significant statistics. However, the average record length for the power spectrum analyses was considerably longer (~73 years) than the period for which τ values were computed (1920-70). Furthermore, the Mann-Kendall statistic is used to identify trends in the data whereas the spectrum may indicate only a very long period oscillation, the exact period of which can not be exactly identified within the resolution of the spectrum. Thus, with a series of 80 points, for example, and using 12 lags, only three spectral estimates are possible at the very low frequency end of the spectrum--at periods of 12, 23, 8, and ∞ years. This lack of low frequency resolution is entirely a function of the short record length and consequent small number of lags leading to few spectral estimates as explained above. Hence, in the example just cited, if a record has a period of >25 years, the statistically significant spectral peak may indicate an <u>apparent</u> frequency of ∞ (trend) but in reality the periodicity may be considerably shorter than the total record length. To investigate this further, all records with statistically significant low frequency spectral peaks were subjected to low pass filtering to examine wavelengths longer than 15 years ($\sigma_G = 2.5$). This had the objective of "smoothing out" higher frequency oscillations while allowing low frequency effects (periods >15 years) to be seen more clearly without appreciable diminution of amplitude. Weights were computed according to the Gaussian ordinate method outlined in WMO (1966). Rather than discuss each record in detail a few brief comments about those of most interest will be made:

<u>Idaho 6542 (winter and summer)</u>. Apart from a minor peak centered on 1908, both records show an upward trend. In

winter this amounts to an increase of approximately one inch (+35%) since 1920. In summer the extraordinary nature of the 1960s is clearly shown; in fact for the entire period 1893-1970, summer precipitation has only exceeded five inches ($\bar{x} + 1.7\sigma$) five times; four of these occasions occurred between 1963 and 1970 (1963, 67, 68 and 70).

Montana 5015 (summer and late summer). These records illustrate a feature which is common to many of the records regardless of season; namely the tendency for precipitation to decrease to a point centered between 1930 and 1935 and then gradually rise thereafter, though at a rate generally less than the decrease to the early 1930s. The winter record for Utah 6357 also illustrates this feature well. Where the record is long enough to identify peaks and troughs, a period of about 60 years is suggested but no great reliance should be placed on this estimate; in other cases the period is evidently extremely low frequency and far exceeds the record length. In view of the relatively short nature of records in the western U.S. this is clearly not the area where one would expect to successfully isolate low frequency oscillations. Furthermore, there are only a few stations exhibiting "significant" long-period periodicities. With $p = 5\%$, 5 "significant" cases in 100 could be expected due to chance alone and this is clearly very close to the actual numbers observed in each set of seasonal records. As there is no favored geographical area for the "significant" station records, little real significance can be attached to these records.

Analysis II

To investigate whether the main spectral estimates isolated were realistic in view of the cumulative problems of short record length and small lags used, further analyses using lags of up to N/4 were carried out on all of the summer and winter records examined in Analysis I (81 records in all). Tables 28 and 29 summarize the main spectral peaks identified in each of the analyses. Although there are a few new peaks identified in the second analysis and of course the spectral resolution is better, the two analyses are very similar suggesting that future analyses would profit from a choice of more lags. This would help to identify the main periodicities, particularly at low frequencies and there is apparently little loss of significance in spite of the lower number of degrees of freedom.

Summary

Power spectrum analysis of seasonal precipitation data has revealed few statistically significant nonrandom features which are characteristic of wide areas or of particular seasons. The most interesting result is the marked 2.0 - 2.2 year cycle of summer and/or late summer precipitation in eastern Utah and western Colorado which was shown to be characteristic of records ranging in length from 63 to 86 years. This supports the general idea of a two-year cycle in the moisture balance of the area (Engelen, 1972a, 1972b) but reveals a new dimension in that summer precipitation had not previously been analyzed directly. Few long-term records (≥65 years) revealed a low frequency oscillation or trend (only 7% of records examined) and other outstanding frequencies were generally only common to a few stations in widely dispersed geographical areas; their real significance is very questionable. Further work is suggested along two main lines. Firstly, the areal extent of the quasi biennial oscillation in summer precipitation centered in the Upper Colorado River Basin needs to be delimited more precisely and, in particular, the years of high and low precipitation should be compared as to the circulation patterns characteristic of each. Other parameters could also shed light on the causes for this phenomenon; in particular, atmospheric pressure at Grand Junction should be examined for a similar periodicity. Secondly, it would be of interest to study the long-term temperature records for the area by power spectral techniques and to compare the precipitation and temperature records by cross-spectral methods. In this regard, a comparison of winter precipitation data from the Northwest and Southwest parts of the region might help to elucidate the inverse relationship between precipitation in those areas already noted (Chapter 2).

CHAPTER 5

DENDROCLIMATOLOGY AND THE CLIMATIC RECORD

As noted in Chapter 1, most studies relating to the climatic history of the western United States have been based on dendroclimatic reconstructions (Douglass, 1928; Schulman, 1956; Fritts, 1965; Glock and Argeter, 1966). In recent years the development of electronic computers has facilitated the use of multivariate statistics to untangle the often complex relationships between climate and tree growth (Fritts *et al.*, 1971; Fritts, 1974). These techniques, coupled with more detailed physiological studies of the relationships between climate and tree growth (Fritts, 1966), have allowed much more meaningful climatic interpretations of past tree growth patterns. An excellent review of the "state of the art" has been provided by Fritts (1971) and detailed discussion of the subject here would be superfluous. However, it should be pointed out that in most dendroclimatic studies published to date there have been very few attempts to verify the inferred climatic information. Some exceptions will be noted here.

Fritts *et al.* (1971) attempted to reconstruct pressure anomaly maps for the years 1700-1899 and selected pentad averages were published. Verification of these reconstructions in the Atlantic sector was restricted by the fact that the reconstructions were for entire seasons whereas comparable historical pressure data (Lamb and Johnson, 1966) were for monthly means. In general the results indicated that the historical position *or* strength of the principal centers of action in the North Atlantic could be assessed but not both position and strength. LaMarche (1974) has inferred a relationship between temperature and tree growth increments of Bristlecone pine near tree line in the White Mountains of California. He shows a synchrony between mean annual temperature in central England (Lamb, 1966) and tree growth at upper tree line in California which he suggests is evidence that "long-term climatic anomalies in these sectors tend to be in phase." Using the tree growth records from both the lower and upper tree line (which relate principally to variations in temperature and precipitation respectively) LaMarche goes on to reconstruct periods of warm-moist, warm-dry, cool-moist and cool-dry conditions and attempts to relate these periods to records of glacial advances in the Sierra Nevada and Rocky Mountain regions. From this study it is inferred that the period from approximately 1870 to 1945 was "warm-moist," and such conditions have only occurred for any length of time once before during the last 1,100 years--between approximately 1055 and 1130 A.D. (820-895 B.P.). The most recent decades were characteristic of a warm-dry period similar to that which followed the earlier warm-moist interval and lasted from approximately 1130 to 1430 A.D. (520-820 B.P.).

Stockton and Fritts (1973) have used tree growth data to reconstruct past levels of Lake Athabasca, northeastern Alberta, where instrumental records only extend from 1935 to 1967. Their results indicate that low lake levels occurred around 1868, 1877, 1887, 1918, 1930, and 1945. Periods of relatively high water level occurred around 1881-1887, 1892-1913, and 1921-1924. They note, however, that the mean lake levels for 1810-1967 are not greatly different from those of the recorded period from 1935 to 1967, though large changes in variance have occurred. Comparison of the reconstructions with historical climatic data (notably the study by Thomas [1965] for the Canadian Prairies) shows more similarities than dissimilarities and it is concluded that "no significant conflict exists between known records of climate and the reconstructed water levels."

Perhaps the most comprehensive attempt at reconstruction and verifying climatic variations over the last 500 years is that of Fritts and Blasing (1973) based on work by Blasing (1975). Although the entire study is not yet published it is clear that reconstructions of seasonal pressure anomaly patterns and their associated temperature and precipitation anomalies are generally good in terms of distribution of the sign of the anomaly, but the actual magnitudes are not always accurate. It was also shown that quite reasonable estimates were possible even for point data. In the opinion of this author, this is the first meaningful attempt to adequately test and verify dendroclimatic reconstructions with independent data, and further work along these lines promises to shed more light on the climatic variability of the past.

Comparison of Historical Precipitation Data and Dendroclimatic Reconstructions

Although this study was not directed at the testing of dendroclimatic reconstructions for the western United States, much of the information presented herein is obviously pertinent to dendroclimatology. Because so few attempts at verification of reconstructions for the western United States have been made, it was felt that a rough indication of the success or otherwise of the reconstructions could be given. However, it must be emphasized that this topic (verification of dendroclimatic reconstruction) demands a detailed study by itself (cf. Blasing, 1975) and so the section which follows must not be considered a definitive critique of the subject. It is merely a general comparison of historical precipitation data and tree growth indices. Furthermore, the study is based on the work of Fritts (1965) in which stepwise multiple regressions were used to obtain climatic information from tree growth increments and hence indicate general climatic conditions from 1501-1940. Clearly more recent work has further clarified the relationships between climate and tree growth but much of this awaits publication.

Although the precise response function of individual trees to climatic variables is unique, a number of studies have indicated that the principal seasons with which tree ring indices correlate more highly are the fall, winter, and spring periods preceding growth (Fritts *et al.*, 1965; Fritts, 1971). This is particularly true for trees in arid and semi-arid areas where soil moisture may be high as a result of relatively wet fall, winter, and spring periods. However, in nonarid regions precipitation during the growing season may be important. A similar relationship has also been seen along an altitudinal gradient: Fritts (1966) notes that "at least for conifers on semiarid sites . . . growth . . . at low elevations relates most closely to the winter component of climate and least closely to the summer component, while growth of conifers at high elevations relates more closely to the climate of late spring and early summer and to that of the previous summer and early autumn." Because of this complexity of tree growth response to climate it is difficult to present a simple comparison between tree growth indices and climate because each chronology may refer to slightly different (and in some cases completely different) periods of time. Furthermore, here we are considering only precipitation and clearly temperature must also play a role (Fritts, 1965). In short, without a detailed study devoted entirely to the verification of dendroclimatic reconstructions we can at this stage only expect to provide a rough evaluation.

The assumptions in this analysis follow the conclusions of Fritts (1965) who notes "it may be concluded that . . . there is a close relationship between the widths of annual rings from trees on semi-arid sites and variations in the aridity of the yearly climates . . . . In general it may be inferred that the *wider the ring, the more moist and cool was the climate*" [emphasis added]. A similar relationship was noted by Fritts (1971) in which mean growth of trees was equated with variations in moisture supply (Fritts, 1971, Figure 13). Thus the question being asked in this analysis can be simply stated as follows: do variations in tree growth increments reflect variations in precipitation as shown by historical climatic data? The following procedure was followed.

Dendroclimatic Data

Using 34 tree ring chronologies (Stokes, Drew and Stockton, 1973) for the ten-state area (Figure 97), the averages and standard deviations for 1951-60 were computed. It should be noted that the chronologies in Stokes *et al.* (1973) are not simple growth increment measurements *but* tree ring indices (growth increments standardized with respect to the natural growth curve of each core) having a mean which approximates to 1 over the entire record length and a homogeneous variance through time (Fritts *et al.*, 1969; Stockton and Fritts, 1971). The averages for *four* five-year periods (1871-1876 to 1887-1892) were then computed, expressed as deviations from the 1951-60 averages and divided by the standard deviations for the 1950s, to provide comparability between records. These data are shown in Figures 98 to 101. Most of the region shows tree growth increments consistently above 1951-60 averages, particularly in southern Wyoming and western Montana. However, three areas are consistently *below* the 1951-60 averages--Colorado and northeastern Utah, southern Arizona and northern California/eastern Oregon (excluding 1877-81). Two chronologies appear to be anomalous (stations 10 and 42 of Stokes *et al.*, 1973). "Station 10" is in southwestern Montana and *is* from limber pine whereas adjacent chronologies are based on Douglas fir (Figure 97) and "station 42" is from pinon pine compared to ponderosa pine or Douglas fir in adjacent chronologies. These are the only two clear anomalies in which are otherwise fairly consistent patterns.

Historical Climatic Data

Historical seasonal precipitation data, expressed as a percentage of the 1951-60 average at analogous stations, were used. Five-year averages were computed from September 1871-August 1876 to September 1886-August 1891, corresponding to the tree ring indices averages for 1872-1876 to 1887-1892. These years were chosen simply because prior to 1870 precipitation data are extremely sparse making meaningful evaluations difficult. Each comparison will be discussed in turn.

1. 1872-1876 cf. Figures 18 to 21 (seasonal precipitation data) and Figure 98 (tree growth indices).

The best correspondence is between tree growth indices and precipitation in spring and winter months which is what one might expect, *a priori*. There are two areas of major disagreement: Colorado and Montana. In Colorado, the few climatic stations suggest that part of the state at least experienced below average winter/spring precipitation if the two periods are considered together; this may account for some of the difference though the growth chronologies are clearly from high elevation sites, not on the plains (Figure 97) where the area of below average precipitation is found. In Montana, precipitation was below average in spring, winter, and fall over most of the state though this is not so at the westernmost station in winter and spring. Here is the thorny problem confronting all attempts to verify dendroclimatic reconstruction surface: How can one group of disparate data points be equated with another equally disparate but spatially distinct group? In such cases it can only be said that a potential disagreement exists but more data points are needed to clarify the degree of conflict or correspondence.

Comparison with the "mean annual" average for September 1871-August 1876 (Figure 102) again shows broad similarities with much of the area being above "average." However, again the broad below average growth zone (encompassing Colorado in particular) is not comparable; the small number of data points in Utah and Oregon make the apparent discrepancies there difficult to evaluate but a potential disagreement does exist.

2. 1877-1881 cf. Figures 22 to 25 (seasonal precipitation data and Figure 99 (tree growth indices).

Tree growth indices show a marked north/south gradient with above average indices north of the southern Wyoming/Idaho/ Oregon line, and below average indices to the south, apart from a zone around the central Arizona/New Mexico border. Only fall precipitation shows a similar pattern and then the chronology for southern Oregon does not correspond and the above average zone in Arizona/New Mexico is absent, perhaps simply due to the few stations in the area. However, when the entire pentad average is considered, correspondence is quite good (Figure 103). In particular, precipitation was indeed above average along a line across southern Idaho and also in parts of Arizona and New Mexico. Again Colorado stations are in disagreement as is the one in southern Arizona. However, considering the differing station distributions, most of the discrepancies could be due to this factor alone.

3. 1882-1886 cf. Figures 26 to 29 (seasonal precipitation data) and Figure 100 (tree growth indices)

Like the previous pentad, tree growth indices correspond best to fall precipitation for 1881-1885, though again eastern Oregon is probably different. When the pentad average is considered (Figure 104), again broad correspondence is seen though many minor differences arise, mainly in areas for which no tree ring data are available. However, once again Colorado stands out as an area of major disagreement.

4. 1887-1891 cf. Figures 30 to 33 (seasonal precipitation data) and Figure 101 (tree growth indices)

As noted in an earlier section, most of the area experienced below average spring, summer, and (except mainly Arizona/New Mexico and eastern Nevada), fall precipitation. Winter precipitation was above average over most of the area with the main exception being Colorado. Tree growth indices do not reflect the dominant below average seasonal patterns but are very similar to winter conditions, with below average growth only over Colorado, southern Arizona and eastern Oregon. The similarities with this winter precipitation pattern are, in fact, quite striking.

Pentad averages (Figure 105) also show quite good correspondence with the below average southern Arizona and Colorado zones identified by dendroclimatic reconstructions.

Discussion

In the above section the approach taken was that the dendroclimatic reconstructions were correct and precipitation data confirming the reconstructions were sought. Thus, in one case precipitation in spring and winter together best fitted the reconstruction whereas in other cases fall, winter, or pentad average precipitation showed better correspondence. This approach could be criticized as being inconsistent. However, in view of the problems of defining concisely for a large number of tree growth chronologies the period to which they all respond, this approach was deemed satisfactory for the present. The results indicate that as expected the period fall to spring showed the best correspondence and these nine months were reflected in the pentad averages which showed reasonable correspondence. However, this comparison has been extremely simplistic; only a comparison of those areas which were above or below normal has received attention. The intensity or degree of correspondence has not been shown and indeed this would be extremely difficult due to the (spatially) quite distinct data sets involved. Also, obviously other meteorological variables (in particular, temperature) have not been considered at this stage and these may improve or negate the correspondence so far established. Also, certain areas stand out as showing very poor agreement with precipitation data. In particular, Colorado is quite consistently "anomalous" and, as almost one-third of the dendrochronologies are from this area, the discrepancy is disturbing. However, again it must be emphasized that with such a simple comparison as this, any differences noted may be entirely related to other variables (such as temperature) not considered in this study. In fact, it has been shown that over a wide area of the western United States monthly precipitation totals and temperature averages are inversely related at the 95% significance level (Leith, 1974) and this inverse relationship is particularly strong in Colorado. Nevertheless, it is clear that despite this simplistic approach there are strong indications that the reconstructions are a fair approximation to reality. On the other hand, it must also be recognized that a consideration of temperature data might not show such good correspondence but any judgment on this matter must await further study.

Perhaps the single most difficult problem associated with verifying dendroclimatic reconstructions is the nature of the two data networks. For example, as mentioned earlier, prior to 1870 there were few climatic stations. In fact, if stations with or more years of record are considered, there were only 25 in the entire ten-state area prior to 1870 and 18 of these were in New Mexico or close to the Oregon/Washington coast. No dendrochronologies exist for the latter area (mainly because the abundant precipitation would make climatic interpretation difficult) and only one published chronology is available for New Mexico and that is in the extreme south. Clearly comparisons based on other stations will be tentative to say the least. As the response function of tree growth is generally defined with respect to a climatic record of 50 to 70 years, independent verification of the reconstructions from a period outside the calibration interval (Fritts, 1974) is obviously confined (in the western United States) to the period between approximately 1870 and 1895. In view of this fact it is suggested that attempts be made to obtain chronologies from areas where historical climatic data are available for this period. In particular, the following locations are recommended: southern and central Arizona, southwestern New Mexico, north-central New Mexico, the Salt Lake City area of Utah, northern Nevada (a transect across the old Pacific Railroad Line), northern Idaho, and eastern Washington. These are all areas where good climatic data exist both for calibration and testing of the reconstructions. In short, what is needed is a sampling network designed to be tested.

CHAPTER 6

SYNOPTIC WEATHER TYPES AND VARIATIONS IN PRECIPITATION

In order to investigate changes in circulation associated with the precipitation variations observed, a catalog of synoptic types was used.[1] The typing scheme was developed by Irving P. Krick and associates at the California Institute of Technology during the late 1930s and 1940s (Elliott, 1943) and is described in depth by Elliott (1949).[2] The classification covers the zone 90°-135°W and recognizes 16 characteristic three-day sequences of synoptic events which form the basis of particular weather types. Each type is identified principally by the positions of the semipermanent elements of circulation (such as the Pacific anticyclone and the Aleutian Low) and the trajectories of polar outbreaks and depression paths, which produce characteristic weather conditions over the western United States. Maps showing the major features and characteristic sequence of events in each type are given in Appendix C. Unfortunately the synoptic catalog is only complete for the period 1928-1958 and summer months from 1956-58 are generally unclassified. However, useful comparisons within this period are possible.

In this section, the frequency of synoptic types associated with relatively high and low decadal precipitation within the catalog period has been examined by season. Fortunately, the three-day catalog is complete for two of the most interesting and contrasting decades in the twentieth century record. Consideration of the data presented in Chapter 3

---

[1] Courtesy of North American Weather Consultants, Santa Barbara, California.

[2] It should be noted that these catalog schemes were derived subjectively with a view to short-term forecasting and its application to synoptic studies of this sort can be questioned. Nevertheless, previous studies (Barry and Bradley, 1972; Bradley and Barry, 1973b) have shown some value in using the catalog in this manner. Further work on the development of an objective catalog of synoptic types for the period 1899-1970, based initially on average types from the C.I.T. catalog is continuing.

(summarized in Table 17) indicates that in all seasons much if not all of the Rocky Mountain area experienced well below average precipitation in the 1930s (see Figures 56, 66, 77 and 87). Only in fall and winter months were there exceptions to this over fairly large areas. This is, of course, the well-known drought period which affected large areas of the country, particularly the mid-West (Lawson *et al.*, 1971) causing disastrous crop failures and economic disaster for thousands. By contrast, the 1940s in the Rockies were extremely wet in all seasons, the principal exceptions occurring in spring months (Table 17). Thus, in general, these two decades provide an excellent contrast in precipitation receipts which presumably reflect fundamental differences in synoptic conditions.

Zonal and Meridional Type Frequencies

Initially, the weather types were grouped into two major units, those characterized by a predominantly zonal circulation and those characteristically meridional (see Footnote 1, Table 30). Figure 106 shows the annual frequency of these types 1928 to 1958 and there is clearly a decrease in zonal and an increase in meridional types throughout the period to at least the early 1950s, after which the trend appears to have been reversed (although there are much data missing in the 1950s, mostly in summer months). Table 30 shows a comparison between zonal and meridional type frequencies in the 1930s and 1940s and it is clear that the decrease in zonal type frequency noted on an annual basis is related mainly to changes in summer and winter seasons (-10% in both seasons). Figure 107 illustrates this further: winter frequencies of zonal types decreased markedly after c. 1941/42 and meridional types showed a complementary increase. Meridional type frequency averaged 79 days per season from 1928/29 to 1941/42 and 90 days from 1942/43 to 1957/58. The corresponding zonal type values were 72 and 62 days respectively.

It is interesting to compare these results with circulation indices for other areas or for the northern hemisphere as a whole. Dzerdzeevski (1966) has described changes in the frequency of zonal and meridional circulation types from 1900-09 to 1950-59 and recognizes two "circulation epochs" the first lasting until ~1921-30 and the second to ~ 1950-59 (Barry and Perry, 1973). Circulation in the former period is predominantly meridional and in the latter zonal which means that "the epoch was more continental and cooler than the second one" (Dzerdzeevski, 1966). However, further consideration of Dzerdzeevski's data (Figure 108) indicates that while the change-over of relative departures from average occurred c. 1921-30, a fundamental change towards less frequent zonal types and more frequent meridional types occurred around the late 1930s/early 1940s. This date is similar to that marking a change in circulation from zonal (late nineteenth century to c. 1940) to meridional (post-1940), as described by Lamb and Johnson (1966). It is also in line with an analysis of the zonal index for 55°-65°N over the entire northern hemisphere by Namias (1957) who shows increasing zonal flow from c. 1915 to the late 1930s/early 1940s and decreasing zonal flow thereafter. In short, all these studies tend to agree that a fundamental change occurred in the circulation regime of the northern hemisphere in the late 1930s or early 1940s, from a predominantly zonal to a predominantly meridional circulation regime.

Circulation Types and Solar Activity

The reason for a change in circulation regime in the late 1930s/early 1940s is not clear but it is of interest to compare the synoptic types described above with values of Baur's solar index[1] which considers both the area of sunspots and of solar faculae (bright spots) on the sun's disc (Baur, 1949, p. 970). This index is thought to be one of the more realistic estimates of variations in the "solar constant" (Lamb, 1972). Figure 109 shows the relationship between the frequency of winter zonal types in the C.I.T. catalog and Baur's index during the same months. A strong positive relationship, statistically significant at 2% level is apparent over the period from 1928/29 to 1953/54. Further study shows that the marked change in Baur's Index is related not to a change in faculae area but to a change in the area of sunspots (Figure 110). Average faculae area from 1900-39 was 1398 millionths of the area of the visible surface of the sun (corrected for foreshortening) and from 1940-53 the area was 1404 millionths. Sunspot areas for the same two periods were 677.5 and 1031 millionths respectively. Thus there is a relationship between increasing sunspot area and decreasing frequency of zonal flow types, or conversely an increase in the frequency of meridional types.

Such a relationship is not entirely unexpected. Willett (1949, 1964) has argued strongly for an inverse relationship between sunspot activity and zonal flow in the northern hemisphere over an 80-year cycle. In the first quarter of

---

[1]Baur's Index is defined thus, $S.I. = 100\,(F/\bar{F} - D/\bar{D})$ where F is the area of solar faculae, D is the area of sunspots and $\bar{F}$ and $\bar{D}$ are the long-term averages of these parameters (from 1900-1949).

the cycle low-latitude zonal flow predominates giving way to higher latitude zonal flow in the second and third quarters and to strong cellular blocking patterns (meridional circulation) in the fourth quarter (reached in the late 1950s in the most recent cycle).

On a shorter time scale, Schuurmans (1969) has investigated the relationship between atmospheric circulation in the northern hemisphere and solar flare activity. He concludes that the ultimate effect of a flare will be the intensification of meridional circulation. In particular, the development of blocking highs in the mid-high latitudes is favored. As flares predominate in periods of high sunspot levels this supports the general hypothesis noted above. Furthermore, Schuurmans has compared the frequency of zonal, half-meridional and meridional circulation types in the Grosswetter catalog of European weather types (Hess and Brezowsky, 1962), in different seasons of the year with different levels of sunspot activity. His results clearly indicate that there is an increase in the frequency of meridional weather types with increased levels of sunspot activity in all seasons of the year but particularly in winter months (Schuurmans, 1969). This is further support for there being a real relationship between solar activity and circulation patterns in the northern hemisphere even though the links forging such a relationship are not understood at present (Mitchell, 1965). Further work on this question using the C.I.T. synoptic catalog and other objective typing schemes (Barry, 1972) is underway.

Individual Synoptic Weather Type Frequencies

A comparison of the individual weather type frequencies for the 1930s and 1940s is given in Table 31. As in Table 30, values are given in percentages of days per season but it should be noted that the seasons are of different length so a 5% change in spring involves an average change of only 3 days per season whereas in winter this would involve a change averaging 8 days per season. Three types show consistent increases--types A, B_{n-a} and B_{n-c} (all meridional types, as one might expect). Collectively these types increased by 5, 9, 6, and 15% in spring and summer, fall and winter months respectively between the 1930s and 1940s.

Type A has a mean upper level trough at 105°-115°W--precisely located over the Rocky Mountain area. Elliott (1949, p. 16) describes the characteristic sequence of events associated with Type A thus:[1] "A surface wave forms along the southeast Alaskan or British Columbian coast and is steered by the upper-level flow rapidly down the coast through Washington and Oregon into the Great Basin area (Nevada, southeastern Oregon, southern Idaho and western Utah). The wave has been occluded and is by now a moderately intense slow-moving cyclone. [In winter] widespread snow flurries are experienced in the Rocky Mountains, and cold air advancing down the West Coast behind the cold front brings frosts as far south as California." This is clearly the sort of synoptic pattern one would expect in a relatively wet period and the fact that in the 1940s it averaged 22% of spring days, 30% of summer days, 25% of fall days and 19% of winter days is very significant.

Type B_{n-c} is also associated with above normal precipitation in the Rocky Mountains and western Great Plains. The characteristic winter sequence is as follows: "an outbreak of cold air from northwestern continental regions moves south then southeast, following a trajectory to the west of the normal path of cold outbreaks. This is responsible for forcing the low entering British Columbia to move more to the south than it might otherwise do. Indeed, in many cases, the low moves as much as 10° of latitude south . . . or a secondary crescent-shaped low of major importance develops in Colorado. Since behind the cold front, east winds blow the cold air upslope to the forward wall of the Rockies and over into the Great Basin, blizzard conditions are experienced in these regions" (Elliott, 1949, p. 42).

Type B_{n-a} does not appear to favor precipitation over the Rocky Mountains. In this type a strong anticyclone over the Great Basin dominates circulation over the Rocky Mountain state and lows generally pass well to the north, across the northern Canadian Rockies and into the northern Prairies. It is thus unlikely that the increased frequency of this type in the 1940s can be related to the changes in precipitation already noted.

Of the types which have decreased in frequency between the 1930s and 1940s three are of interest. Type B_{n-b}, like B_{n-a} is characterized by a strong and persistent Great Basin

---

[1]It should be noted that all quotes from Elliott (1949) describing North American Weather Types pertain specifically to conditions in winter months. However, the basic sequences of events in each type are the same in all months and differ only in intensity (i.e., in the actual temperature characteristics, type of precipitation--rain or snow, etc.).

anticyclone which diverts Pacific storms northward as they enter the North American continent. The type results in "below normal [precipitation] . . . in the Rocky Mountain region as far north as the 50th parallel" (Elliott, 1949, p. 42). If this type and type B_{n-a} are considered together (they are both sub-types of type B_n) then the overall change of the two types is a decrease in frequency in all seasons except winter (Table 31) and this is consistent with the observed changes in precipitation.

Type B, which decreased considerably in frequency only during summer months (when it is normally most frequent anyway) is also associated with near or below normal precipitation in the Rockies. "At the surface, the low center apparently marches eastward from the Gulf of Alaska through northern Canada, with a frontal zone trailing southward from the center across southern Canada and the United States. As the front advances into the continent, precipitation occurs from Alaska to Oregon, but as it advances further east into the Great Plains region, precipitation becomes negligible except in northern and central Canada . . . a polar-Pacific air mass following the frontal zone forms a temporary high in the Great Basin" (Elliott, 1949). Once again the synoptic conditions associated with this type are such that one would expect their frequency to diminish in periods of heavy precipitation. However, it should be noted that this type did increase in frequency in fall and particularly spring months. This points out a general problem involved in attempting to relate synoptic types on a monthly or seasonal basis to monthly or seasonal precipitation totals, viz: how does one know that the precipitation events were associated with a particular type or types? The short answer is clearly that one does not know without considering the particular precipitation receipts associated with each type sequence on a daily basis. However, one of the valuable assets of this particular typing scheme is that it was designed for the purpose of short-term forecasting and thus the usual precipitation patterns and temperature conditions associated with the types are quite well established. Hence it is reasonable to assume that the increased frequency of particular types normally characterized by above normal precipitation is probably causing the observed pattern of increased seasonal precipitation, and _vice versa_. This could not be said of an objectively typed scheme without first looking at daily precipitation receipts associated with particular types and this is in itself no small task.

Finally, another type which decreased in frequency in spring, fall and (only slightly) summer months is Type C_L. This is also characterized by below average precipitation in the Rockies due to a persistent anticyclone over the Great Basin with cyclones forced either far to the north (over the

Northwest Territories of Canada) or to the south over southern California and southern Arizona. A higher frequency of this type in the 1930s compared to the 1940s is thus consistent with precipitation conditions already noted for these decades.

A comparison of precipitation in the 1950s with that in the 1930s and 1940s indicates that the most marked contrasts are between the 1940s and 1950s in summer and fall seasons. In both seasons, the 1950s were well below average at almost all stations and in fall the majority of stations recorded period minima (Figure 81). It is thus of interest to compare the frequency of synoptic types in these decades to see if the observed precipitation differences can be accounted for synoptically. Unfortunately, summer days from 1954 to 1958 are generally unclassified and so such a comparison is not possible for this season. Fall types have been compared (Table 32); however, it should be noted that the period cataloged only extends to 1958. Of particular interest is the marked decrease in frequency of Type A, which, as noted above, is generally associated with above normal precipitation. In the 1940s this type occurs on 25% of fall days and is in fact the most common type. In the 1950s, however, it occurs on only 11% of days and ranks fourth most frequent. Also of note is the 10% increase in Type B_{n-b} which is associated with below average precipitation in the area, as noted above. In the 1940s this type was relatively infrequent (7% of days) whereas in the 1950s it was the dominant type (17% of days). Other type frequency changes are not so clearly in line with the observed precipitation conditions. Type B (associated with relatively dry conditions) decreased from the 1940s to the 1950s whereas Type B_{n-c} (associated with wet conditions) increased in frequency. However, Type C_L (also associated with below average precipitation) increased in frequency. Overall, then, the major changes in frequency between the 1940s and 1950s were associated with decreases in the main types normally bringing relatively heavy precipitation to the area and increases in the frequency of types favoring relatively dry conditions. Once again, this is consistent with the observed precipitation data.

Summary

In conclusion it can be stated that the synoptic weather type catalog has proved useful in characterizing those synoptic events associated with and responsible for the contrasting precipitation regimes of the 1930s, 1940s, and in some seasons, 1950s. In particular, changes in the frequencies of five main types--A, B, B_{n-a}, B_{n-b}, and B_{n-c} were shown to be of major significance to precipitation variations in these three decades. The lower frequency of zonal types after 1940 is in line with other circulation studies in the northern hemisphere and there are indications that variations in solar activity may be in some way responsible for this change. However, it must be stated that this is at present only a recognized covariation and of course must await a meaningful explanation of the mechanisms linking solar variations with the atmospheric circulation. Until accurate long-term measurements of the "solar constant" over all wavelengths are made, such a goal will probably prove to be extremely elusive.

CHAPTER 7

SUMMARY AND CONCLUSIONS

In the preceding six chapters, secular changes of precipitation on a seasonal basis have been documented and compared with dendroclimatic reconstructions, long-term records have been examined for evidence of nonrandomness, and synoptic weather types have been considered in relation to the observed variations of precipitation. The following main points have arisen from these discussions.

1. Although no temperature data were analyzed in this study, a survey of literature for the western United States shows most studies in general agreement that the area was warmer in the mid-nineteenth century than it was during 1931-60. Temperatures gradually fell to the late 1920s or early 1930s and rose rapidly during the 1930s and 1940s. The 1960s have been warmer than 1931-60 averages, particularly in winter months. This is the inverse of the generally observed global pattern of warming from c. 1880 to 1930-1940 and cooling thereafter.

2. Comparison of precipitation data from the latter half of the nineteenth century with 1951-60 averages at comparable stations indicates that from at least 1865 to 1890, winter and spring precipitation was above average over much of the western United States (see Chapter 2 and Appendix B for relative magnitudes). In summer, however, the period 1865-1890 was drier than in the 1950s, though prior to 1865 this may not have been the case. Fall precipitation is more complex than other seasons but quite large areas were relatively dry for much of the late nineteenth century, compared to averages in the 1950s. As many stations in the Rocky Mountain area (at least) recorded lowest decadal fall precipitation amounts this century in the 1950s, it seems likely that the latter half of the nineteenth century was exceptionally dry over large areas.

3. Nineteenth-century winter precipitation records in the Southwest show marked peaks at four- to six-year intervals which are paralleled by equally marked troughs in the records of stations in the Northwest. This inverse relationship reflects changes in latitudinal position of

depressions crossing the west coast of the United States in different years. Such a pattern has been recognized by others in monthly precipitation for the western United States and also in tree growth increments since 1700. However, the reason for a four- to six-year periodicity is not clear.

4. Careful screening of long-term precipitation records in the Rocky Mountain states indicated a large number of potentially useful records of 50 years or more in length. Tests on the homogeneity of these data showed that most unhomogeneous records were found in winter and adjacent seasons while relatively few summer and late summer records were unhomogeneous, presumably reflecting snowfall-recording problems in winter months.

5. Variability of precipitation varies more or less inversely with precipitation amounts in all seasons though the relationship is not simply linear in all seasons. Coefficients of variation are highest in fall and lowest in winter months.

6. In general terms, precipitation in the Rocky Mountain states was relatively high in the 1890s, 1910s and/or 1920s, 1940s and 1960s. Other decades were relatively dry, particularly the 1930s.

   Spring. The most recent 30-year period was wetter than the 1911-40 period over most of the area particularly the northern Rockies. However, the 1890s and 1900s were relatively wet and it is likely that the period 1881-1910 was wetter than 1941-70.

   Summer. The period 1941-70 was probably the most anomalously wet period for at least 110 years (and perhaps a very much longer period). Over much of the Rockies, the 1960s were extraordinarily wet.

   Fall. As in other seasons the 1940s and 1960s were relatively wet resulting in a relatively high "normal." However, precipitation amounts have been higher in several earlier decades particularly the 1890s, 1910s, and 1920s. Over almost the entire area twentieth-century decadal minima were recorded in the 1950s.

   Winter. In all areas except Idaho, precipitation in recent decades has been considerably less than was characteristic of the late nineteenth and early twentieth centuries. Recurrent anomaly patterns with negative departures centered over southern Idaho are characteristic of the winter record.

7. Consideration of the data outlined in points 1, 2, and 6 suggest that the recent "normal" period is indeed anomalous when placed in the context of the last 100-120 years. In particular, precipitation totals in summer months have been unusually high and in winter months unusually low. One hundred years ago the climate of the Rockies was characterized by wetter winters and springs but much drier summer and (in some areas) much drier falls compared to the post-1940 climate of the region. Temperatures a century ago were generally warmer particularly in summer months. The recent anomalous "normal" is related to a change towards a more meridional circulation (see 10 below) which has intensified in the 1960s suggesting that an early return to climatic conditions characteristic of the early twentieth century is unlikely.

8. Power spectrum analysis of over 200 seasonal precipitation records indicated few nonrandom features which could be considered statistically significant. Only one periodicity was of major interest--that at 2.0-2.2 years. This is characteristic of summer and late summer records in Utah and western Colorado only.

   Some other records (with no coherent geographical distribution) showed long-period trends but records were generally too short to isolate their frequency with any degree of accuracy. However, several of these records, when smoothed, showed a characteristic low "pivot-point" around 1930-35 (precipitation falling to 1930-35 and increasing thereafter).

9. Comparison of dendroclimatic reconstructions and historical precipitation data (1870-1890, i.e., a period outside the dendroclimatic calibration period) showed quite good correspondence despite the crude analysis undertaken.

10. Observed precipitation variations in the Rockies during the 1930s, 1940s, and 1950s were shown to be related to changes in the frequency of certain synoptic weather types. In particular, there has been a marked increase in meridional circulation types in the post-1940 period. This concurs with results from other parts of the northern hemisphere. A relationship between these changes and variations in solar activity (particularly increases in sunspot area) during the 1930s and 1940s was demonstrated.

11. It is interesting to note that, whereas the most recent 30 year "normal" is indeed anomalous where compared to records of the last 100-120 years, the entire period considered in this study may itself be anomalous when

viewed in context of the last millenium. If LaMarche's dendroclimatic inferences are correct (LaMarche, 1974), then conditions similar to the most recent "normal" have not prevailed since 1330 A.D. when a 200-year "warm-dry" spell ended, replaced by "cool-moist" conditions. Furthermore, the "warm-moist" conditions of the period c. 1870-1945 may only have occurred once in the last 1100 years--from c. 1055 and 1130 A.D. (820-895 B.P.).

In conclusion, it can be stated that seasonal precipitation totals over the Rocky Mountains and adjacent western states have undergone very significant changes over the last 120 years. Secular changes in precipitation throughout the world have received little attention in the literature, mainly due to a general skepticism regarding the quality and accuracy of precipitation recording procedures. Such skepticism is even more prevalent regarding nineteenth-century data. However, this study has demonstrated that such an attitude is unwarranted. Early records were well kept enabling broad-scale regional patterns to be discerned even in this topographically diverse region. Furthermore, even though recording precipitation (particularly snow) in mountainous regions is a problem and topoclimatic variations may be great, it is clear that by the judicious use of carefully screened long-term data, coherent and meaningful regional changes can be identified without resorting to techniques which artificially remove the effects of topography. Topography is clearly important to the total amount of precipitation received at any one point (and indeed to the steering of some small-scale systems along broad basins such as those of the Snake and Upper Colorado rivers) but in terms of secular changes in precipitation, it appears to be of much less importance than the regional synoptic controls which determine the broad-scale changes observed. This author thus concurs with Lamb's (1966) remark (made after studying precipitation changes in different parts of the world) that

> It is gratifying to find that there are such simple trends of rainfall, changing nearly uniformly over wide areas and intelligibly related to the large-scale atmospheric circulation. This might not have been expected in view of the characteristically complex patterns that local rainfall amounts present, varying with all the intricate details of topography.

Table 1

Earliest Instrumental Meteorological Stations in the Western United States

State	Station	Lat.	Long.	Elevation (ft)	Month of 1st obs. Temp.	Month of 1st obs. Precip.	Source of data (see Appendix A)
Arizona	Fort Defiance	35°43'	109°10'	6500	Dec.1851	May 1852	Greely 1891a
California	Yerba Buena (San Francisco)	?	?	?	Nov.1826		Beechey (1831)++
	Monterey Barracks	36°36'	121°52;	140		May 1847	Lawson 1855
Colorado	Fort Massachusetts	37°32'	105°23'	8365	Sep.1852	Oct.1852	Lawson 1855
Idaho*	Cantonment Loring (Fort Hall)	43°04'	112°27'	4800	Aug.1849		Lawson 1855
	Fort Lapwai					Jan.1864	USWB Bull W
Montana	Fort Benton	47°52'	110°43'	2630	Oct.1862	Oct.1862	Microfilm (National Archives Record T907, Box 293)
Nevada	Fort Churchill	39°20'	119°05'	4284	Oct.1860	Dec.1860	Greely 1891a
New Mexico	Ft. Marcy(Santa Fe)	35°41'	105°57'	7026	Jan.1849		Greely 1891a
	Cebolleta					Dec.1849	Lawson 1855
Oregon*	Fort Dalles	45°22'	120°50'	350	Sep.1850	Jul.1850	Lawson 1855
	(Dalles of Columbia)	36'	55'?				Greely 1889
	Ft. George (Oregon?*)	46°18'	123°00'	?	Jun.1821		Scoulder 1827+

* See also discussion in the text, p. 19.

\+ Scouler, Mr. 1827; On the temperature of the northwest coast of America. Edinburgh Journal of Science, Vol. 6, 251-53.

++ Beechey, F.W., 1831. Narrative of a Voyage to the Pacific and Bering's Strait to Cooperate with the Polar Expeditions in the years 1825, 26, 27, 28. London, H. Colburn and R. Bentley (4th Ed.).

Table 1 (continued)

Earliest Instrumental Meteorological Stations in the Western United States

State	Station	Lat.	Long.	Elevation (ft)	Month of 1st obs. Temp.	Month of 1st obs. Precip.	Source of data (see Appendix A)
Utah	Gt. Salt Lake City	40°46'	111°54'	4354	Jan.1850		Lawson 1855
	Salt Lake City	40°46'	112°06'	4361		Feb.1857	Greely 1891a
Washington (western)	Fort Steilacoom	47°11'	122°34'	300	Nov.1849	Nov.1849	Greely 1889
Washington (eastern)	Fort Walla Walla	46°03'	118°20'	1396	Jan.1857	Jan.1857	Greely 1889
Wyoming	Fort Laramie	42°12'	104°47'	4519	Sep.1849	Sep.1879	Lawson 1855

Table 2

Dates of Establishment of

National Weather Service Regional Offices

State	City	Began	1st Weather-Crop Bulletin	Number of Stations in July 1 1895
Arizona	Tucson	Oct. 1891	April 1892	42
California	Sacramento	Sept 1891	May 2, 1890	335
Colorado	Denver*	Jan. 1885	Mar. 1891	95
Idaho	Idaho Falls	July 1892	May 1893	31
Montana	Helena	Nov. 1891	Apr. 1892	43
New Mexico	Santa Fe	July 1891	Apr. 1892	31
Nevada	Carson City	Feb. 1887	Sept 1893	63
Oregon	Portland	Mar. 1887	Mar. 1889	60
Utah	Salt Lake City	Sept 1891	Apr. 1892	35
Washington	Seattle	Summer 1891	Spring 1892	47
Wyoming	Cheyenne	Fall 1891	Apr. 1892	15

* "Colorado Meteorological Association."

Table 3

Summary of Seasonal Precipitation Totals Relative to 1951-60 Averages at Analog Stations

Note: Refer to text and Figures 2 to 33 for station distribution and discussion of individual pentads

	Springs	Summers	Falls	Winters
1886-90	Generally $<\bar{x}$ over most of the area	Predominantly $<\bar{x}$, except in Washington, S. Oregon, S. Arizona & E. Colorado.	$>\bar{x}$ in most of Arizona and New Mexico, Utah & E.Nevada & E.Central Montana. Below $\bar{x}$ elsewhere.	$>\bar{x}$ except NW coast, Colorado & E. Wyoming (?) area.
1881-85	Generally $>\bar{x}$ except S.Arizona/S.New Mexico, central Nevada & most of Montana.	Mixed; N&E central $>\bar{x}$; elsewhere $<\bar{x}$.	Well above $\bar{x}$ N of a line from N.W.Nevada to N.Colorado, & in E.Arizona & most of New Mexico. $<\bar{x}$ elsewhere.	$>\bar{x}$ except NW and E margins.
1876-80	Generally $>\bar{x}$ except N.Nevada/N.Utah/S. Idaho & S.Arizona, New Mexico & E. Colorado.	Well below $\bar{x}$ in most areas except SE & parts of Montana & NW	$>\bar{x}$ in N (except S. Oregon & NW coast) & in S. Arizona. $\bar{x}$ elsewhere.	Generally $>\bar{x}$, except in zone from N.Utah/N.E. Nevada to C.Arizona.
1871-75	$>\bar{x}$ over most of area except N.Nevada S. Oregon & zone NE across Idaho.	$<\bar{x}$ over most of the area but above in eastern margins & extreme SW	$>\bar{x}$ in E and S $<\bar{x}$ in W and N	Markedly $>\bar{x}$ in almost all areas.

Table 3 (continued)

	Springs	Summers	Falls	Winters
1866-70	$>\bar{x}$ in N. states generally $>\bar{x}$ in Arizona, New Mexico (excluding SE) & Colorado	$>\bar{x}$ at almost all stations	Generally $<\bar{x}$ except perhaps in central part of region.	Generally $>\bar{x}$
1861-65	?	$>x$ near NW coast but less inland along Columbia River	Well below $\bar{x}$ in NW	?
1856-60	Well below $\bar{x}$ in N. Mexico; elsewhere = ?	$>\bar{x}$ in New Mexico & away from NW coast	$>\bar{x}$ in New Mexico & at most stations in NW	$>\bar{x}$ in NW and SE; elsewhere = ?
1851-55 (New Mexico only)	Slightly $>\bar{x}$	$>\bar{x}$ (147%)	Markedly $>\bar{x}$ ($\sim$174%)	$>\bar{x}$ ($\simeq$134%)

Table 4

Winter Precipitation at Oregon Stations[1]

	1873	1883	1884	1888	Average for Period	
Portland	28.37	25.43	22.66	17.31	34.55	1871-89
	5	3	2	1		
Eola	26.15	22.41	20.69	13.46	27.37	1871-89
	6	4	2	1		
Roseburg		18.67	18.54	12.98	24.06	1877-89
		4	3	1		
Albany		27.15	20.34	15.60	30.02	1879-89
		3	2	1		
Astoria	49.25		28.06	39.00	54.31	1856-75 and
	8		1	2		1884-89

[1]Figures in row 2 for each station indicate the rank of winter precipitation within the period given in the final column. Hence, 1 indicates precipitation was the period minimum, 2 the second lowest, etc.

Table 5

Winter Precipitation at Washington Stations[1]
(Values in Inches)

	1873	1883	1884	1888	Average for Period	
San Juan Island	9.30				15.20	1861-73
	2					
Fort Colville	2.71				9.65	1866-78
	1					
Cathlamet	37.50*				52.05	1872-75
	1					
Port Townsend	6.46	9.59	6.76	7.24	11.20	1876-89
	1	7	2	3		
Cape Disappointment	38.78	28.04	28.99	33.29	44.91	1871-89
	5	1	2	3		
Walla Walla	5.30	8.43	9.91	5.71	8.63	1873-89
	2	7	11	3		
Olympia		21.45	21.59	20.52	36.60	1877-89
		2	3	1		
Ediz Hook Lighthouse		18.23	17.52	13.59	19.87	1878-89
		3	2	1		
Dayton		16.49	11.43		15.50	1881-84
		3	1			
Port Blakely		21.66	24.37		29.76	1878-86
		1	3			
Neah Bay		prob. < 50.00	54.75		75.45	1878-80 and 1884-86
		1?	2			
New Tacoma			17.08		25.02	1884-86
			1			
Spokane Falls		8.56	9.39		10.63	1881-86
		1	2			
Tatoosh Lighthouse		48.05	49.36		60.80	1883-86
		1	2			
Kennewick			2.34		4.09	1884-86
			1			

* 2 months interpolated (Greely, 1889)

---

[1] Figures in row 2 for each station indicate the rank of winter precipitation within the period given in the final column. Hence, 1 indicates precipitation was the period minimum, 2 the second lowest, etc.

Table 6

Matrix of Mann-Kendall Values Identified as Statistically Significant between Stations Indicated

A

Index Number	0364	1044	1202	1318	2112	2173	2409	3489	4985	6157	7286	8501
0392			X									
0432					X							
1202						X						X
1737								X				
2409	X			X								
2609							X					
3013									X			
3489	X											
3984											X	
4038							X					
6157							X					
8597							X			X		

B

Index Number	2112	4985	6157	7286
0432	X			
3013		X		
3984				X
9587			X	

Table 7

Tests of Homogeneity of Precipitation Records

Colorado

Station Index Number	Station Name	Elevation (feet)	Year Record Begins	Seasons* 1	2	3	4	5
5722	Montrose No. 2	5830	1885	S				
2192	Delta	5055	1888					R
3488	Grand Junction	4855	1892				R	R
2432	Durango	6550	1894	S	S			
1741**	Collbran	6130	1892		S	S		R
3359	Glenwood Sprs	5823	1901	R	S	S		R
3146	Fruita	4507	1902					
7656	Silverton	9322	1906	S				R
7936	Steamboat Springs	6770	1908					
2281	Dillon	9065	1909			S		
6513	Pitkin	9200	1909				S	
7017	Rico	8842	1909	S				R
3951	Hermit	9001	1909		S	R		S
3113	Fraser	8560	1909					
1959	Crested Butte	8855	1909					
7618	Shoshone	5933	1910			S		R
3016	Fort Lewis	7595	1911					
0228	Ames	8701	1913		S		R	
4250	Ignacio	6424	1914		S			

* 1 = Spring (A,M); 2 = Summer (J,J,A); 3 = Late Summer (J,A,S); 4 = Fall (S,O): 5 = Winter (N,D,J,F,M).

** Records end 1966.

Table 8

Tests of Homogeneity of Precipitation Records

Idaho

Station Index Number	Station Name	Elevation (feet)	Year Record Begins	Season* 1	2	3	4	5
1022	Boise	2838	1864	S			R	R
7264	Porthill	1800	1889	S		S		
6891	Payette	2150	1890	S				
6152	Moscow	2660	1892					
6542	Oakley	4600	1893	R				
1408	Cambridge	2650	1894					
0915	Blackfoot	4487	1896			R	R	
0470	Ashton	5220	1897	S		S		S
6388	New Meadows	3870	1903					S
6681	Orofino	2150	1903		S	S	S	
4457	Idaho Falls	4730	1904	R		S	R	R
1380	Caldwell	2370	1904	S				
4831	Kellogg	2312	1905		S		R	S
2676	Driggs	6097	1907	R				R
3732	Grace	5400	1907					
8818	Sugar	4890	1907					
9498	Wallace Wood-land Park	2935	1907					R
5011	Kooskia	1260	1908		S	S	S	
5462	Mackay	5897	1908					S
3631	Glen's Ferry	2580	1909				R	R
9294	Twin Falls	3770	1909	S			S	R
8137	Sand Point	2100	1910	S	S			S
7386	Priest River	2380	1911				R	
4295	Hollister	4550	1912	S			S	
4442	Idaho City	3965	1913	S				
0010	Aberdeen	4405	1914					
6053	Montpelier	5960	1914		S			
4670	Jerome	3785	1915					
1663	Challis	5175	1915					
2187	Council	2950	1915					S
0448	Arrow Rock Dam	3275	1916					
5708	McCall	5025	1916					
3448	Garden Valley	3212	1916					

* 1 = Spring (A,M); 2 = Summer (J,J,A); 3 = Late Summer (J,A,S); 4 = Fall (S,O); 5 = Winter (N,D,J,F,M).

Table 8 (Continued)

Tests of Homogeneity of Precipitation Records

Idaho

Station Index Number	Station Name	Elevation (feet)	Year Record Begins	Season* 1	2	3	4	5
1002	Bliss	3265	1917				R	R
4140	Hazleton	4060	1917					R
5544	Malad	4420	1917		R			
3942	Hailey	5328	1918					S
5275	Lifton	5926	1918		R			R

* 1 = Spring (A,M); 2 = Summer (J,J,A); 3 = Late Summer (J,A,S); 4 = Fall (S,O); 5 = Winter (N,D,J,F,M).

Table 9

Tests of Homogeneity of Precipitation Records

Montana

Station Index Number	Station Name	Elevation (feet)	Year Record Begins	Season* 1	2	3	4	5
5685	Miles City	2360	1877					R
4055	Helena	3828	1880				R	
3581	Glendive	2076	1889					
1044	Bozeman	4856	1894					R
1318	Butte	5526	1894			R		R
2409	Dillon	5218	1895	R	R	R	R	R
5015	Libby	2080	1895					
4985	Lewistown	4145	1896	S				S
2112	Crow Agency	3030	1898	S				R
0364	Augusta	4070	1901			R	R	S
3994**	Havre	2584	1892					
1297	Busby	3440	1903					
1722	Chinook	2350	1903				A	
0802	Billings	3567	1905					R
2604	East Anaconda	5511	1905					
2689	Ekalaka	3425	1905					R
7382	Savage	1985	1905					
3139	Fortine	3000	1906					
1552	Cascade	3390	1907					
6157	Norris	4800	1907	S		R		
1102	Bridger	3680	1908					
3984	Haugen	3100	1908	R				
7286	Saint Ignatius	2900	1908	R			R	
4241	Holter	3487	1908					S
2173	Cut Bank	3838	1908					
0780	Big Timber	4100	1909					
5337	Malta	2255	1909					
5666	Mildred	2407	1909					R
8501	Valier	3805	1911					
8927	White Sulphur Springs	5157	1911				R	S
1080	Brady Aznoe	3329	1912					A
1202	Browning	4355	1912	S			S	S
3489	Gibson Dam	4590	1913	R	S		S	R

* 1 = Spring (A,M); 2 = Summer (J,J,A); 3 = Late Summer (J,A,S); 4 = Fall (S,O); 5 = Winter (N,D,J,F,M)

** Records end 1960 A = too much data absent.

Table 9 (Continued)

Tests of Homogeneity of Precipitation Records

Montana

Station Index Number	Station Name	Elevation (feet)	Year Record Begins	Season* 1	2	3	4	5
4038	Hebgen Dam	6489	1913					
0392	Babb	4300	1914					
3013	Flat Willow	3138	1915	S				S
8597	Virginia City	5776	1916	S		R	R	
1737	Choteau	3945	1918		R		S	S
0432	Ballantine	3000	1919	S				

* 1 = Spring (A,M); 2 = Summer (J,J,A); 3 = Late Summer (J,A,S); 4 = Fall (S,O); 5 = Winter (N,D,J,F,M).

Table 10

Tests of Homogeneity of Precipitation Records

Utah

Station Index Number	Station Name	Elevation (feet)	Year Record Begins	Season* 1	2	3	4	5
2996	Fort Duchesne	4990	1887		S			
5065	Levan	5300	1889					
5186	Logan	4785	1891	S	S	S	S	R
2828	Fillmore	5160	1892				R	R
3809	Heber	5580	1893			S		S
5402	Manti	5585	1894	R			R	R
7714	Scipio	5306	1894					R
8771	Tooele	4820	1896					S
2101	Deseret	4585	1899					
4856	Laketown	59-8	1900				R	
5752	Modena	5460	1901	S			S	R
2484	Emery	6200	1901					S
2418	Elberta	4690	1902	S				
5826	Morgan	5070	1903					R
8973	Utah Lake	4497	1904			S	S	R
0738	Blanding	6036	1905					S
6357	Oak City	5075	1905					
2253	Duchesne	5510	1906					S
4527	Kanosh	5015	1907					
5837	Moroni	5525	1908	S				S
8119	Spanish Fork	4711	1910				S	
6534	Orderville	5460	1910					
3611	Hanksville	4308	1910				S	
0061	Alpine	4935	1910					
6658	Park Valley	5570	1911					
5610	Midvale	4342	1911			S		
7271	Richmond	4680	1911		S			S
9382	Wendover	4237	1911					S
6897**	Piute Dam	5900	1911					
7686	Santaquin	5100	1914	S		R		
7909	Snake Creek	5950	1914			S		
7318	Riverdale	4390	1914	S		S	S	
7846	Silver Lake Brighton	8740	1915					S

* 1 = Spring (A,M); 2 = Summer (J,J,A); 3 = Late Summer (J,A,S); 4 = Fall (S,O); 5 = Winter (N,D,J,F,M).

** Records end 1969

Table 10 (Continued)

Tests of Homogeneity of Precipitation Records

Utah

Station Index Number	Station Name	Elevation (feet)	Year Record Begins	Season* 1	2	3	4	5
0086	Alton	7040	1915	S			S	
5969	Myton	5030	1915		S			S
3896	Hiawatha	7230	1916	S				S
1759	Cottonwood Weir	4950	1917					
8705	Thompson	5150	1918				S	

* 1 = Spring (A,M); 2 = Summer (J,J,A); 3 = Late Summer (J,A,S); 4 = Fall (S,O); 5 = Winter (N,D,J,F,M).

Table 11

Tests of Homogeneity of Precipitation Records

Wyoming

Station Index Number	Station Name	Elevation (feet)	Year Record Begins	Season* 1	2	3	4	5
1675	Cheyenne	6126	1871	S				S
9905	Yellowstone Park	6230	1889					
5830	Lusk	5000	1889			R	R	S
3490	Fort Laramie	4760	1894	S	S			R
0540	Basin	3837	1898		A	A	A	
3100	Evanston	6780	1899	S	R	S	S	R
5410**	Laramie	7266	1890	S	S		S	R
7105	Pathfinder Dam	5930	1899	S				
7235	Pine Bluffs	5047	1900				A	A
1730	Chugwater	5282	1900	S				
0915	Border	6120	1902		S	S		
4065	Green River	6089	1904	S		S	S	S
5345	Lake Yellowstone	7762	1904	A			A	A
2995	Elk Mountain	7270	1905	S	S			
7990	Saratoga	6790	1906					A
2715	Dubois	6917	1906					S
7380	Powell	4378	1907	S				
8385	South Pass City	7875	1909				S	S
0140	Alta	6431	1909	S			S	S
5770	Lovell	3837	1909					
5170+	Kirtley	5200	1904					
3630	Foxpark	9045	1910	A		A		A
4440	Hecla	6800	1912	A			S	

* 1 = Spring (A,M); 2 = Summer (J,J,A); 3 = Late Summer (J,A,S);

4 = Fall (S,O); 5 = Winter (N,D,J,F,M). A = Too much data absent

** Records end 1961 + Records end 1964

Table 11 (Continued)

Tests of Homogeneity of Precipitation Records

Wyoming

Station Index Number	Station Name	Elevation (feet)	Year Record Begins	Season* 1	2	3	4	5
1610++	Centennial	8060	1911					
1840++	Cody	4990	1911					
1905	Colony	3553	1914		S		S	R
3855	Gillette	4556	1915		A			
2415	Deaver	4105	1916	S				
8160	Sheridan Field Station	3800	1917					
6660	Newcastle	4265	1918					
7760	Riverton	4954	1918				S	S
7115	Pavillion	5400	1920					
4910	Jackson	6244	1920	S			S	S
2595	Diversion Dam	5574	1920					

* 1 = Spring (A,M); 2 = Summer (J,J,A); 3 = Late Summer (J,A,S); 4 = Fall (S,O); 5 = Winter (N,D,J,F,M).

++ Records end 1968

A = Too much data absent

Table 12

Percentage of Seasonal Precipitation Records Rejected, or Suspected of Being Inhomogeneous

State	Spring	Summer	Late Summer	Fall	Winter
Colorado	26	32	26	11	47
Idaho	32	18	16	29	42
Montana	28	8	13	25	43
Utah	24	11	18	29	47
Wyoming	41	19	13	32	40
Average	30	18	17	25	44

Table 13

Summary of Elevation Size Categories* for Long-Term Precipitation Stations in the Study Area

Elevations (feet)	Number of stations in:						
	Colorado	Idaho	Montana	Utah	Wyoming	Σ	%
1001 - 2000		2	1			3	2
2001 - 3000		12	10			22	13
3001 - 4000		7	14		4	25	15
4001 - 5000	2	8	8	14	8	40	23
5001 - 6000	4	8	5	19	6	42	25
6001 - 7000	4	1	1	2	10	18	11
7001 - 8000	1			2	4	7	4
8001 - 9000	4			1	1	6	4
9001 - 10,000	4				1	5	3
Total	19	38	39	38	34	168	100
Average Elevation by State (feet)	7,211	3,729	3,715	5,399	5,790		

Overall Average Elevation: 4,866

* Elevations from station indices in State Annual Summaries of Climatological Data for 1970 (U.S. Weather Bureau).

Table 14

Hypsometric Division of Study Area

State	Area (square miles)	Percent of State at an Elevation (in thousands of ft) between 1 and 2	2 and 5	5 and 10	>10
Colorado*	104,247	0	2.5	78.5	19
Idaho	83,557	0.5	43.5	54	2
Montana	147,138	2	73	24	1
Utah	84,916	0	31	66	3
Wyoming	97,914	0	19	77	4
Entire Study Area	517,772	0.6	37	57	6
Percent of long-term precipitation stations in elevation zones		2	51	47	0

* West of 106°W only.

Table 15

Highest Elevation Weather Bureau Stations with Precipitation Normals for 1941-70

State	Index No.	Station	Latitude	Longitude	Elevation (ft)
Colorado	4884	Leadville	39°15'	106°18'	10,158*
Idaho	4598	Island Park Dam	44°25'	111°24'	6,300
Montana	1995	Cooke City	45°36'	107°27'	7,553**
Utah	7846	Silver Lake Brighton	40°36'	111°35'	8,740+
Wyoming	3630	Foxpark	41°05'	106°09'	9,045+

*Only station normal at an elevation $>$10,000 feet.

**Only station normal in the state at an elevation $>$7,000 feet.

+Only station normal in the state at an elevation $>$8,000 feet.

Table 16

Precipitation Records with Statistically

Significant τ Values (Mann-Kendall Statistic), 1920-70

Spring	Tau + or -	Significance level %	Summer	Tau + or -	Significance level %
Co2281	-	5	Co3488	-	10
Co3113	-	10	Id1002	+	5
Co7618	+	10	Id1380	+	10
Id3732	-	10	Id2676	+	10
Id6152	+	10	Id3448	+	5
Id8818	+	5	Id4295	+	5
Mo5015	+	5	Id4670	+	5
Mo7382	+	10	Id6542	+	10
Wy7760	-	5	Id7264	+	10
			Id7386	+	10
			Id8818	+	1
			Mo4241	+	10
			Mo5337	-	5
			Mo7286	+	10
			Mo7382	+	10
			Mo8597	+	1
			Ut7318	+	5
			Wy0140	+	10
			Wy7380	+	10
			Wy9905	+	10

Late Summer	Tau + or -	Significance level %	Fall	Tau + or -	Significance level %
Co0228	+	5	Wy3490	-	10
Co3488	-	5	Wy6660	-	5
Id4295	+	5	Wy7105	-	10
Mo5337	-	5	Wy7115	-	5
Ut2253	-	10			
Ut4856	-	10			
Ut5969	-	5			
Wy3490	-	5			
Wy4910	-	10			
Wy7105	-	5			
Wy7760	-	10			

Table 16 (Continued)

Precipitation Records with Statistically

Significant τ Values (Mann-Kendall Statistic), 1920-70

Winter	Tau + or -	Significance level %
Co1959	+	5
Co2281	-	1
Id6542	+	5
Id7264	+	5
Id7386	+	10
Mo0780	-	10
Mo1297	+	10
Mo2173	-	10
Mo5337	-	10
Ut4527	-	10
Ut5610	-	10
Wy0540	-	0.1
Wy1610	-	1
Wy1840	-	10
Wy2415	-	5
Wy2995	-	10
Wy5-70	-	1
Wy6660	-	5
Wy8160	-	5

Table 17

Summary of the Seasonal Decadal Precipitation, 1890-1970 in Terms of the 1941-70 Normal

	Spring	Summer	Fall	Winter
1960s	$\bar{x}$ except in SE, S.Idaho/N.Utah,W. Idaho & a band N/S through W.Montana.	$>\bar{x}$ in all areas except N.Idaho & N.Montana.	Below $\bar{x}$ except in N.Montana, N&C.Idaho, C&E.Wyoming & S.Utah.	Predominantly $<\bar{x}$ except in S.Utah, W.Idaho, & parts of S.Montana.
1950s	Almost the inverse of the 1940s; $>\bar{x}$ in NW half, $<\bar{x}$ in SE half.	Below $\bar{x}$ in all areas except NW Idaho.	Markedly $<\bar{x}$ in all areas.	$<\bar{x}$ in a zone across Idaho, S.Montana & N. Wyoming, $>\bar{x}$ elsewhere.
1940s	$<\bar{x}$ in NW & W except N.Idaho; $>\bar{x}$ in C & SE.	Above $\bar{x}$ in all areas except S.Idaho, S. Utah, & S&C.Colorado.	Predominantly $>\bar{x}$ except in N.W. Colorado & W.C.Idaho.	Predominantly $<\bar{x}$ except in N.Idaho, C.Utah & parts of W.Colorado.
1930s	Markedly $<\bar{x}$ particularly in SW. Montana, W.Idaho & E.Wyoming.	Predominantly $<\bar{x}$.	Generally $<\bar{x}$ but above in N.C. Montana, S.Idaho, Utah border areas & parts of W. Colorado.	Generally $<\bar{x}$ except in N.E. Montana, C.Wyoming & Upper Colorado River area in Utah & Colorado.
1920s	$<\bar{x}$ in most areas except S.Idaho, E&S. Utah & parts of C.Wyoming & W. Colorado.	Above $\bar{x}$ in Utah, W. Colorado & most of Wyoming. Below $\bar{x}$ elsewhere.	Above $\bar{x}$ in all areas except W.Idaho & NW. Montana.	$<\bar{x}$ in W (W&S.Idaho & most of Utah) & $>\bar{x}$ elsewhere.

Table 17 (continued)

	Spring	Summer	Fall	Winter
1910s	Below $\bar{x}$ in C.Idaho & along a band through C.Utah to SW Colorado. Also $<\bar{x}$ in parts of N&SC. Montana. $>\bar{x}$ elsewhere.	Below $\bar{x}$ in all areas except S.Utah & S. Colorado.	Above average in all areas except N. Utah.	$<\bar{x}$ in NW, $>\bar{x}$ in S and E.
1900s	Predominantly $>\bar{x}$ except in W.Idaho.	Below $\bar{x}$ in nearly all areas.	$<\bar{x}$ in Idaho, C. Colorado, Montana & W.Wyoming; $\bar{x}$ in E.Wyoming, Utah & Colorado.	Generally $>\bar{x}$ except in C. Wyoming.
1890s	Generally slightly $>\bar{x}$ except in S where precipitation is $<\bar{x}$.	Below $\bar{x}$ in all areas.	Above $\bar{x}$ (?) except in S & C. Montana	Mainly $>\bar{x}$ except in W. Idaho & the Utah/Colorado border area.

Table 18

STATION RECORDS ANALYZED BY POWER SPECTRUM ANALYSIS, (LAGS $\sim \frac{N}{7}$)

All Records $\geq$ 65 Years, and not Suspect or Rejected

Colorado

Spring	Summer	Late Summer	Fall	Winter
1741	2192	2192	1741	2432
2192	3146	2432	2192	3146
3146	3488	3146	2432	5722
3488	5722	3488	3146	
		5722	3359	
			5722	

Idaho

Spring	Summer	Late Summer	Fall	Winter
0915	0470	1380	0470	0915
1408	0915	1408	1380	1380
4831	1380	4831	1408	1408
6152	1408	6152	6152	6152
6388	4457	6388	6388	6542
6681	6152	6542	6542	6681
	6388	6891	6891	6891
	6542		7264	7264
	6891			
	7264			

Montana

Spring	Summer	Late Summer	Fall	Winter
0364	0364	0802	0802	0364
0802	0802	1044	1044	1297
1044	1044	1297	1297	2604
1297	1297	2112	1318	3581
1318	1318	2604	2112	4055
2604	2112	2689	2604	4985
2689	2604	3581	2689	5015
3581	2689	3994	3581	7382
3994	3581	4055	3994	
4055	3994	4985	4985	
5015	4055	5015	5015	
7382	4985	7382	7382	
	5015			
	7382			

Utah

Spring	Summer	Late Summer	Fall	Winter
0738	0738	0738	0738	2101
2101	2101	2101	2101	2418
2484	2418	2418	2418	2996
2828	2484	2484	2484	4856
2996	2828	2828	2996	5065
3809	3809	2996	3809	6357
4856	4856	4865	5065	
5065	5065	5065	5826	
5826	5402	5402	6357	
6357	5752	5752	7714	
7714	5826	5826	8771	
8771	6357	6357		
8973	7714	7714		
	8771	8771		
	8973			

Wyoming

Spring	Summer	Late Summer	Fall	Winter
0540	1730	1730	0915	0540
0915	4065	2995	1730	0915
5830	5345	3490	2995	1730
7235	5830	5345	3490	2995
9905	7105	5410	9905	7105
	7235	7105		9905
	9905	7235		
		9905		

Table 19

SUMMARY OF SEASONAL RECORDS EXAMINED

	Spring	Summer	Late Summer	Fall	Winter	Average Record Length (years)
Colorado	4	4	5	6	3	77
Idaho	6	10	7	8	8	74
Montana	12	14	12	12	8	73
Utah	13	15	14	11	6	73
Wyoming	5	7	8	6	6	73
	40	50	46	43	31	(210) Total

Note. Seasonal records are not necessarily all from different stations. In all, the 210 seasonal records came from 63 stations.

Table 20

SEASONAL RECORDS EXHIBITING STATISTICALLY SIGNIFICANT (>95%)

LAG 1 AUTOCORRELATION COEFFICIENTS (r_1)

State	Number	Record Length (years)	Statistically Significant r_1 Value in: Spring	Summer	Late Summer	Fall	Winter
Colorado	2192	83			-0.348*		
	2432	76			-0.266		
	3016+	59		-0.478**			
	3146	68		-0.338		+0.209	
	3488	79		-0.280			
	5722	86			-0.258		
Idaho	0470	74		+0.218			
	0915	75		+0.207			+0.244
	1408	77			+0.184		
	4457	66		+0.236			
	6891	81			+0.200		
	7264	81		+0.297*			+0.238
Montana	1297	67					+0.196
	1318	77				+0.270*	
	3581	81					+0.524**
	4055	90					+0.349**
	4985	74					+0.184
	5015	75					+0.264*
Utah	2484	69				+0.255	
	3896+	54		-0.469*			
	4856	70					+0.336*
	5826	68				+0.190	
	8705+	53		-0.361			
	8771	75				+0.277	
Wyoming	0540	72					+0.196
	2995	65				+0.273	+0.313*
	5830	82	-0.314*				

* Significant at > 99% level

** Significant at > 99.9% level

+ Stations analyzed in a subsequent study of the Colorado River Basin area. For this, records did not have to be ≥ 65 years.

Table 21

FREQUENCIES OF PRINCIPAL SPECTRAL PEAKS AND GAPS[1]: COLORADO

	Significance Level (%)						
	Peaks			Gaps			
	99	97.5	95	1	2.5	5	% Significance of r_1
Spring		- None -			- None -		
Summer							
3146	0.45(2.2) 0.5(2.0)					0.0(∞)	-, 5%
3488	0.45(2.2) 0.5(2.0)						-, 5%
5722		0.5(2.0)	0.462(2.16)				
Late Summer							
2192	0.5(2.0)	0.458(2.18)		0.125(8)			-, 1%
2432		0.5(2.0)	0.455(2.2)				-, 5%
3146	0.5(2.0)	0.45(2.16)			0.35(2.9)		
3488			0.5(2.0)				
5722		0.462(2.16) 0.5(2.0)					-, 5%
Fall							
2432						0.0(∞)	
Winter							
2432			0.0(∞)				
5722					0.5(2)		

[1]Periods corresponding to the frequencies given in parentheses (in years).

Table 22

FREQUENCIES OF PRINCIPAL SPECTRAL PEAKS AND GAPS[1]: IDAHO

	Significance Level (%)					
	Peaks			Gaps		
	99	97.5	95	1	2.5	5
Spring						
		- None -			- None -	
Summer						
1380						0.5(2.0)
6542		0.0(∞)				
Late Summer						
1380			0.2(5.0)			
Fall						
1408					0.5(2.0)	
6152		0.167(6.0)			0.0(∞) 0.042(23.8)	
6891				0.0(∞)	0.5(2.0)	
Winter						
6152	0.3(3.3)	0.292(3.4)		0.0(∞)		
6542		0.0(∞)				
6681					0.0(∞)	

[1]Periods corresponding to the frequencies given in parentheses (in years).

Table 23

FREQUENCIES OF PRINCIPAL SPECTRAL PEAKS AND GAPS[1]: MONTANA

	Significance Level (%)					
	Peaks			Gaps		
	99	97.5	95	1	2.5	5
Spring						
0802			0.1(10.0)			
1297				0.25(4.0)		0.3(3.3)
2689			0.4(2.5)			
4055		0.0(∞)				
Summer						
1044					0.5(2.0)	
2112			0.273(3.7)			
3581			{0.25(4.0) 0.292(3.4)}			
4055		0.154(6.5)	0.308(3.2)			0.5(2.0)
5015	0.0(∞)					0.136(7.35)
Late Summer						
0802						0.5(2.0)
1044						0.409(2.4)
2112			0.227(4.4)			0.0(∞)
		{0.273(3.66)			0.5(2.0)	0.455(2.2)
2604			{0.3(3.3) 0.35(2.86)}			
2689					0.5(2.0)	
4055					0.5(2.0)	
5015	0.0(∞)					0.5(2.0)
Fall						
1044			0.5(2.0)			0.364(2.75)
2604			0.5(2.0)			
2689			0.2(5.0)			
3581			0.25(4.0)			
7382		0.25(4.0)				
Winter						
2604			0.5(2.0)			

[1]Periods corresponding to the frequencies given in parentheses (in years)

Table 24

FREQUENCIES OF PRINCIPAL SPECTRAL PEAKS AND GAPS[1]: UTAH

	Significance Level (%)					
	Peaks			Gaps		
	99	97.5	95	1	2.5	5
Spring						
0738						0.0(∞)
2484				0.0(∞)		
2828			0.33(3.0) 0.375(2.6)			
2996			0.083(12.0)			
5065			0.33(3.0)			
5826			0.0(∞)			0.25(4.0)
6357		0.35(2.86)				
8973		0.35(2.86)				
Summer						
0738			0.5(2.0)			
2418		0.5(2.0)			0.35(2.9)	
2484		0.5(2.0)				
3809		0.458(2.18)				
4856	0.5(2.0)					0.364(2.7)
5402			0.455(2.2)			
5826			0.05(20.0)			
7714			0.455(2.2)			
8973		0.5(2.0)				
Late Summer						
2418			0.5(2.0)			
5402		0.0(∞)			0.182(5.5) 0.227(4.4)	
5826					0.15(6.7) 0.2(5.0)	
Fall						
2101		0.0(∞)				
2996			0.375(2.6)		- None -	
6357			0.0(∞)			
Winter						
2101		0.0(∞)				
5065				0.125(8.0)		0.167(6.0)
6357			0.0(∞)	0.1(10.0)		

[1]Periods corresponding to the frequencies given in parentheses (in years).

Table 25

FREQUENCIES OF PRINCIPAL SPECTRAL PEAKS AND GAPS[1]: WYOMING

	Significance Level (%)					
	Peaks			Gaps		
	99	97.5	95	1	2.5	5
Spring						
0540			0.318(3.14)	0.0(∞)	0.045(22.2)	
0915		0.35(2.9)				
5830		0.417(2.4)	0.375(2.7)		0.25(4.0)	0.208(4.8)
					0.0(∞)	
Summer						
4065			0.2(5.0)		0.35(2.9)	
7105			0.182(5.5)			0.455(2.2)
7235			0.273(3.7)			
9905		0.292(3.4)				
Late Summer						
1730		0.35(2.9)		0.0(∞)		
2995	0.0(∞)					
3490			0.364(2.8)			
5410			0.273(3.7)			
9905				0.5(2.0)		
Fall						
1730			0.2(5.0)			
2995						0.35(2.9)
						0.4(2.5)
7105			0.182(5.5)			0.0(∞)
Winter						
		- None -			- None -	

[1]Periods corresponding to the frequencies given in parentheses (in years).

Table 26

SUMMARY OF PRINCIPAL PERIODICITIES (IN YEARS) IDENTIFIED BY

POWER SPECTRUM ANALYSIS[1]

	Spring	Summer	Late Summer	Fall	Winter
Colorado		2.0 (1) 2.2 (3)	2.16 (1) 2.18 (1) 2.0 (5) 2.2 (2)		∞ (1)
Idaho		∞ (1)	5.0 (1)	6.0 (1)	3.3 (1) 3.4 (1) ∞ (1)
Montana	2.5 (1) 10.0 (1) ∞ (1)	3.2 (1) 3.4 (1) 3.7 (1) 4.0 (1) 6.5 (1) ∞ (1)	2.86 (1) 3.31 (1) 3.66 (1) 4.41 (1) ∞ (1)	2.0 (3) 4.0 (2) 5.0 (1)	2.0 (1)
Utah	2.6 (1) 2.86 (2) 3.0 (2) 12.0 (1) ∞ (1)	2.0 (5) 2.2 (4) 20 (1)	2.0 (1) ∞ (1)	2.6 (1) ∞ (2)	∞ (2)
Wyoming	2.4 (1) 2.7 (1) 2.9 (1) 3.14 (1)	3.4 (1) 3.7 (1) 5.0 (1) 5.5 (1)	2.8 (1) 2.9 (1) 3.7 (1) ∞ (1)	5.0 (1) 5.5 (1)	

[1]Numbers in parentheses indicate the number of records exhibiting this periodicity. Note that peaks are not necessarily all from different stations (see Tables 21 to 25).

Table 27

Additional Summer Records Analyzed in Eastern Utah and Western Colorado: Significant Spectral Peaks[1]

State	Years	99	97.5	95
Colorado				
1959	62	No significant peaks		
2281	62	0.5(2.0)		
3016	59			
3113	61	0.5(2.0)		0.44(2.25)
6513	62	No significant peaks		
7017	61	No significant peaks		
7618	61	0.5(2.0)		0.44(2.25)
7656	64	0.5(2.0)	0.45(2.2)	
7936	62		0.5 (2.0)	
Utah				
0061	60		0.5 (2.0)	
0086	55		0.5 (2.0)	
1759	54		0.5 (2.0)	
3611	61			0.056(17.86)
3896	54			
4527	63			0.5 (2.0)
5610	59		0.5 (2.0)	
5837	63	No significant peaks		
6534	61		0.5 (2.0)	
6658	59	No significant peaks		
6897	58	0.5(2.0)		
7318	57		0.5 (2.0)	
7846	55	0.5(2.0)		
7909	57	0.5(2.0)		0.44(2.25)
8119	61	0.5(2.0)		0.44(2.25)
8705	53			
9382	59	No significant peaks		

[1]Periods corresponding to frequencies given in parentheses.

Table 28

Comparison of Power Spectrum Analyses with

Lags $\sim \frac{N}{7}$ and Lags $\sim \frac{N}{4}$

Summer

State	Station	Main Spectral Peaks $\frac{N}{7}$	Main Spectral Peaks $\frac{N}{4}$
Colorado	3146	2.2, 2.0	2.3, 2.1, 2.0
	3488	2.2, 2.0	2.2, 2.1, 2.0
	5722	2.2, 2.0	2.0
Idaho	0470	No significant peaks	
	0915	No significant peaks	
	1380	No significant peaks	
	4457	No significant peaks	
	6542	∞	∞, 40.0
	7264	No significant peaks	
Montana	1044	No significant peaks	
	2112	3.7	3.8
	3581	4.0, 3.4	3.8, 3.5
	4055	6.5, 3.2	6.6, 5.7, 3.5, 3.3
	5015	∞	∞, 38.5
Utah	0738	2.0	2.0
	2418	2.0	2.0
	2484	2.0	2.0
	3809	2.2	2.35, 2.2
	4856	2.0	2.1, 2.0
	5402	2.2	2.2
	5826	20.0	16.95
	7714	2.2	2.2
	8973	2.0	2.0

Table 28 (Continued)

Comparison of Power Spectrum Analyses with

Lags $\sim \frac{N}{7}$ and Lags $\sim \frac{N}{4}$

Summer

State	Station	Main Spectral Power $\frac{N}{7}$	Main Spectral Power $\frac{N}{4}$
Wyoming	4065	5.0	5.7
	7105	5.5	5.15
	7235	3.7	3.6
	9905	3.4	{3.5 {3.2

Table 29

Comparison of Power Spectrum Analyses with Maximum

Lags up to $\sim\frac{N}{7}$ and up to $\sim\frac{N}{4}$

Winter

State	Station	Main Spectral Peaks	
		$\frac{N}{7}$	$\frac{N}{4}$
Colorado	2432	∞	38.5, 2.7
	5722		11.0
Idaho	0915	No significant peaks	
	1408		2.2
	6152	3.3, 3.4	3.3, 3.1, 2.9
	6542	∞	∞, 40.0
	6681	No significant peaks	
	6891	No significant peaks	
	7264	No significant peaks	
Montana	1297	No significant peaks	
	2604	2.0	2.3, 2.1
	3581	No significant peaks	
	4055	No significant peaks	
	4985	No significant peaks	
	5015	No significant peaks	
Utah	2101	∞	∞
	4856	No significant peaks	
	5065	No significant peaks	∞, 2.2
	6357	∞	∞
Wyoming	0540	No significant peaks	
	2995	No significant peaks	

Table 30

Zonal and Meridional Weather Type Frequencies[1] in the 1930s and 1940s (Percentage of Days per Season)[2]

		Spring (A,M)	Summer (J,J,A)	Fall (S,O)	Winter (N,D,J,D,M)
a) 1931-40	Zonal	46	31	30	49
	Meridional	53	69	70	50
b) 1941-50	Zonal	47	21	33	39
	Meridional	52	76	65	60

	Zonal Type Frequencies (b-a)[2]	Meridional Type Frequencies (b-a)[2]
Spring	+1%	-1%
Summer	-10%	+8%
Fall	+3%	-5%
Winter	-10%	+10%

[1]Zonal types = types B, B_S, B_{S-B}, E_M, E_H, E_L, E_N

Meridional types = A, B_{n-a}, B_{n-b}, B_{n-c}, C_H, C_L, D, E_J

For discussion of these types see Elliott, 1949; for idealized maps, see Appendix C.

[2]Discrepancies between decades are due to differences in amounts of missing data per decade and to rounding errors.

Table 31

FREQUENCIES OF THREE-DAY SYNOPTIC TYPES IN THE 1930s AND 1940s (PERCENTAGE OF DAYS PER SEASON)[1]

Types (Elliott, 1949):	A	B	B_{n-a}	B_{n-b}	B_{n-c}	B_s	B_{SB}	C_H	C_L	D	E_M	E_H	E_L	E_N
Spring														
1931-40	19	11	3	11	10	1	1	1	7	3	13	1	14	5
1941-50	22	19	4	7	11	0	0	1	3	4	11	0	16	1
Summer														
1931-40	27	24	20	5	9	0	0	0	1	8	3	0	4	<1
1941-50	30	16	23	1	12	0	0	0	<1	10	1	0	3	<1
Fall														
1931-40	19	15	10	10	13	0	3	0	9	9	3	0	6	2
1941-50	25	18	10	7	13	2	2	0	5	4	4	0	5	2
Winter														
1931/2-1939/40	12	9	7	9	4	3	3	1	12	4	10	4	11	8
1941/2-1949/50	19	8	12	7	7	1	2	1	12	3	9	1	11	6

[1]Discrepancies between decades are due to differences in amounts of missing data per decade, and to rounding errors.

Table 31

Change in Type Frequency from the 1930s to the 1940s[1]

	A	B	B_{n-a}	B_{n-b}	B_{n-c}	B_S	B_{SB}	C_H	C_L	D	E_M	E_H	E_L	E_N
Spring	+3	+8	+1	-4	+1	-1	-1	0	-4	+1	-2	-1	+2	-4
Summer	+3	-8	+3	-4	+3	0	0	0	-1	+2	-2	0	-1	0
Fall	+6	+3	0	-3	0	+2	-1	0	-4	-5	+1	0	-1	0
Winter	+7	-1	+5	-2	+3	-2	-1	0	0	-1	-1	-3	0	-2

[1]Discrepancies between decades are due to differences in amounts of missing data per decade, and to rounding errors.

Table 32

FREQUENCIES OF THREE-DAY SYNOPTIC TYPES, 1951-58[1] (PERCENTAGE OF DAYS PER SEASON)

	A	B	B_{n-a}	B_{n-b}	B_{n-c}	B_S	B_{S-B}	C_H	C_L	D	E_M	E_H	E_L	E_N	Absent
Spring	14	19	8	6	14	1	0	0	11	7	7	1	1	4	10
Summer	(insufficient data)														
Fall	11	13	9	17	16	1	1	0	9	4	6	0	5	4	4
Winter (51/2-57/58)	16	8	6	11	10	4	0	1	6	7	10	2	9	9	1
Changes in Type Frequencies from the 1930s to the 1950s[1]															
Spring 50s-30s	-5	+8	+5	-5	+4	0	-1	-1	+4	+4	-6	0	-13	-1	
Fall 50s-30s	-8	-2	-1	+7	+3	+1	-2	0	0	-5	+3	0	-1	+2	
Winter 50s-30s	+4	-1	-1	+2	+6	+1	-3	0	-6	+3	0	-2	-2	+1	
Changes in Type Frequencies from the 1940s to the 1950s[1]															
Spring	-8	0	+4	-1	+3	-1	0	-1	+8	+3	-4	1	-15	+3	
Fall	-14	-5	-1	+10	+3	-1	-1	0	+4	0	+2	0	0	+2	
Winter	-3	0	-6	+4	+3	+3	-2	0	-6	+4	+1	+1	-2	+3	

[1]Discrepancies between decades are due to differences in amounts of missing data per decade and to rounding errors.

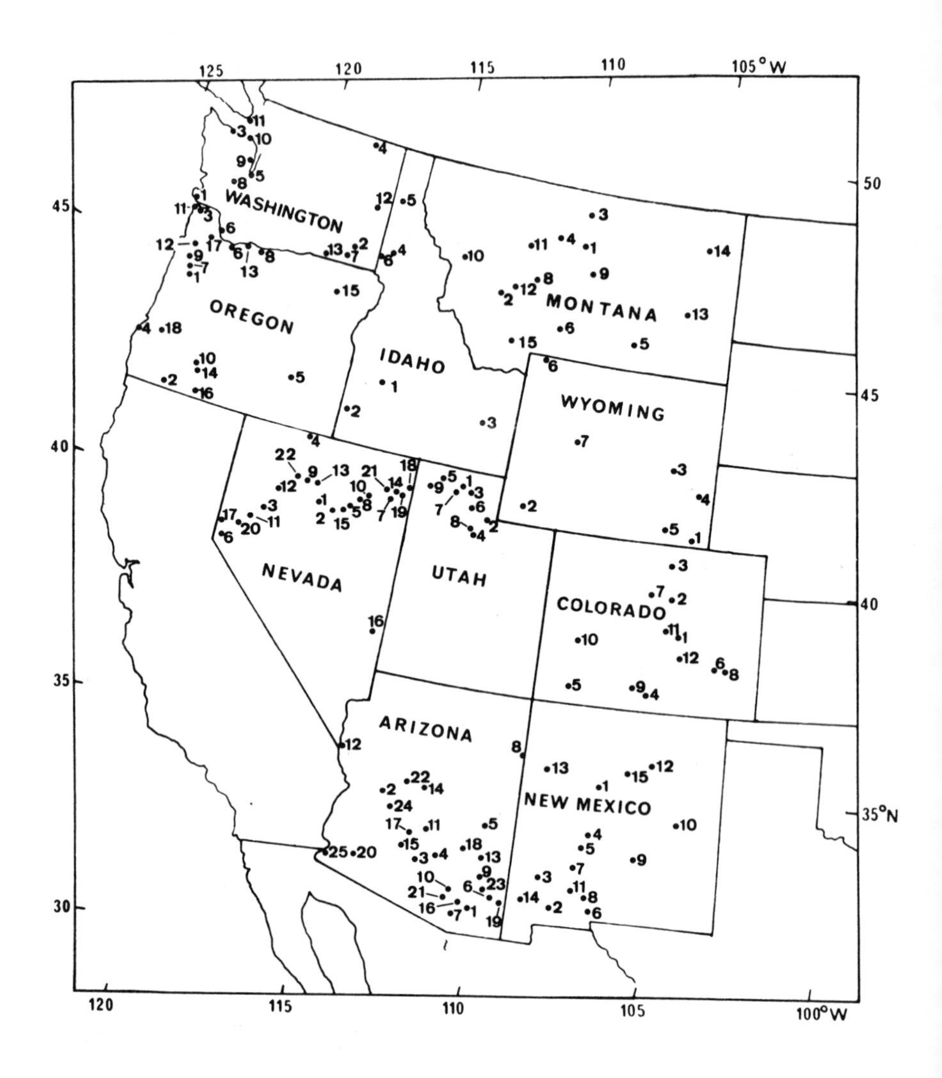

Figure 1: Precipitation stations operating for more than four seasons, 1851-1890.

KEY TO FIGURE 1

Arizona

1. Benson
2. Camp Date Creek
3. Casa Grande
4. Florence
5. Fort Apache
6. Fort Bowie
7. Fort Buchanan
8. Fort Defiance
9. Fort Grant
10. Fort Lowell
11. Fort McDowell
12. Fort Mojave
13. Fort Thomas
14. Fort Verde
15. Maricopa
16. Pantano
17. Phoenix
18. San Carlos
19. San Simon
20. Texas Hill
21. Tucson
22. Whipple Barracks
23. Willcox
24. Wickenburg
25. Yuma

Colorado

1. Colorado Springs
2. Denver
3. Fort Collins
4. Fort Garland
5. Fort Lewis
6. Fort Lyon
7. Georgetown
8. Las Animas
9. Monte Vista
10. Montrose
11. Pike's Peak
12. South Pueblo

Idaho

1. Boise
2. Camp Three Forks
3. Fort Hall
4. Fort Lapwai
5. Fort Sherman
6. Lewiston

Montana

1. Camp Cooke
2. Deer Lodge
3. Fort Assiniboine
4. Fort Benton
5. Fort Custer
6. Fort Ellis
7. Fort Keogh
8. Fort Logan
9. Fort Maginnis
10. Fort Missoula
11. Fort Shaw
12. Helena
13. Miles City
14. Poplar
15. Virginia City

Nevada

1. Battle Mountain
2. Beowawe
3. Brown's
4. Camp McDermit
5. Carlin
6. Carson City
7. Cedar Pass
8. Elko
9. Golconda
10. Halleck
11. Hot Springs
12. Humboldt
13. Iron Point
14. Otego
15. Palisade
16. Pioche
17. Reno
18. Tecoma
19. Toano
20. Wadsworth
21. Wells
22. Winnemucca

New Mexico

1. Albuquerque
2. Deming
3. Fort Bayard
4. Fort Conrad
5. Fort Craig
6. Fort Fillmore
7. Fort McRae
8. Fort Selden
9. Fort Stanton
10. Fort Sumner
11. Fort Thorn
12. Fort Union
13. Fort Wingate
14. Lordsburg
15. Santa Fe

Oregon

1. Albany
2. Ashland
3. Astoria
4. Bandon
5. Camp Harney
6. Cascade Locks
7. Eola
8. Fort Dalles
9. Fort Hoskins
10. Fort Klamath
11. Fort Stevens
12. Fort Yamhill
13. Hood River
14. Klamath Agency
15. La Grande
16. Linkville
17. Portland
18. Roseburg

Utah

1. Blue Creek
2. Coalville
3. Corrine
4. Douglas
5. Kelton
6. Ogden
7. Promontory
8. Salt Lake City
9. Terrace

Washington

1. Cape Disappointment
2. Dayton
3. Ediz Hook Lighthouse
4. Fort Colville
5. Fort Steilacoom
6. Fort Vancouver
7. Fort Walla Walla
8. Olympia
9. Port Blakely
10. Port Townsend
11. San Juan Island
12. Spokane Falls
13. Walla Walla

Wyoming

1. Cheyenne
2. Fort Bridger
3. Fort Fetterman
4. Fort Laramie
5. Fort Sanders
6. Fort Sheridan
7. Fort Washakie

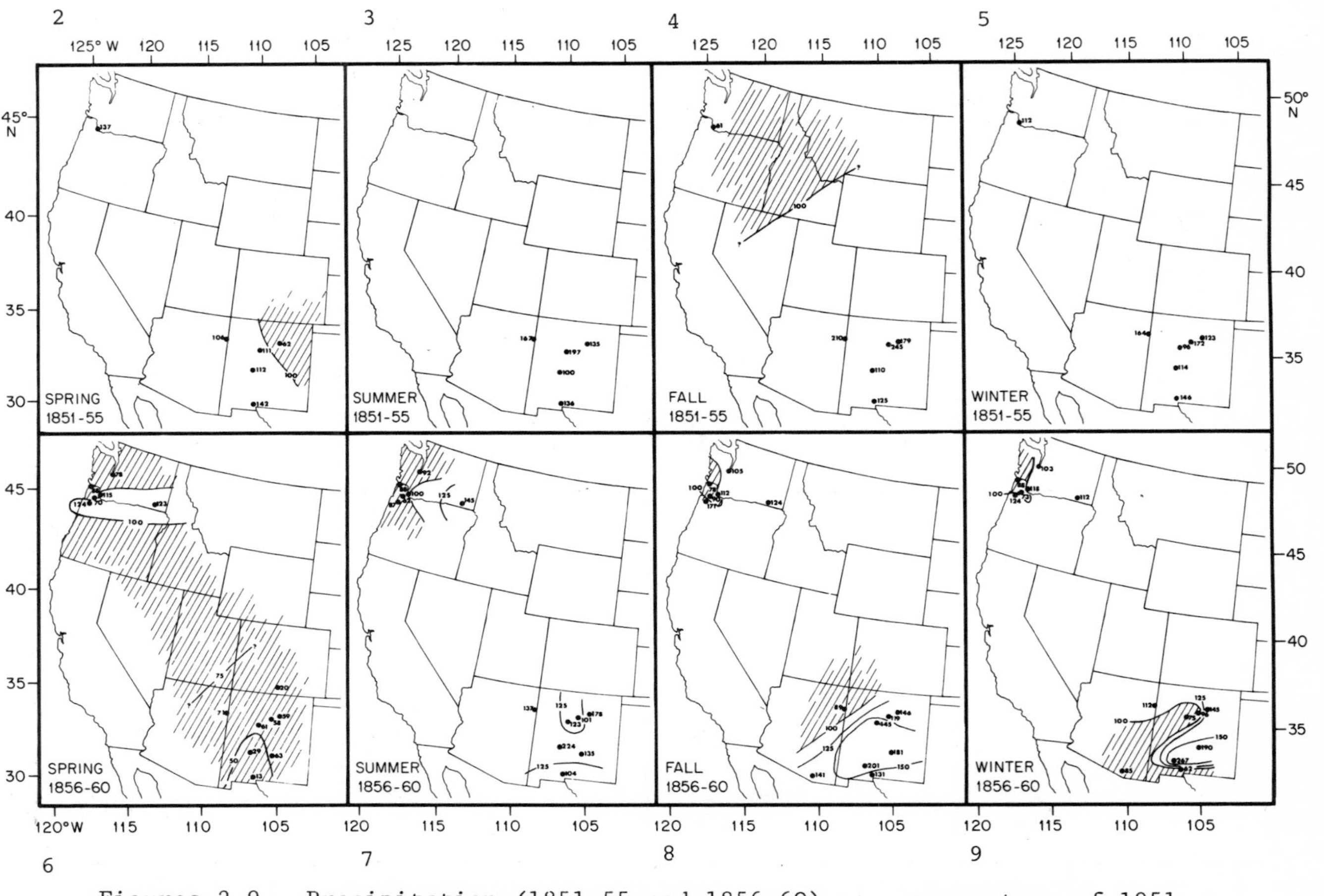

Figures 2-9: Precipitation (1851-55 and 1856-60) as a percentage of 1951-60 averages at analog stations.

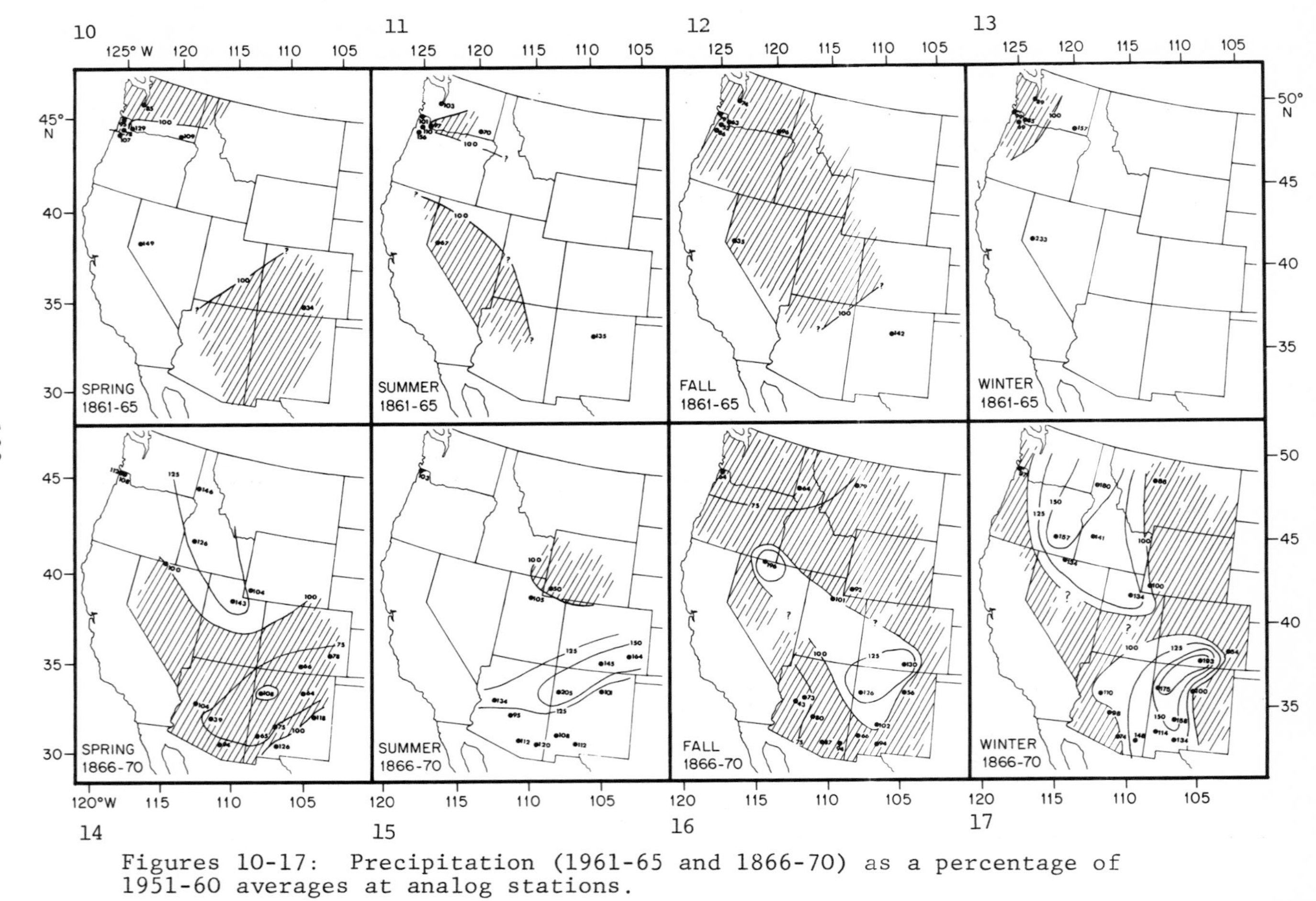

Figures 10-17: Precipitation (1961-65 and 1866-70) as a percentage of 1951-60 averages at analog stations.

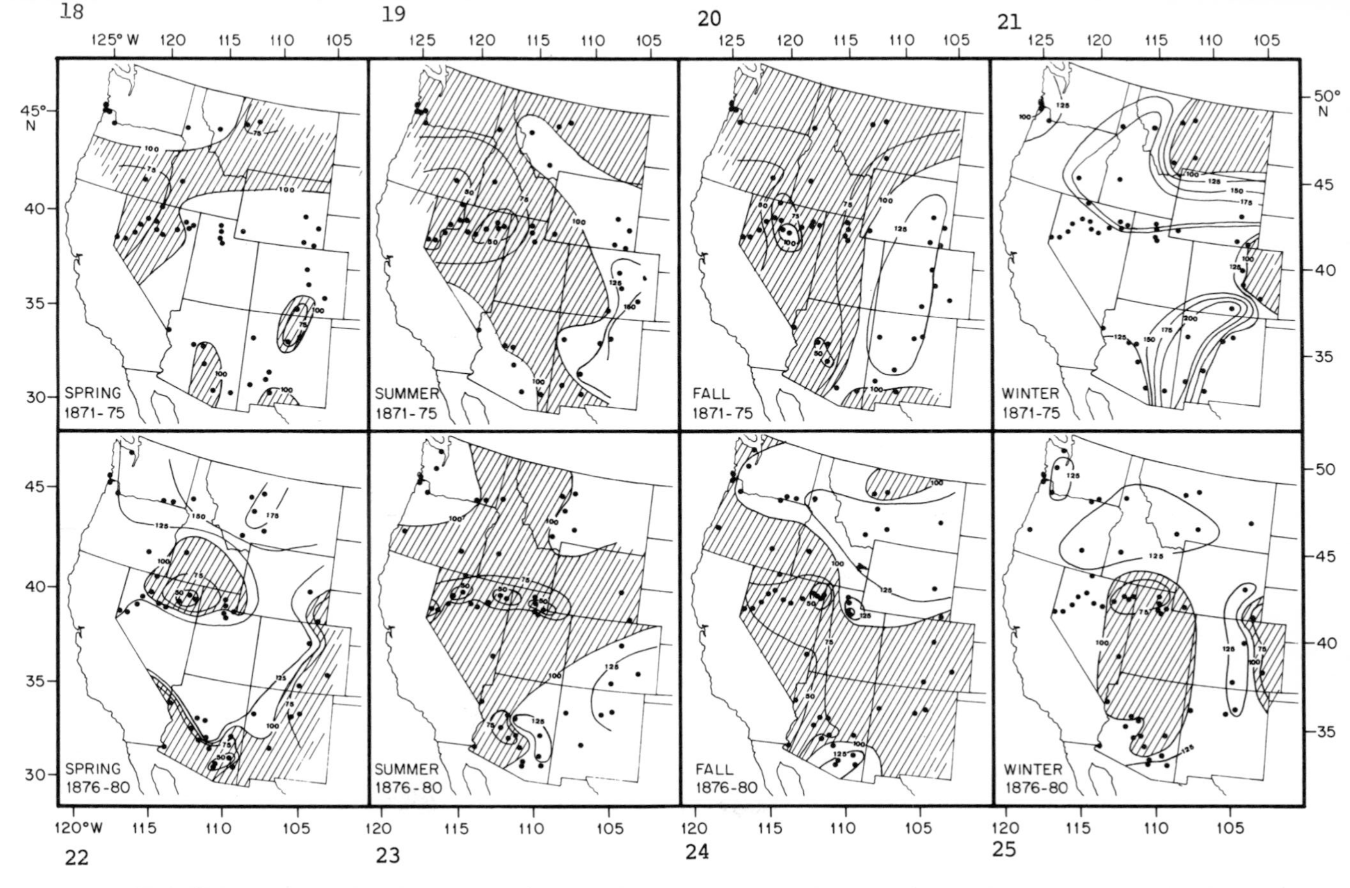

Figures 18-25. Precipitation (1971-75 and 1876-80) as a percentage of 1961-60 averages at analog stations.

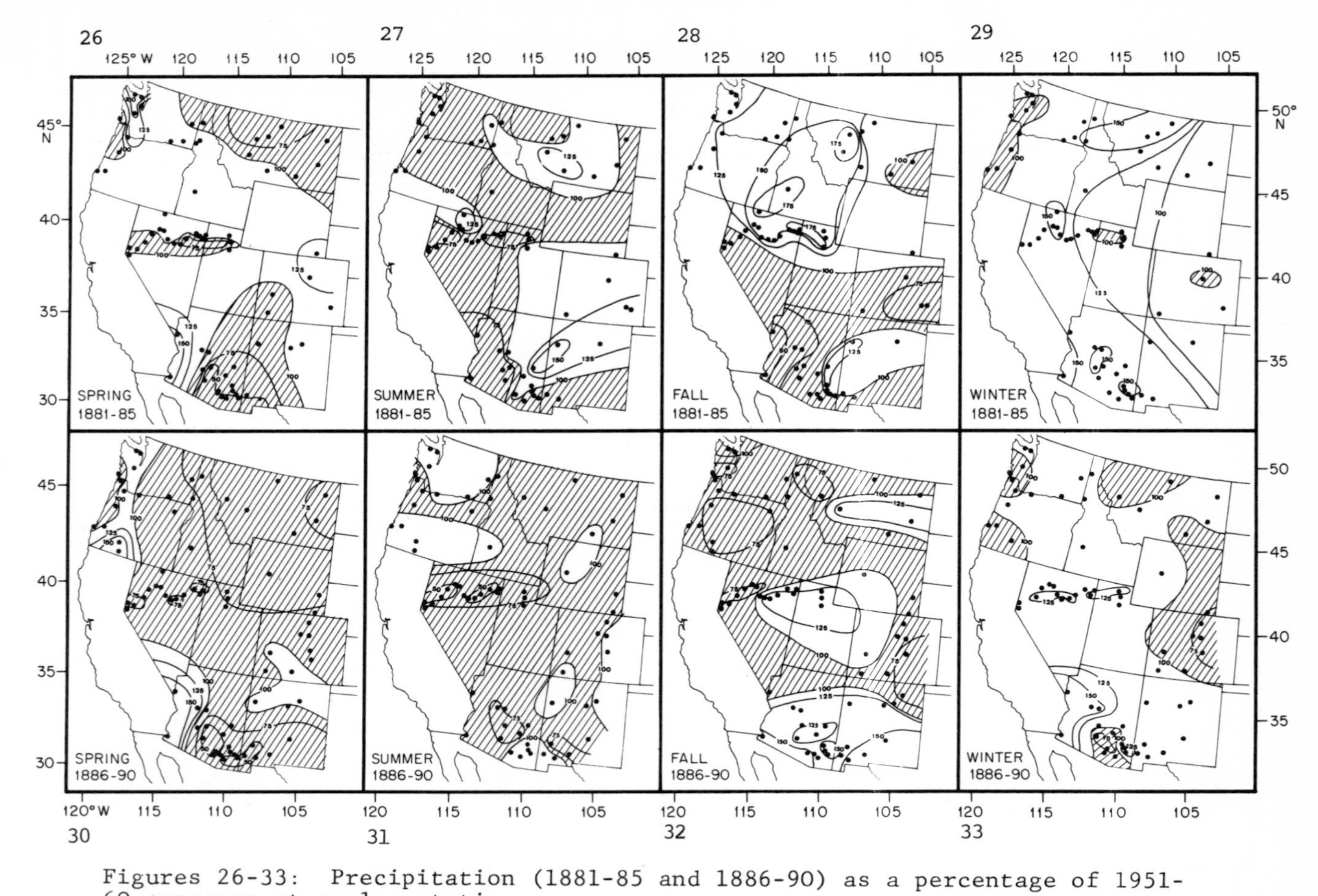

Figures 26-33: Precipitation (1881-85 and 1886-90) as a percentage of 1951-60 averages at analog stations.

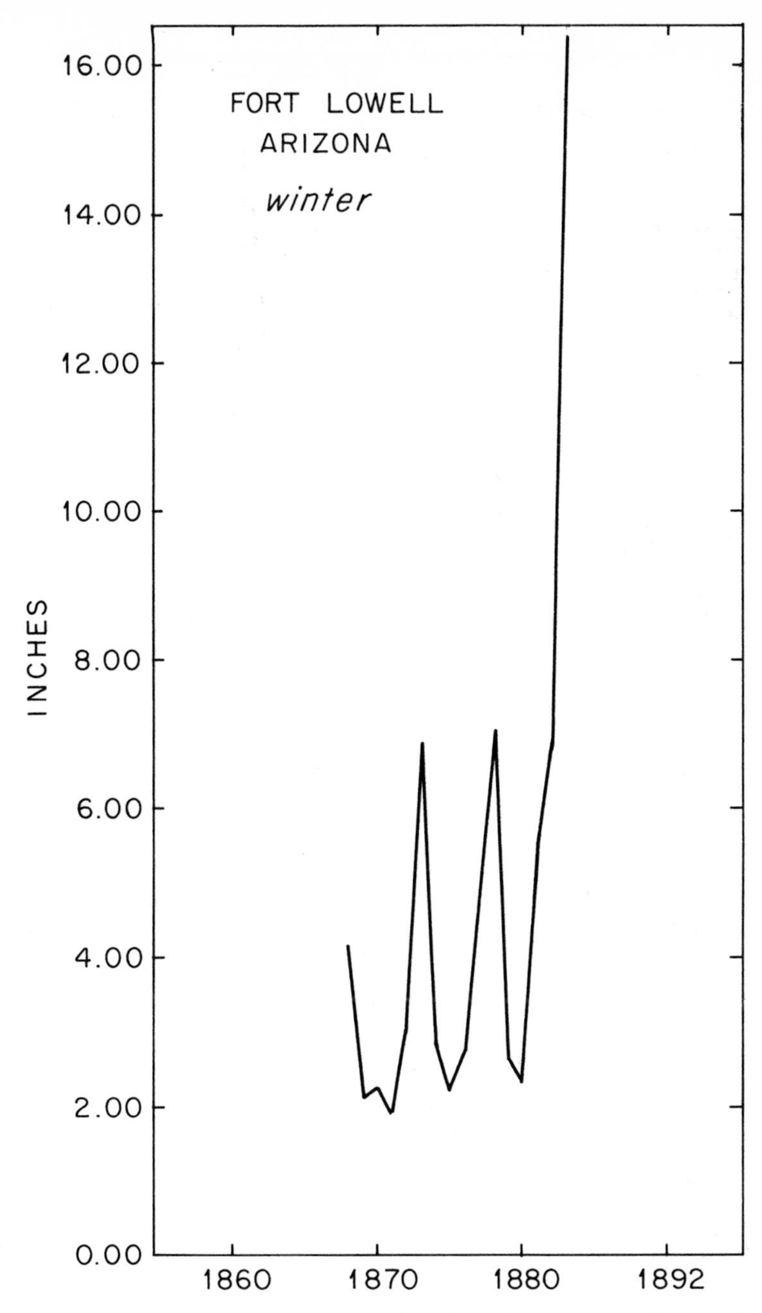

Figure 34: Recurrent peaks in winter precipitation at Fort Lowell, Arizona.

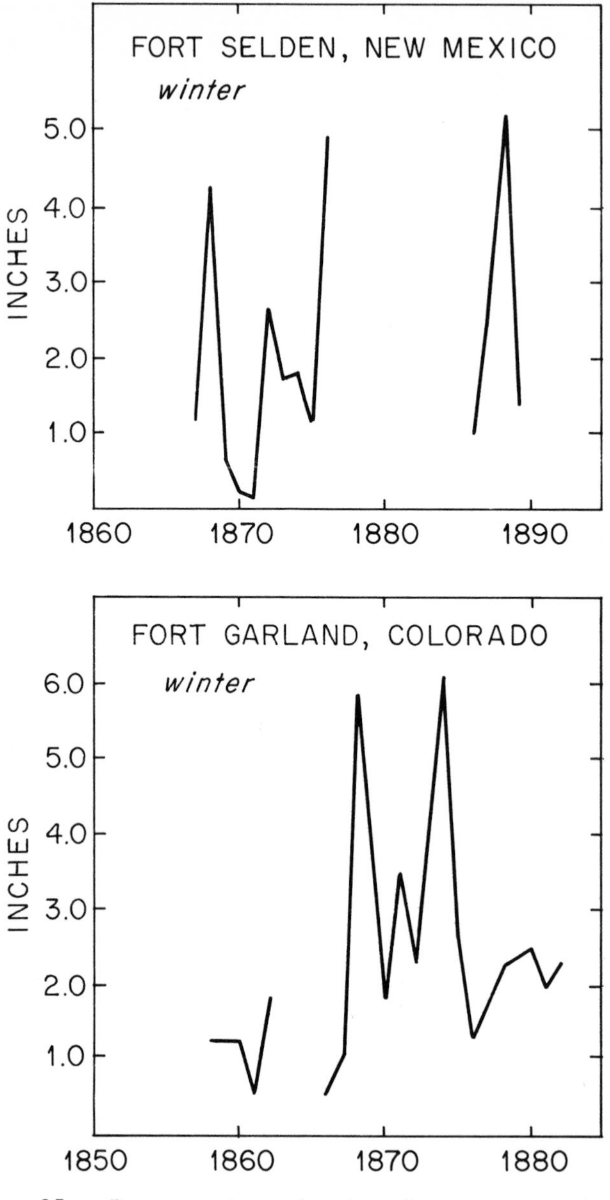

Figure 35: Recurrent peaks in winter precipitation at Fort Selden, New Mexico and Fort Garland, Colorado.

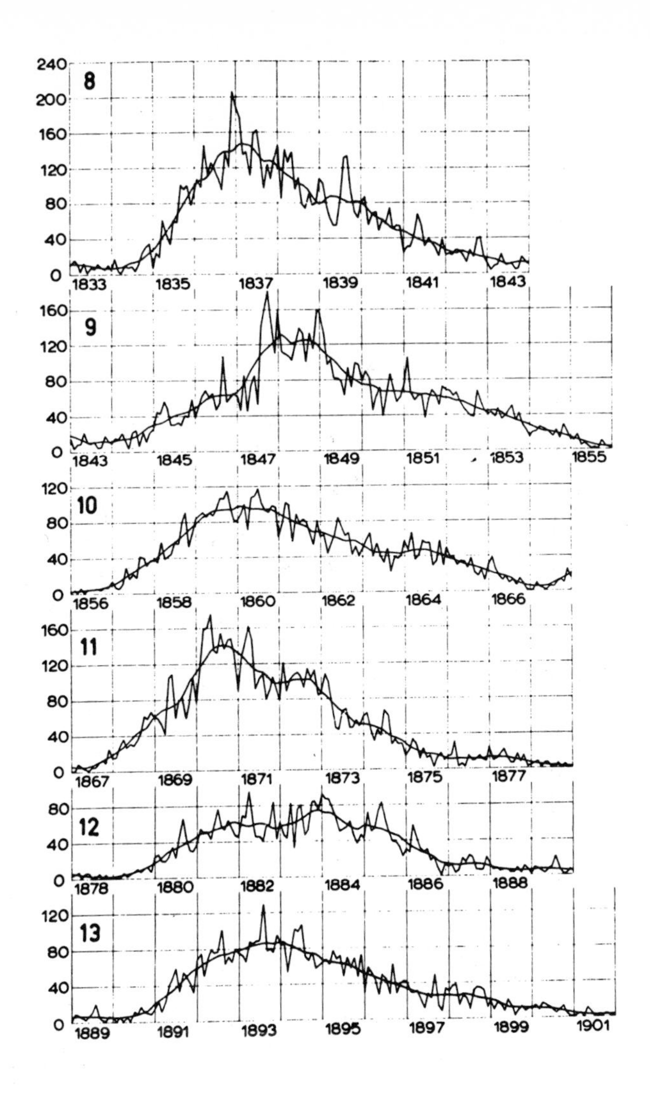

Figure 36: Wolf relative sunspot numbers in the nineteenth century (from Waldmeier, 1961).

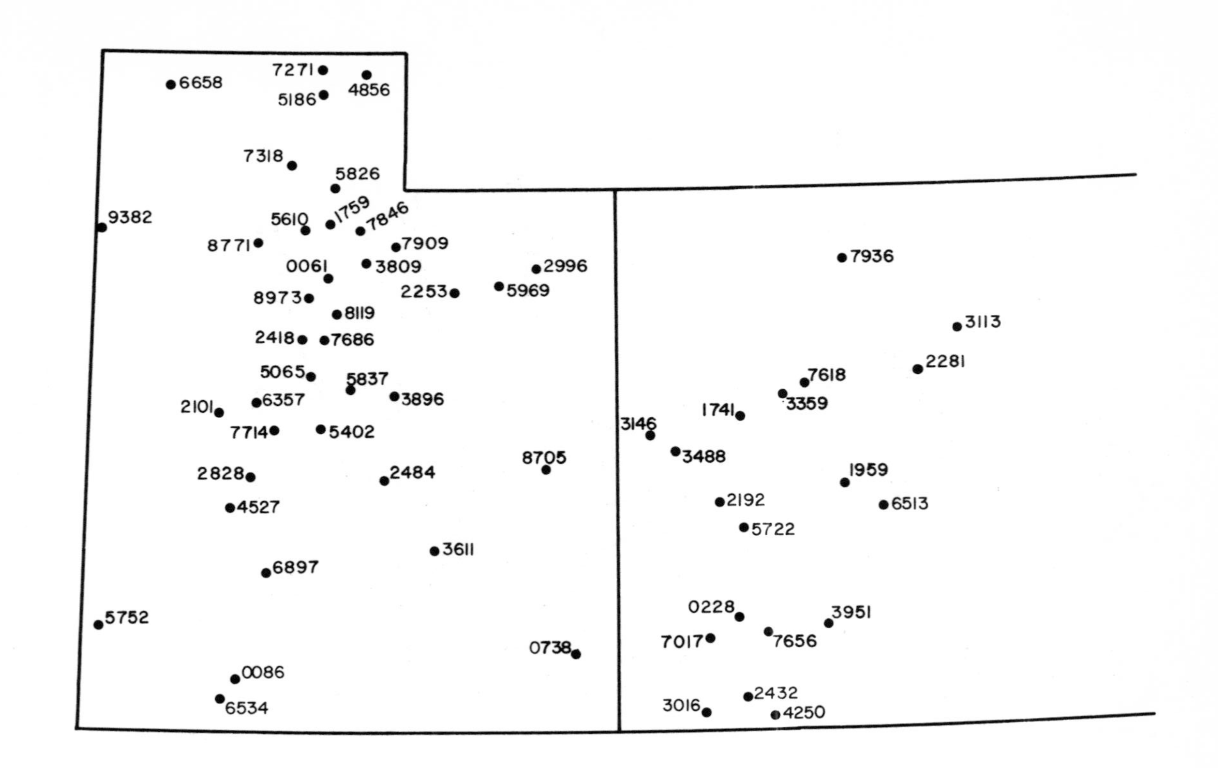

Figure 37: Continuous precipitation data: western Colorado and Utah.

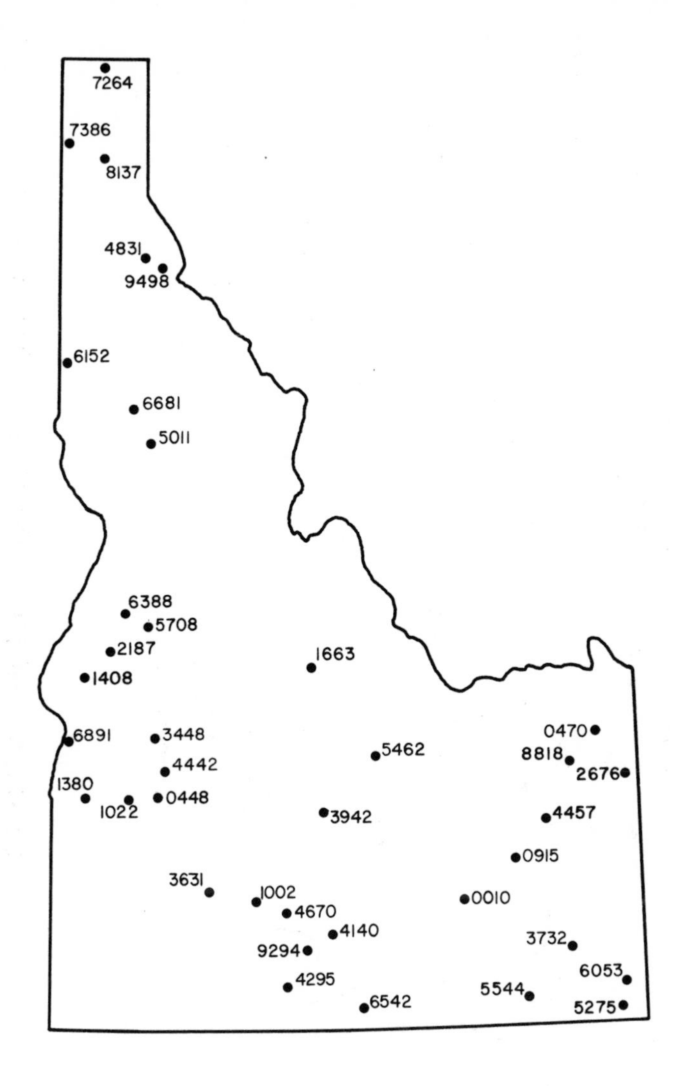

Figure 38: Continuous precipitation data: Idaho.

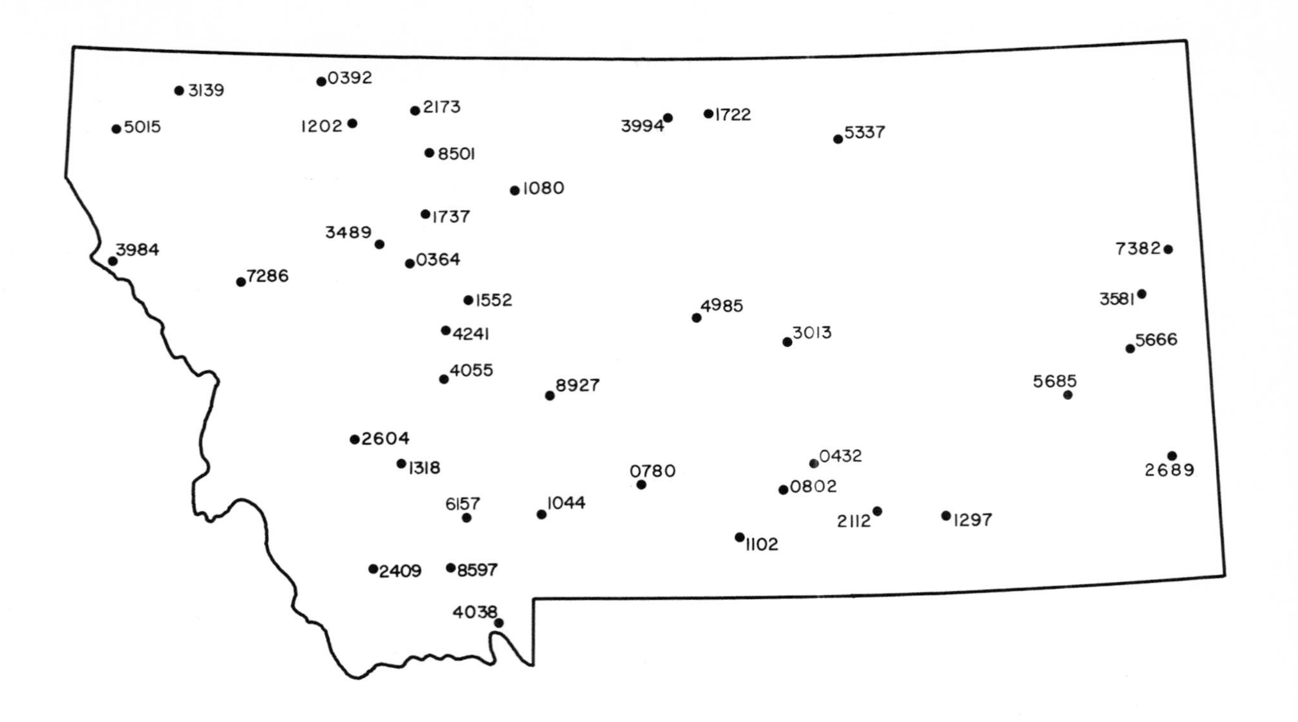

STATION INDEX NUMBERS — MONTANA

Figure 39: Continuous precipitation data: Montana.

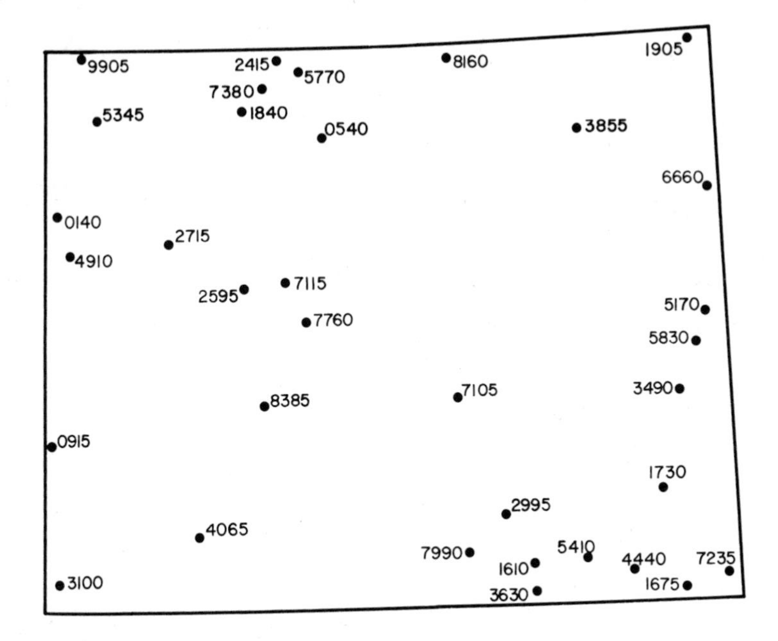

STATION INDEX NUMBERS — WYOMING

Figure 40: Continuous precipitation data: Wyoming.

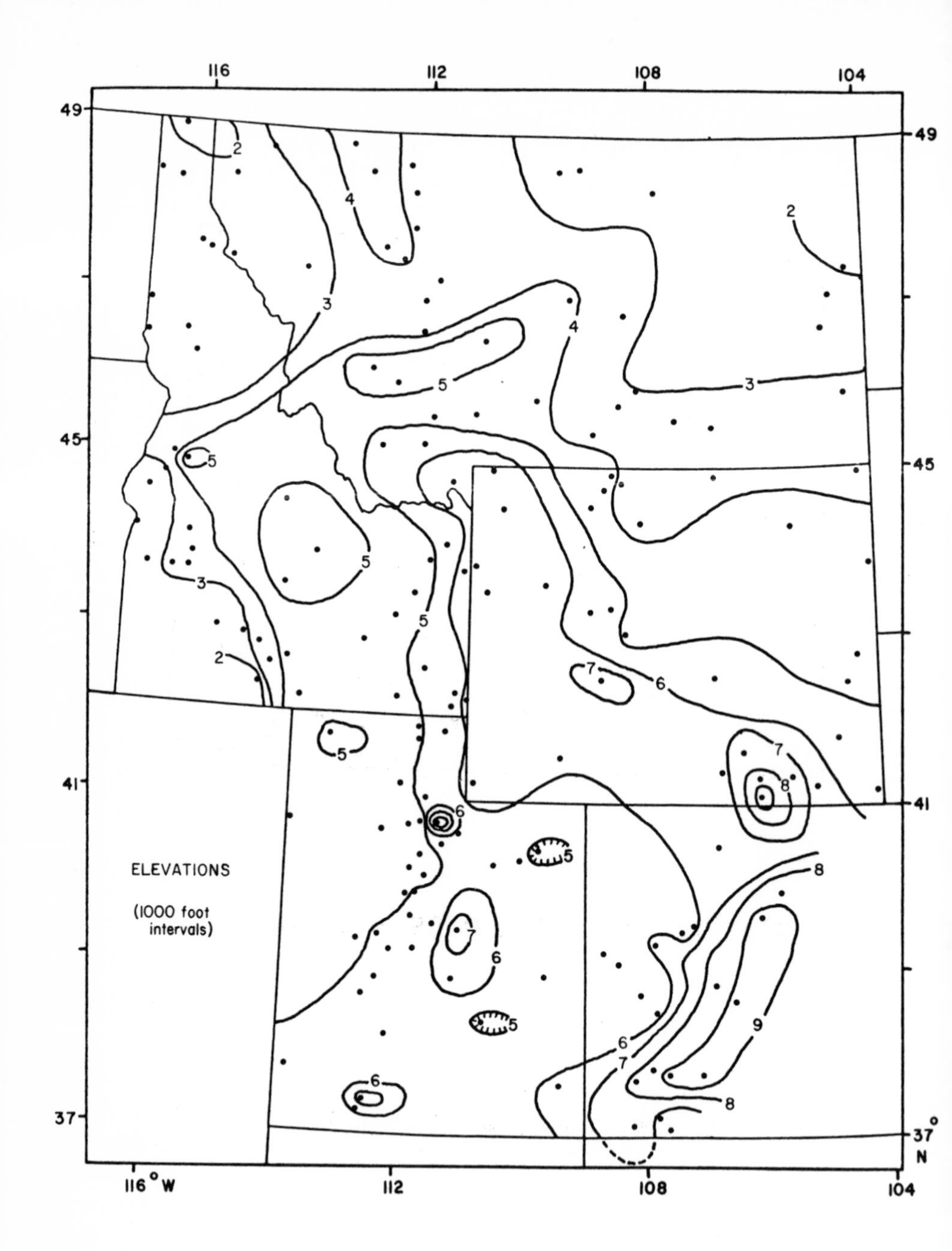

Figure 41: Generalized contours of station elevation.

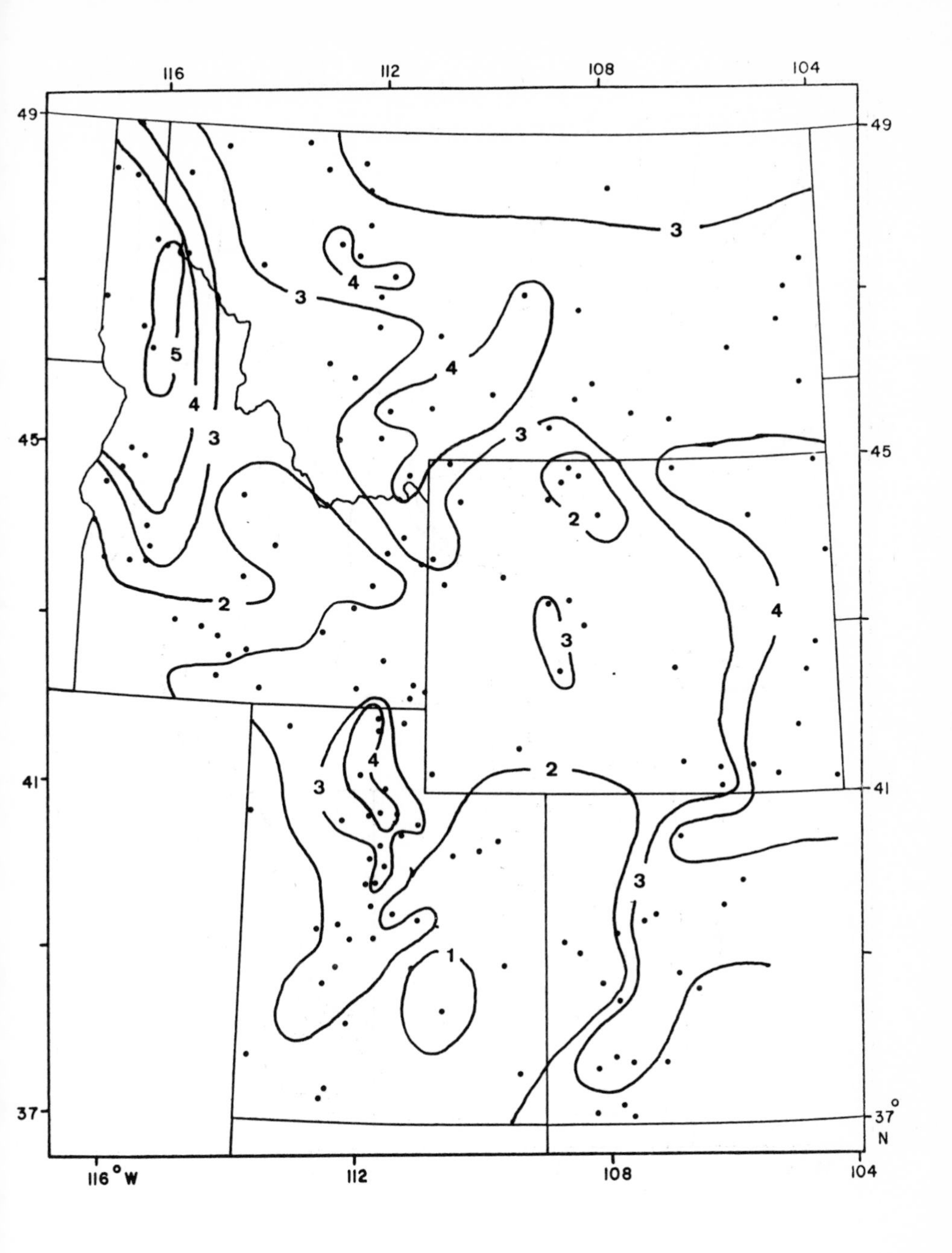

Figure 42: Average spring precipitation, 1941-70 (inches).

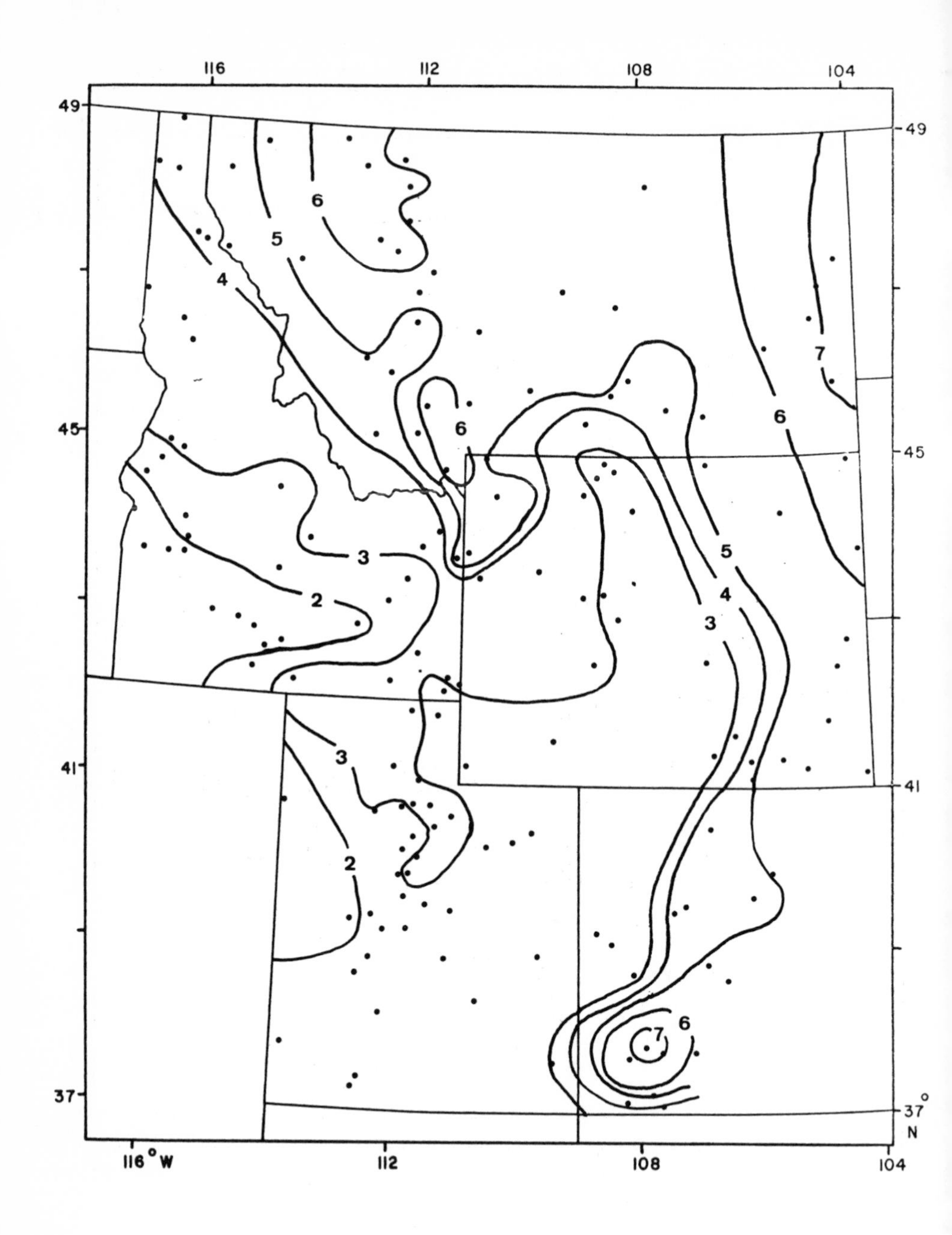

Figure 43: Average summer precipitation, 1941-70 (inches).

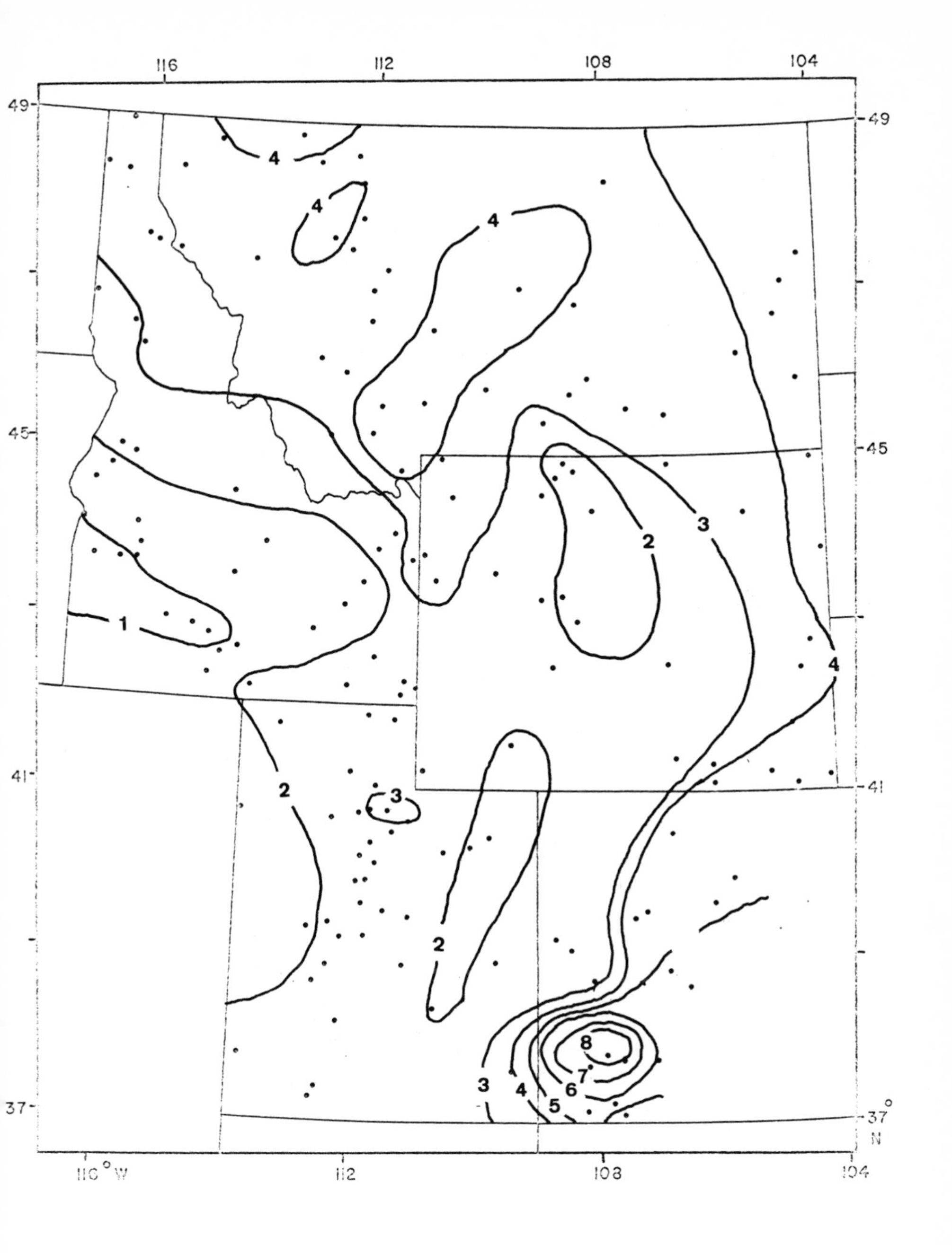

Figure 44: Average late summer precipitation, 1941-70 (inches).

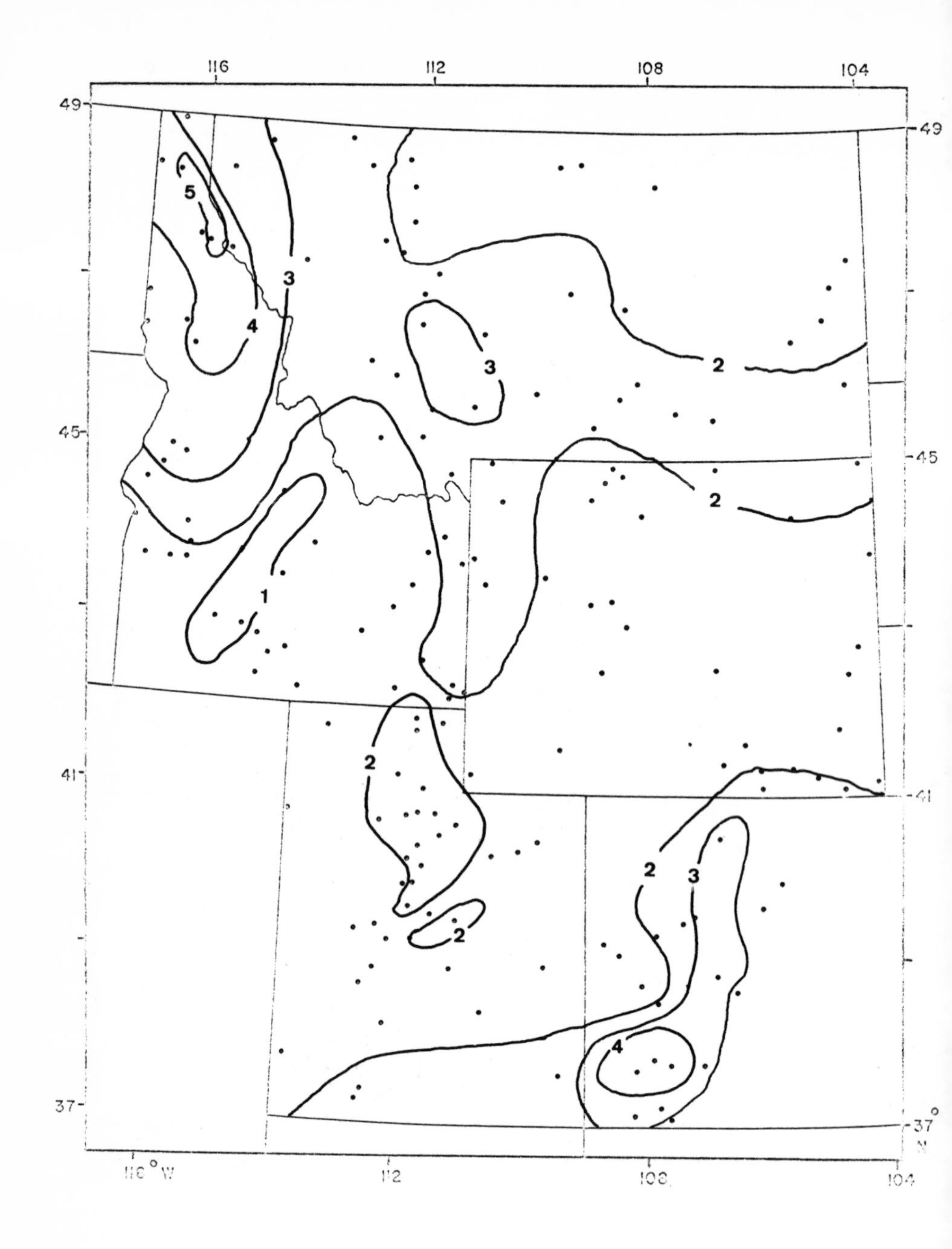

Figure 45: Average fall precipitation, 1941-70 (inches).

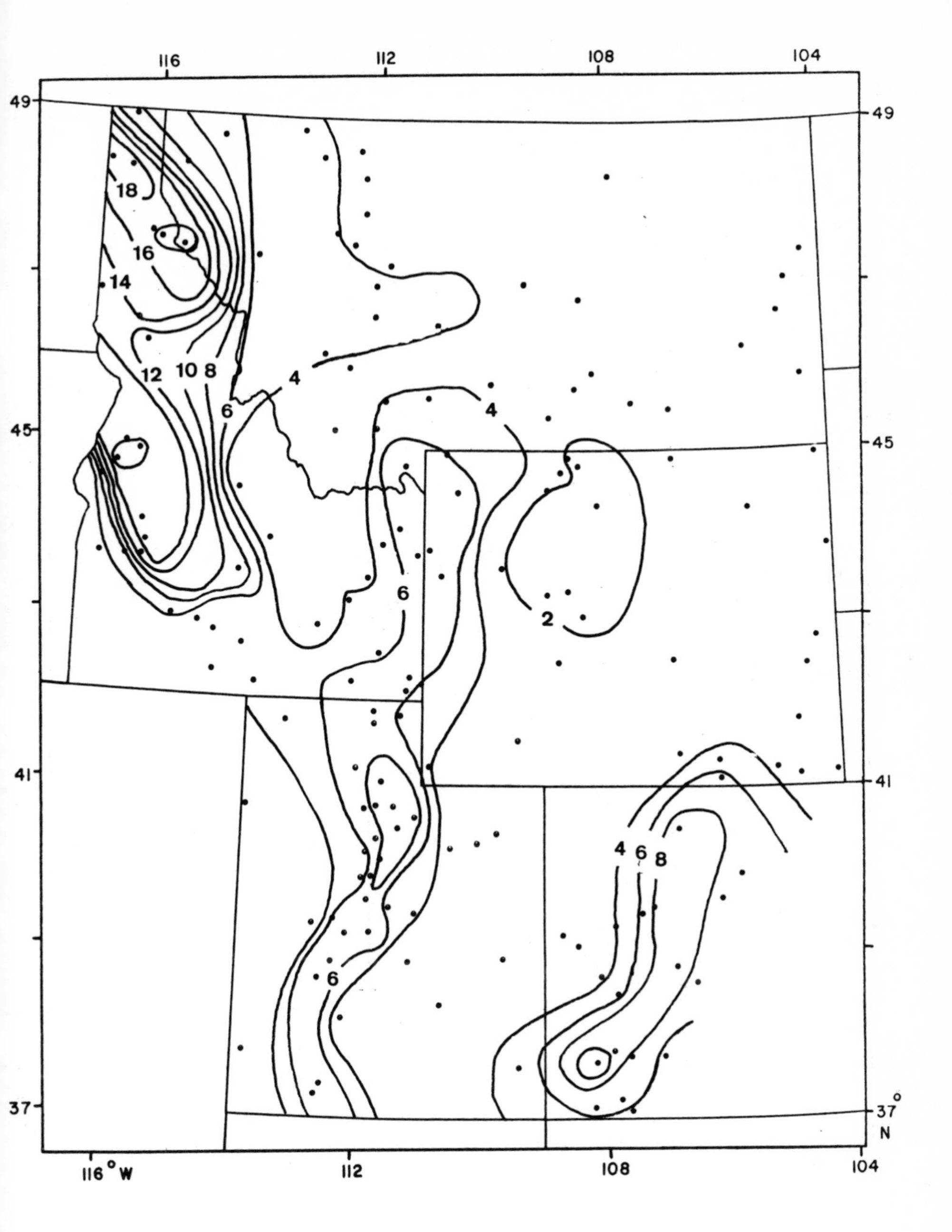

Figure 46: Average winter precipitation, 1941-70 (inches)

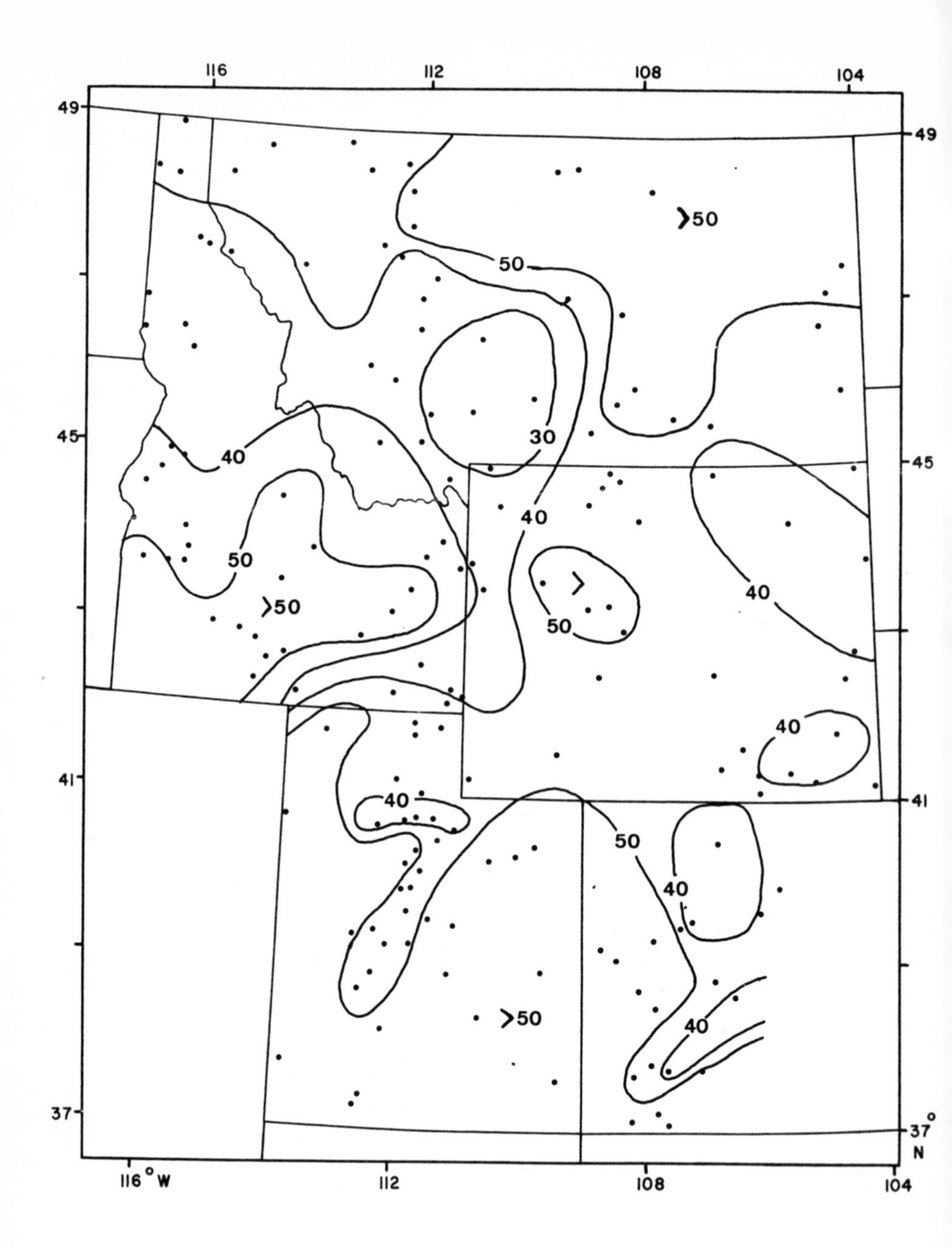

Figure 47: Coefficient of variation (%) of spring precipitation, 1941-70.

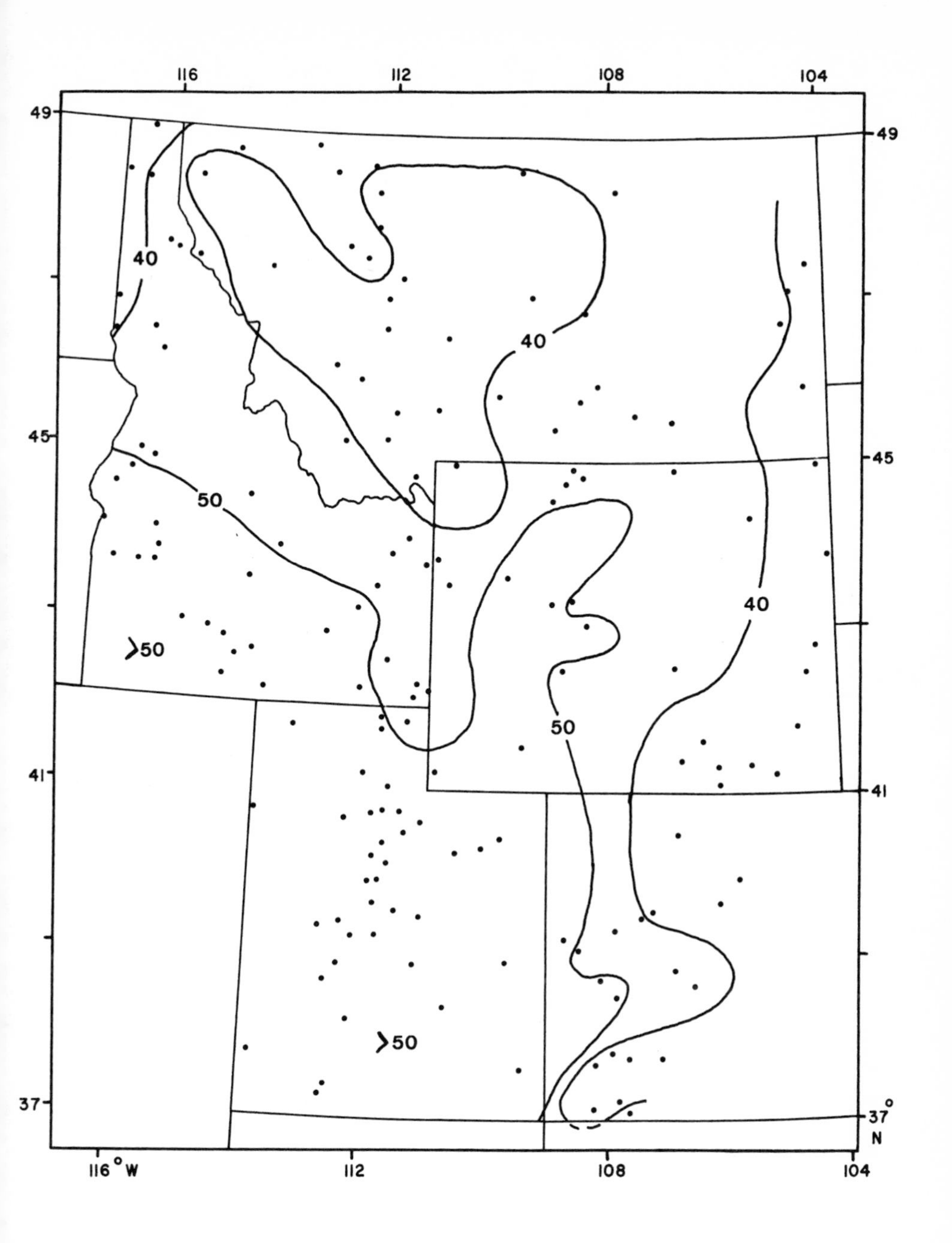

Figure 48: Coefficient of variation (%) summer precipitation, 1941-70.

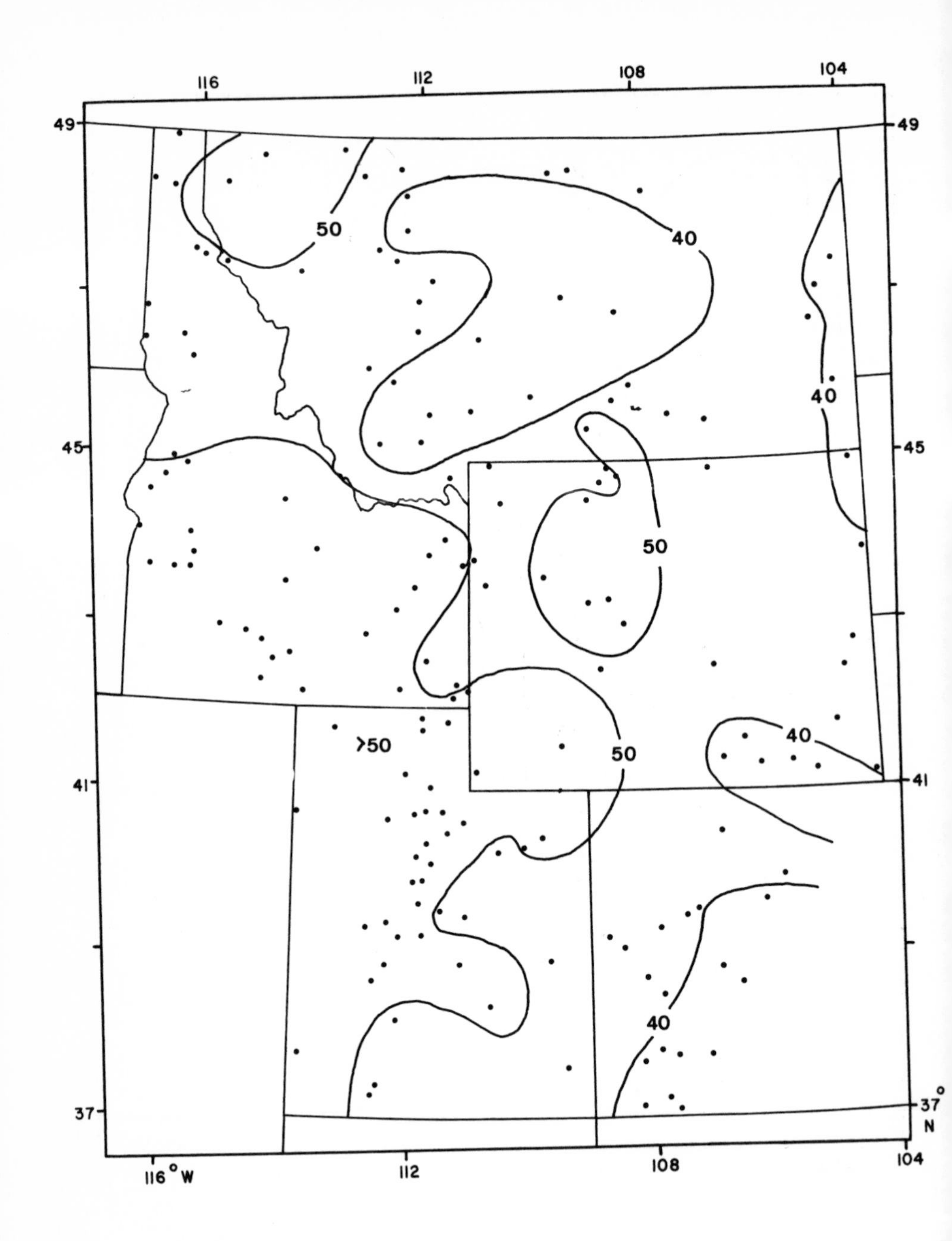

Figure 49: Coefficient of variation (%) late summer precipitation, 1941-70.

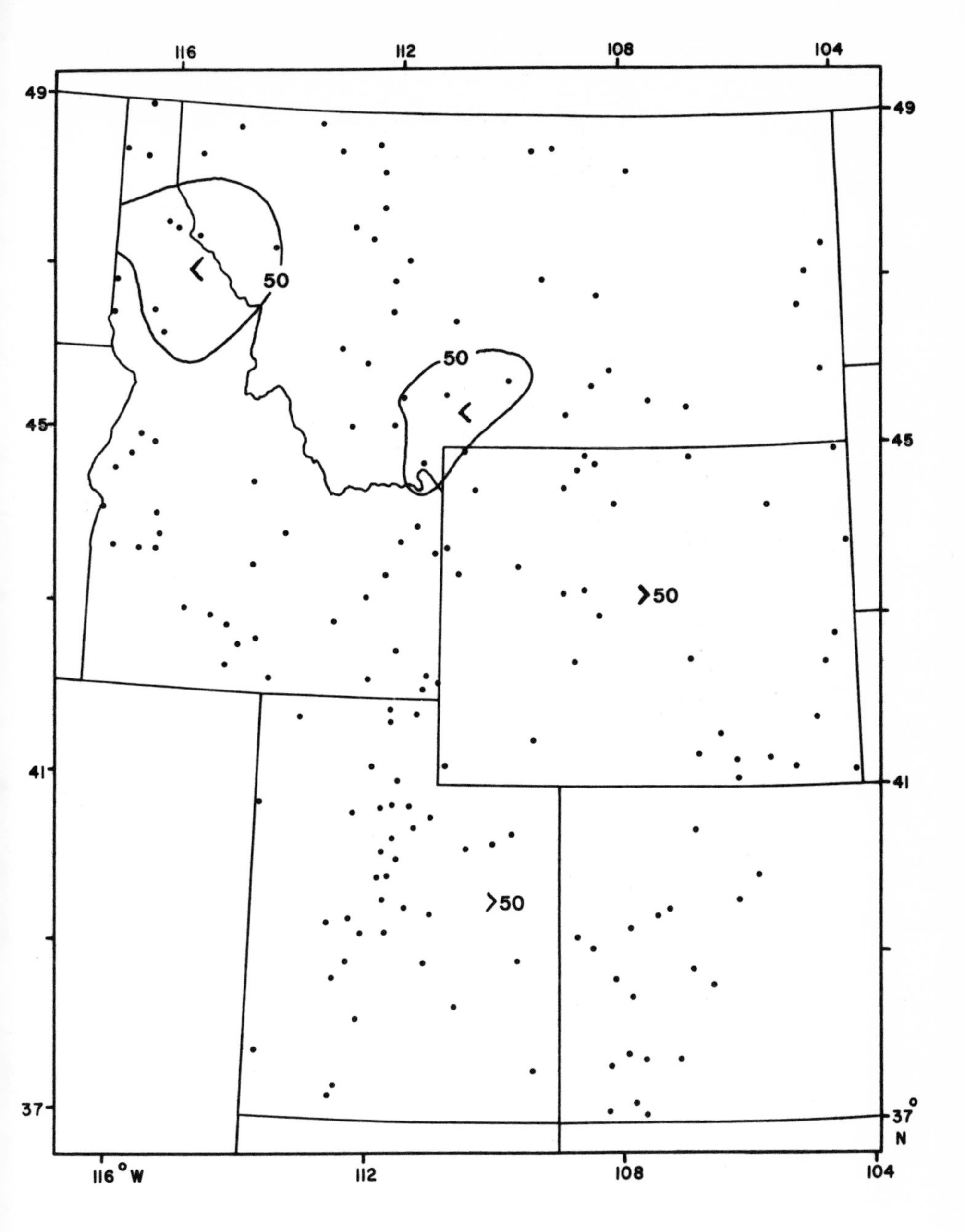

Figure 50. Coefficient of variation (%) fall precipitation, 1941-70.

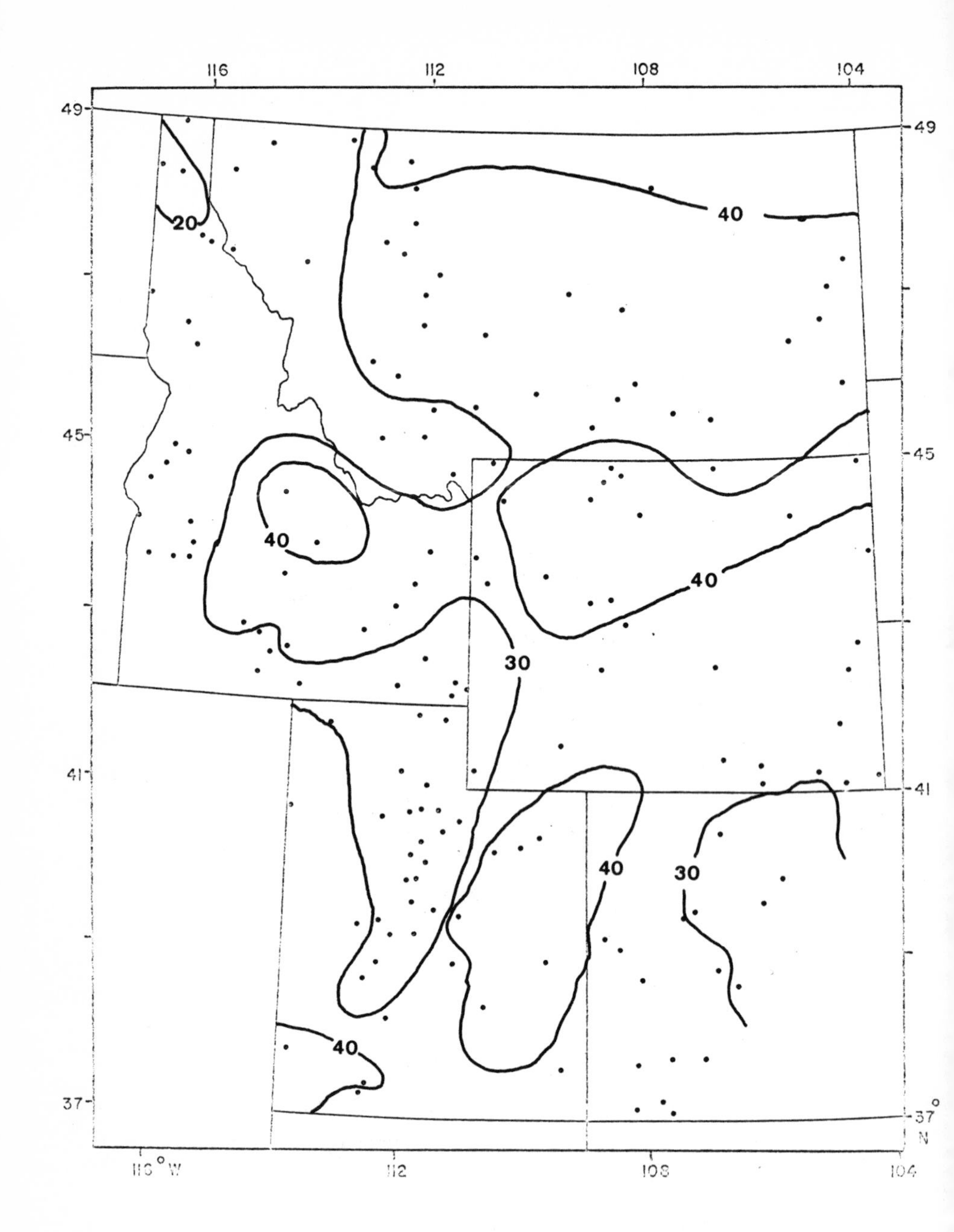

Figure 51: Coefficient of variation (%) winter precipitation, 1941-70.

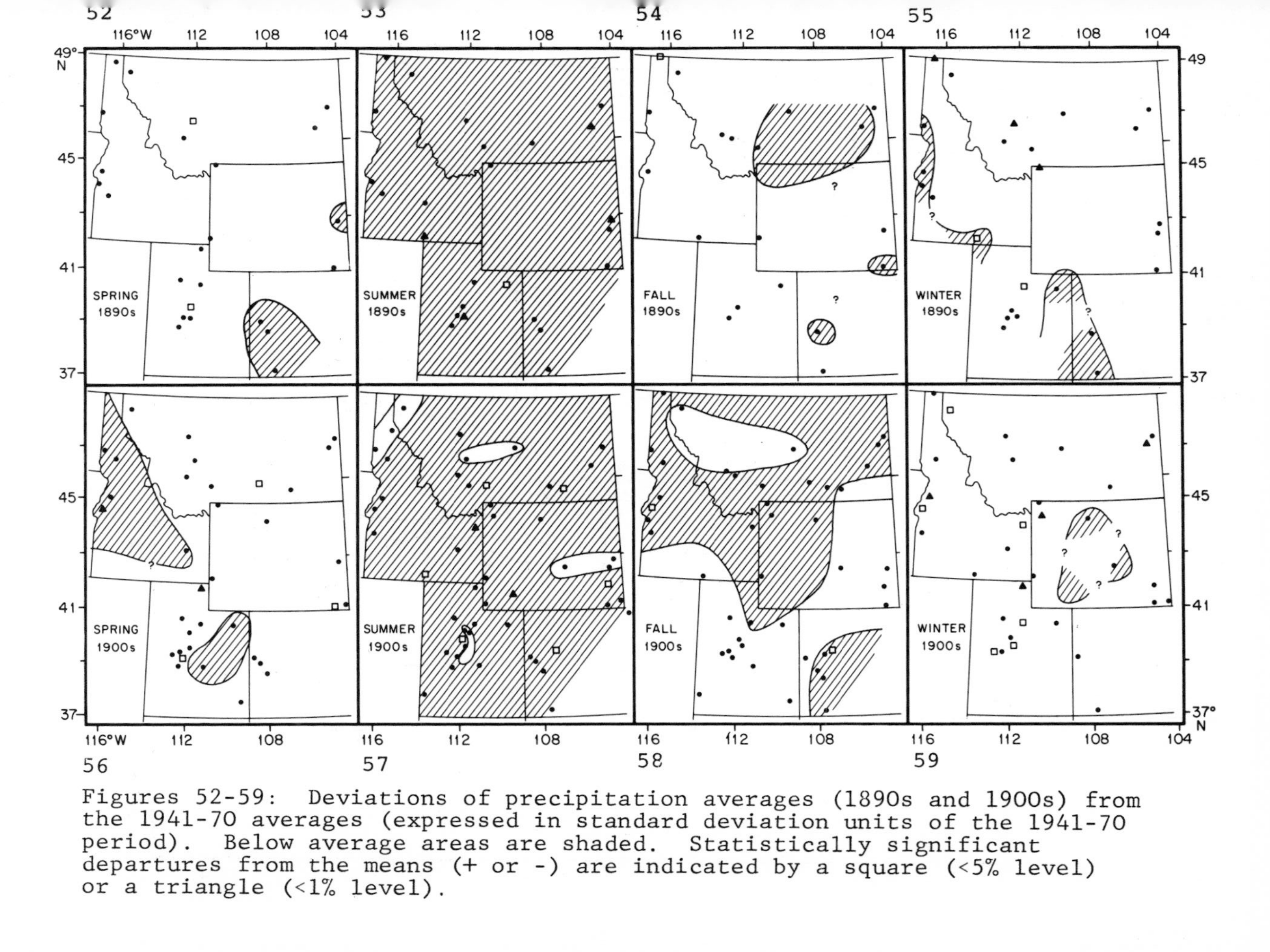

Figures 52-59: Deviations of precipitation averages (1890s and 1900s) from the 1941-70 averages (expressed in standard deviation units of the 1941-70 period). Below average areas are shaded. Statistically significant departures from the means (+ or -) are indicated by a square (<5% level) or a triangle (<1% level).

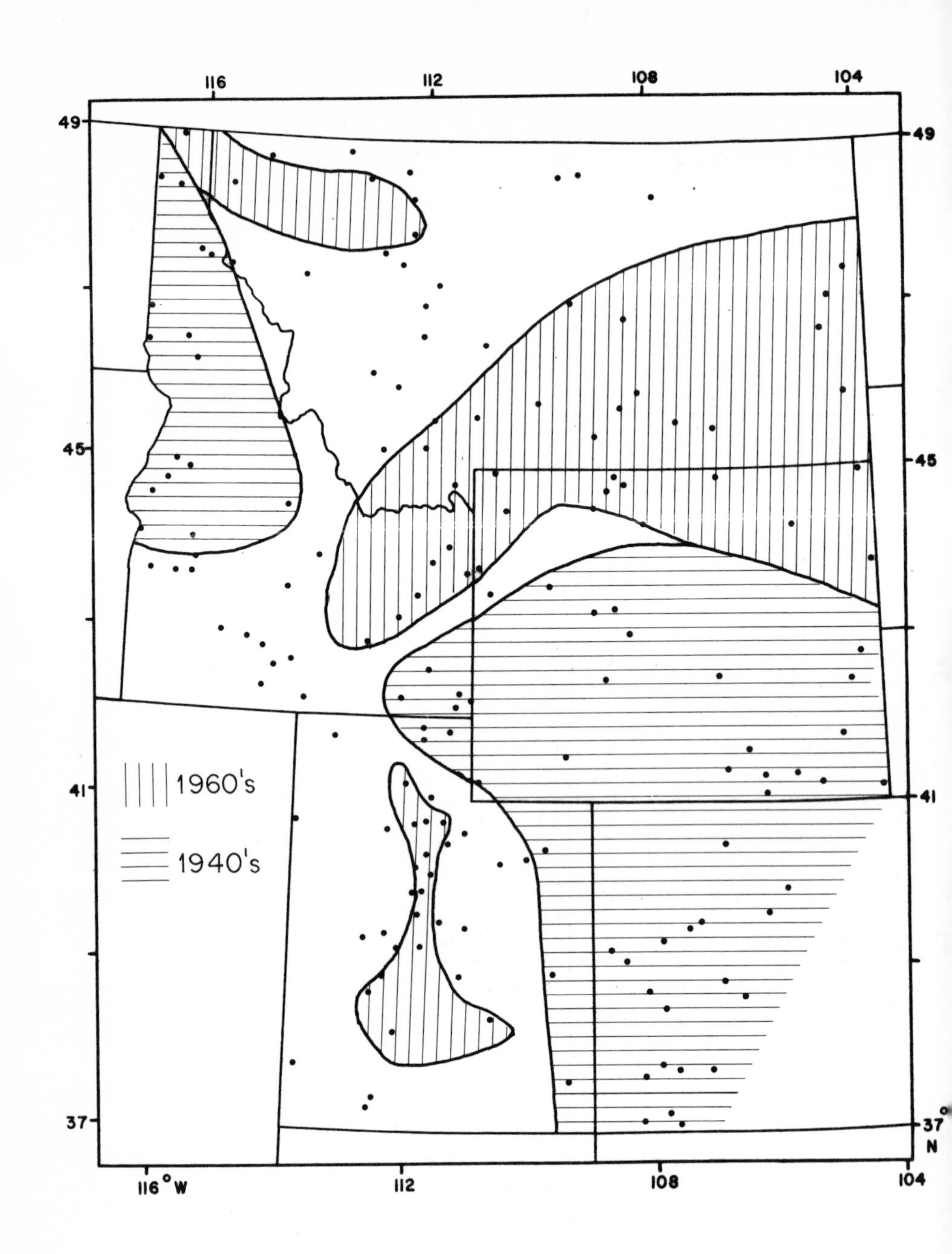

Figure 60: Periods of decadal precipitation maxima (1911-70) in spring months.

Figure 61: Tau values of spring precipitation, 1920-1970.

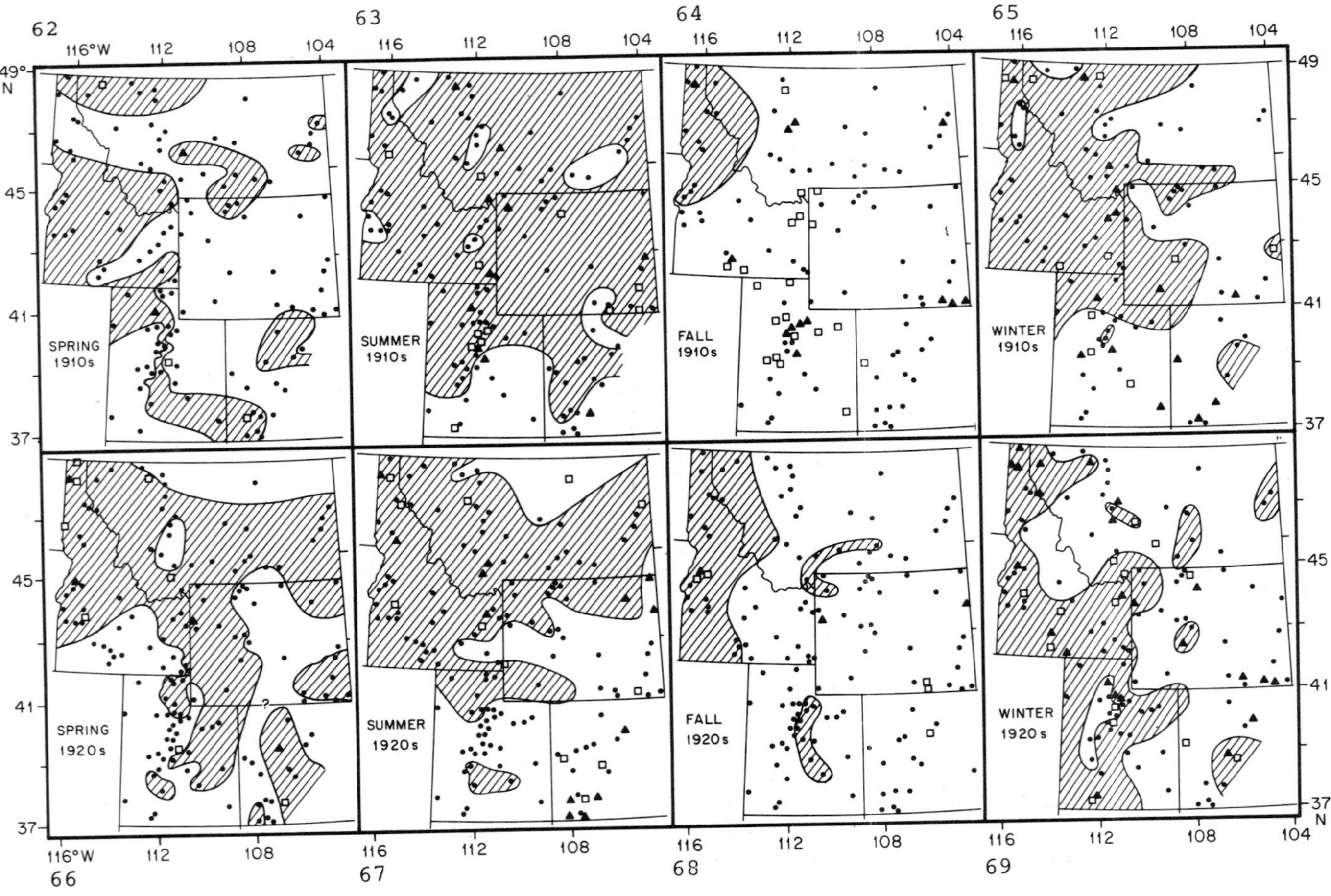

Figures 62-69: Deviations of precipitation averages for 1910s and 1920s from the 1941-70 averages (expressed in standard deviation units of the

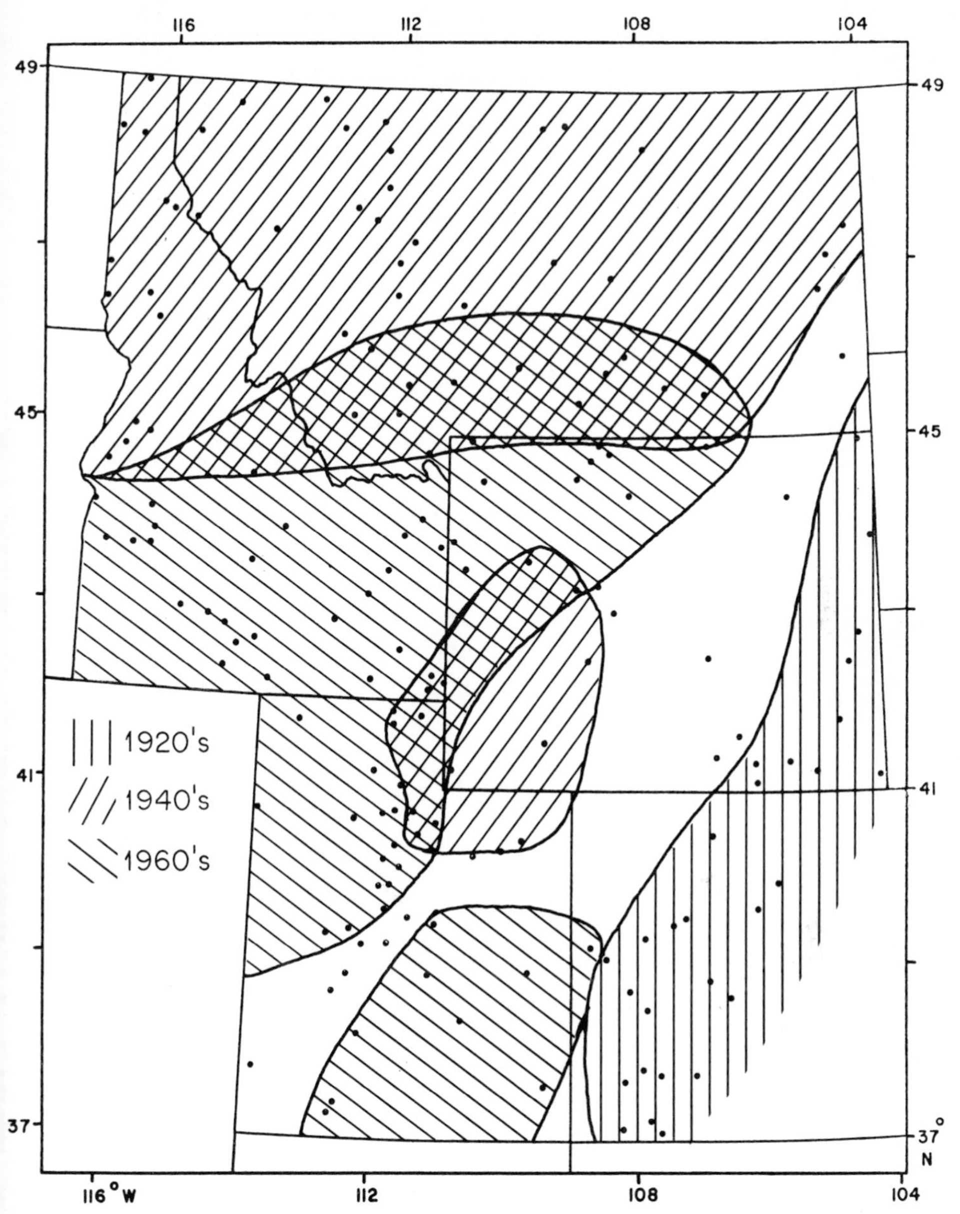

Figure 70: Periods of decadal precipitation maxima in summer months. Areas of overlap indicate regions where decadal maxima are found in one decade at approximately half the stations and in another decade at the remaining stations.

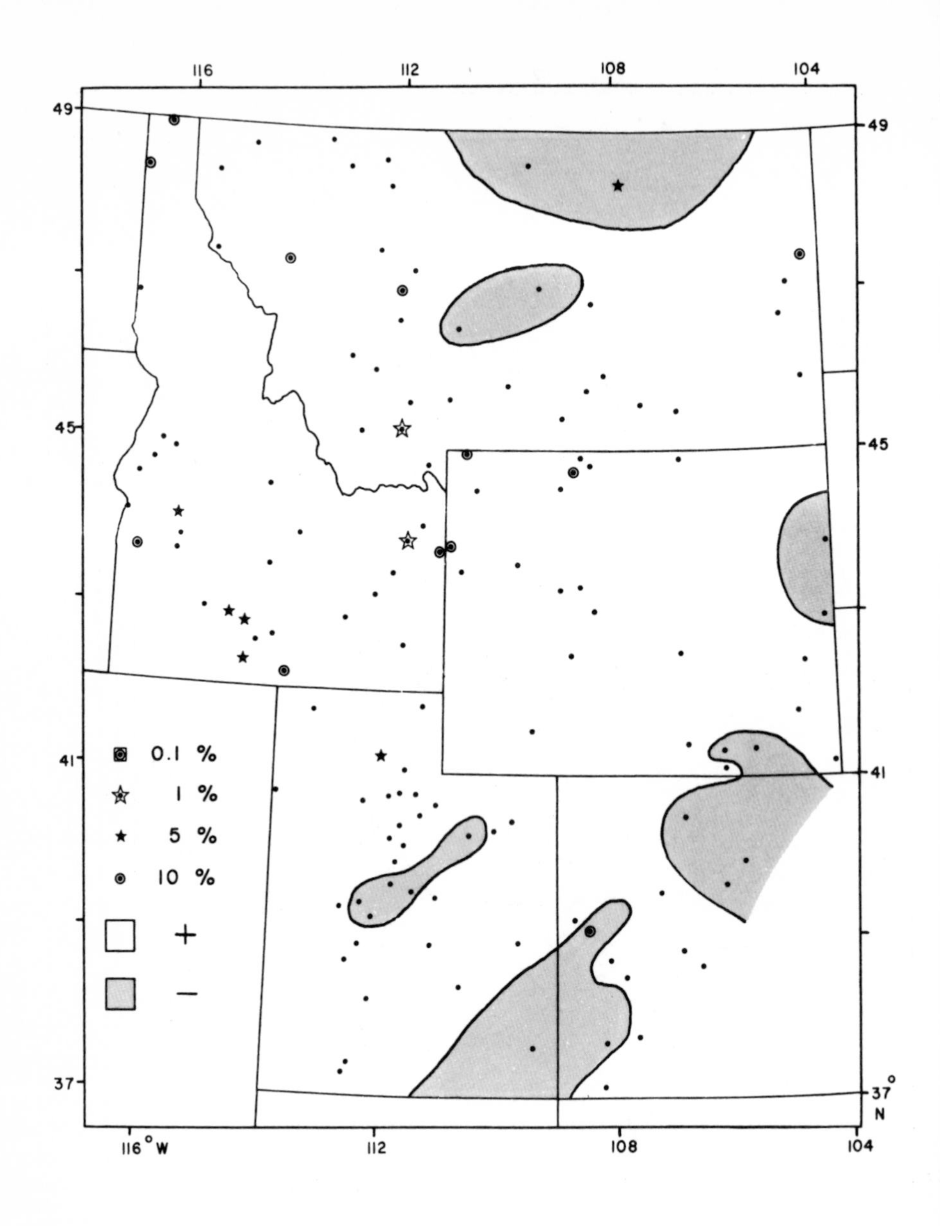

Figure 71: Tau values of summer precipitation, 1920-1970.

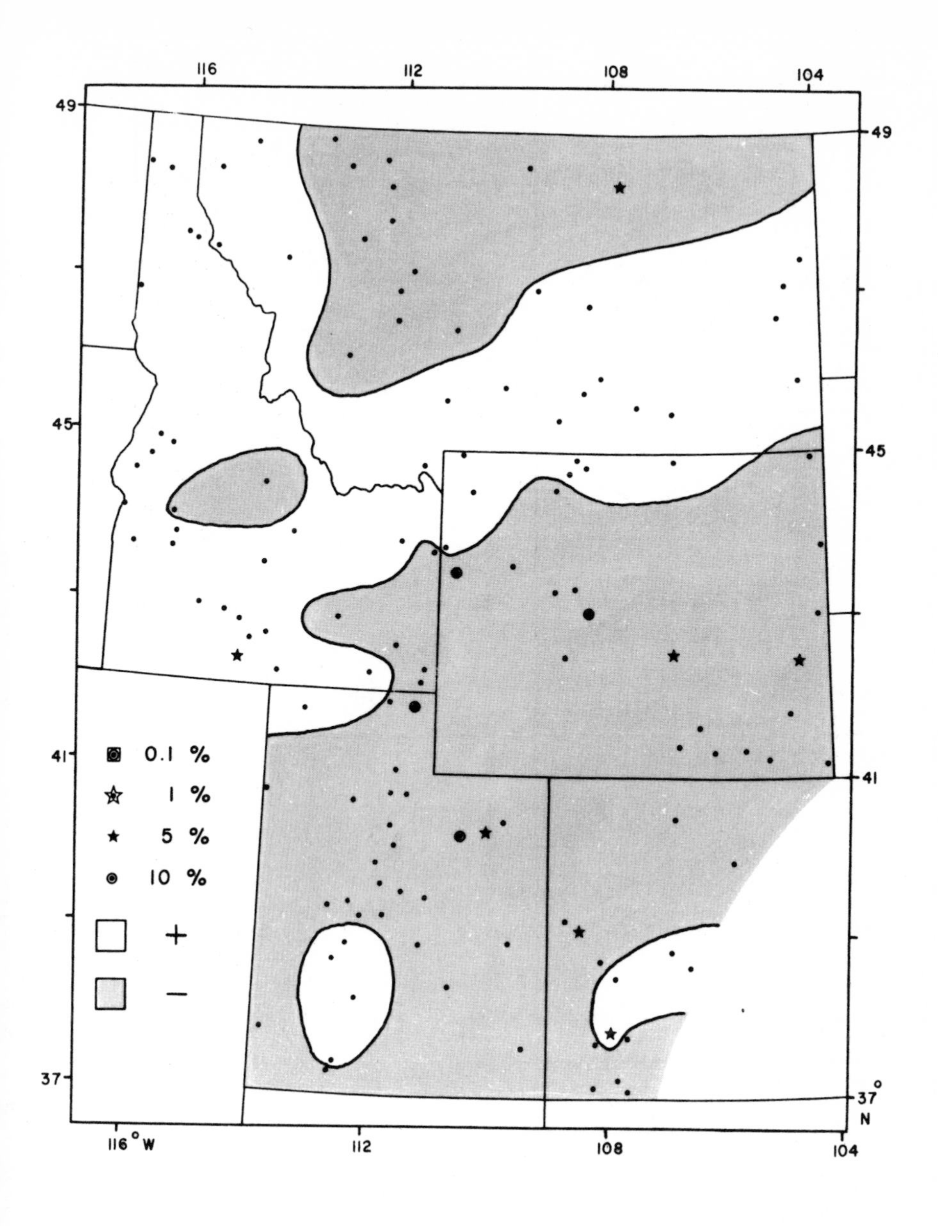

Figure 72: Tau values of late summer precipitation, 1920-1970.

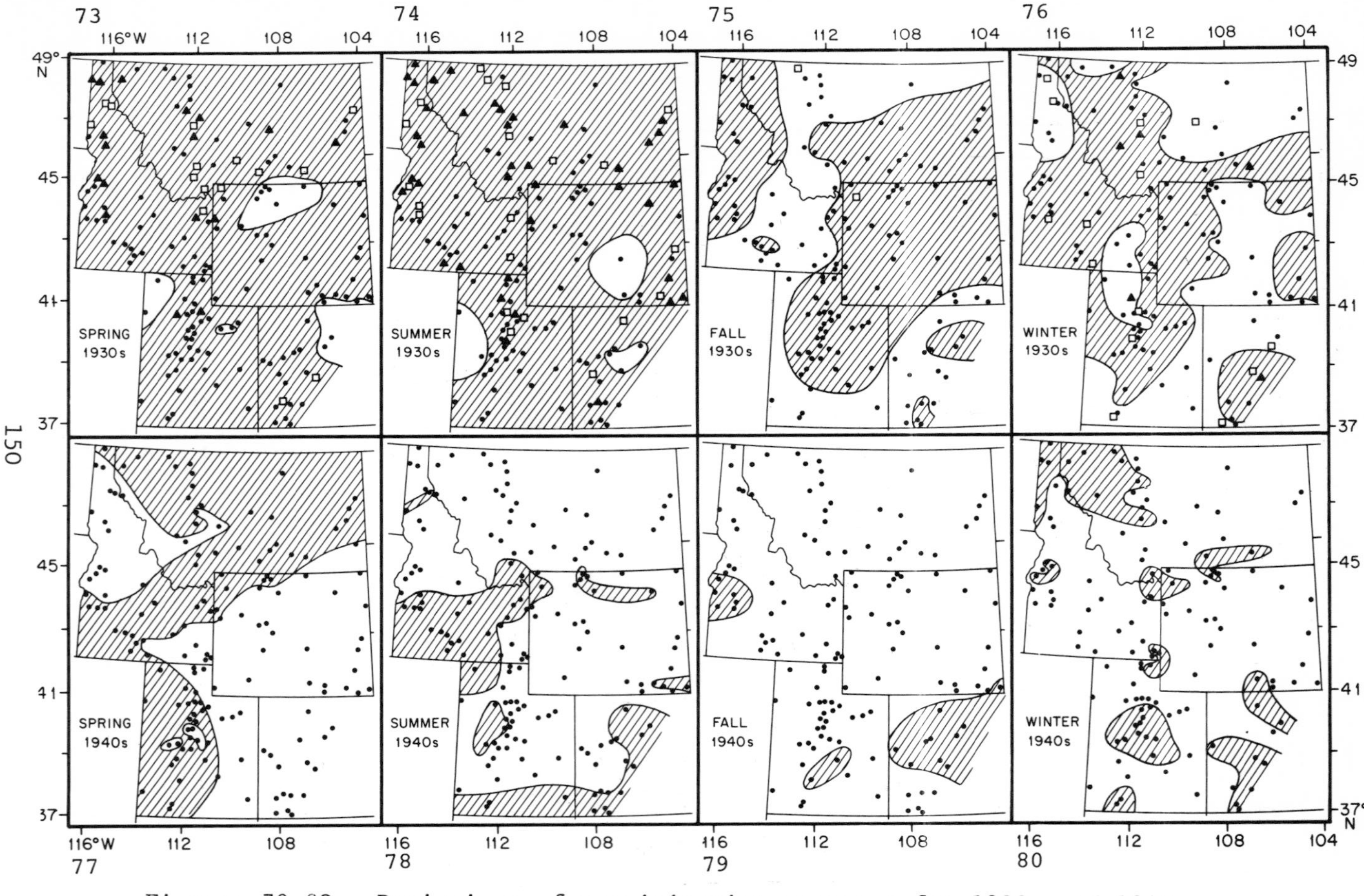

Figures 73-80. Deviations of precipitation averages for 1930s and 1940s from the 1941-70 averages (expressed in standard deviation units of the 1941-70 period). See Figures 52-59 for explanation.

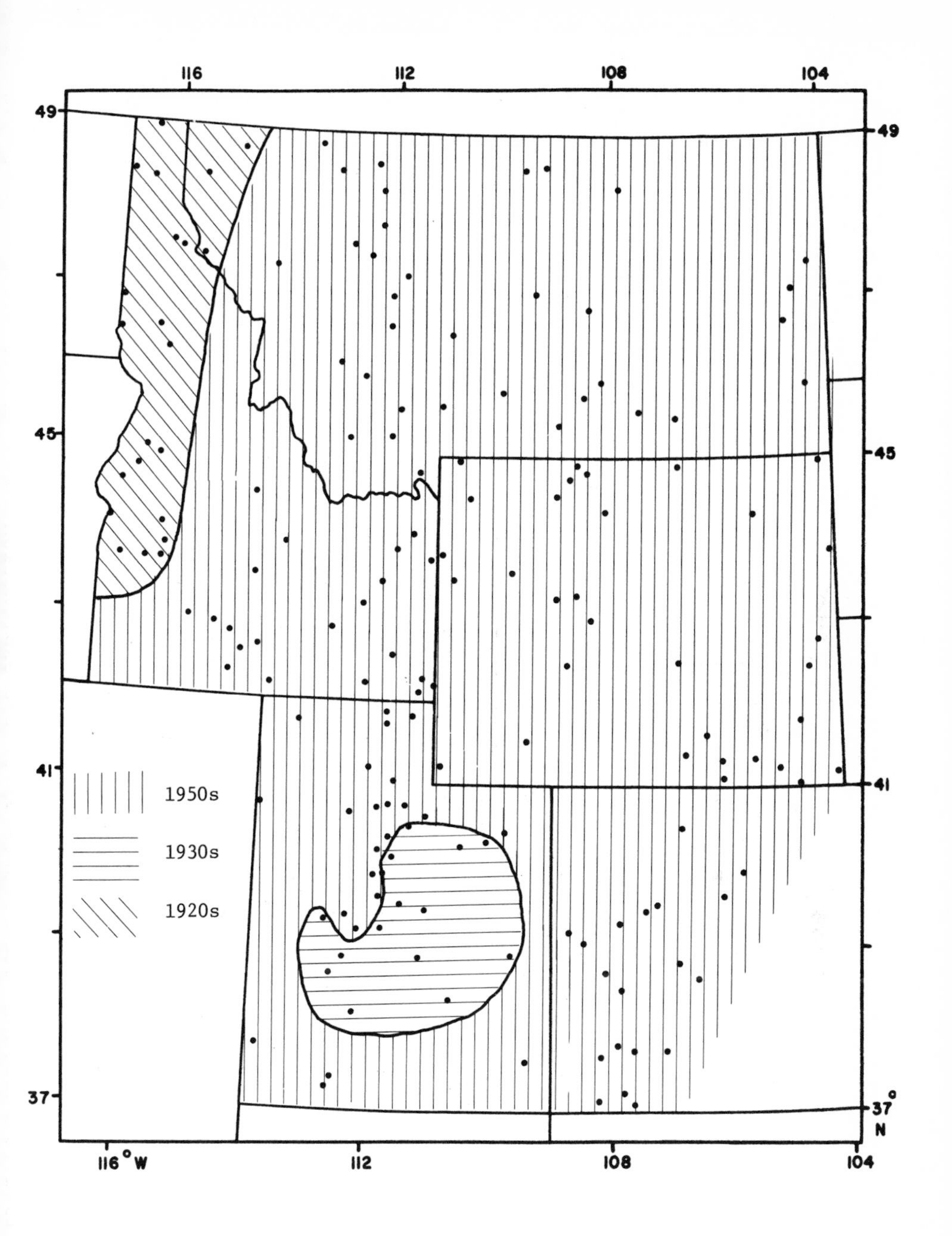

Figure 81. Periods of decadal precipitation minima in fall months.

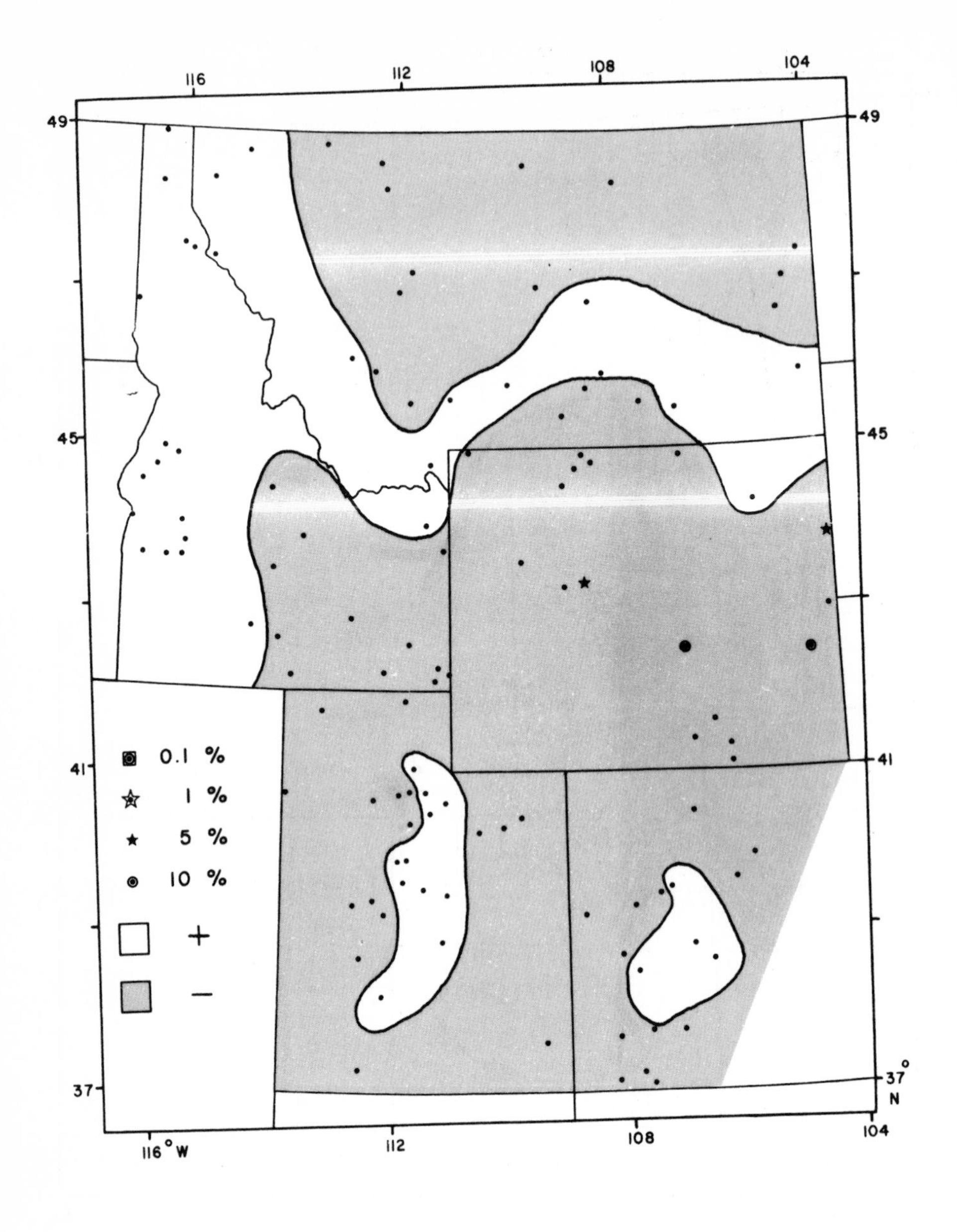

Figure 82: Tau values of fall precipitation, 1920-1970.

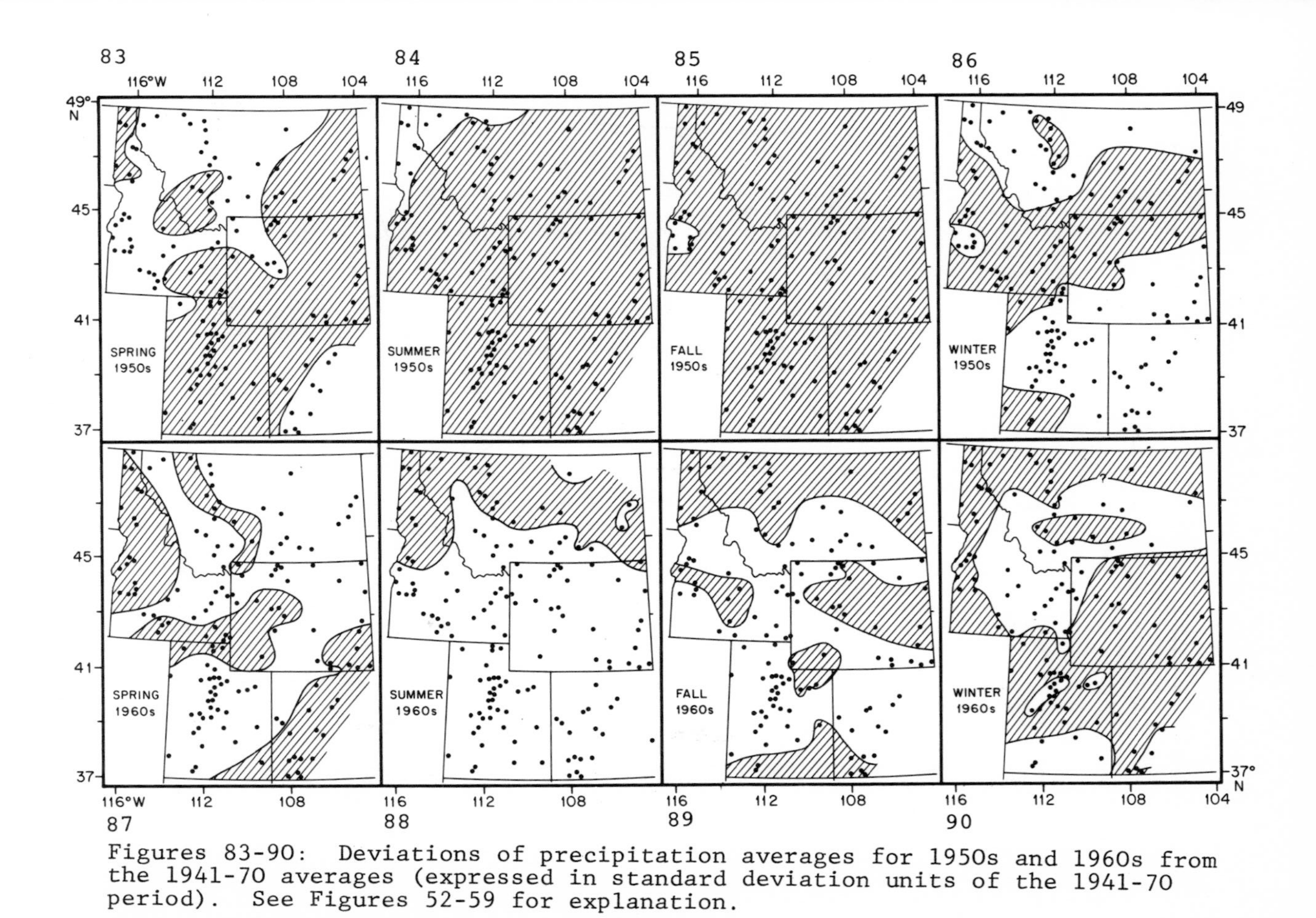

Figures 83-90: Deviations of precipitation averages for 1950s and 1960s from the 1941-70 averages (expressed in standard deviation units of the 1941-70 period). See Figures 52-59 for explanation.

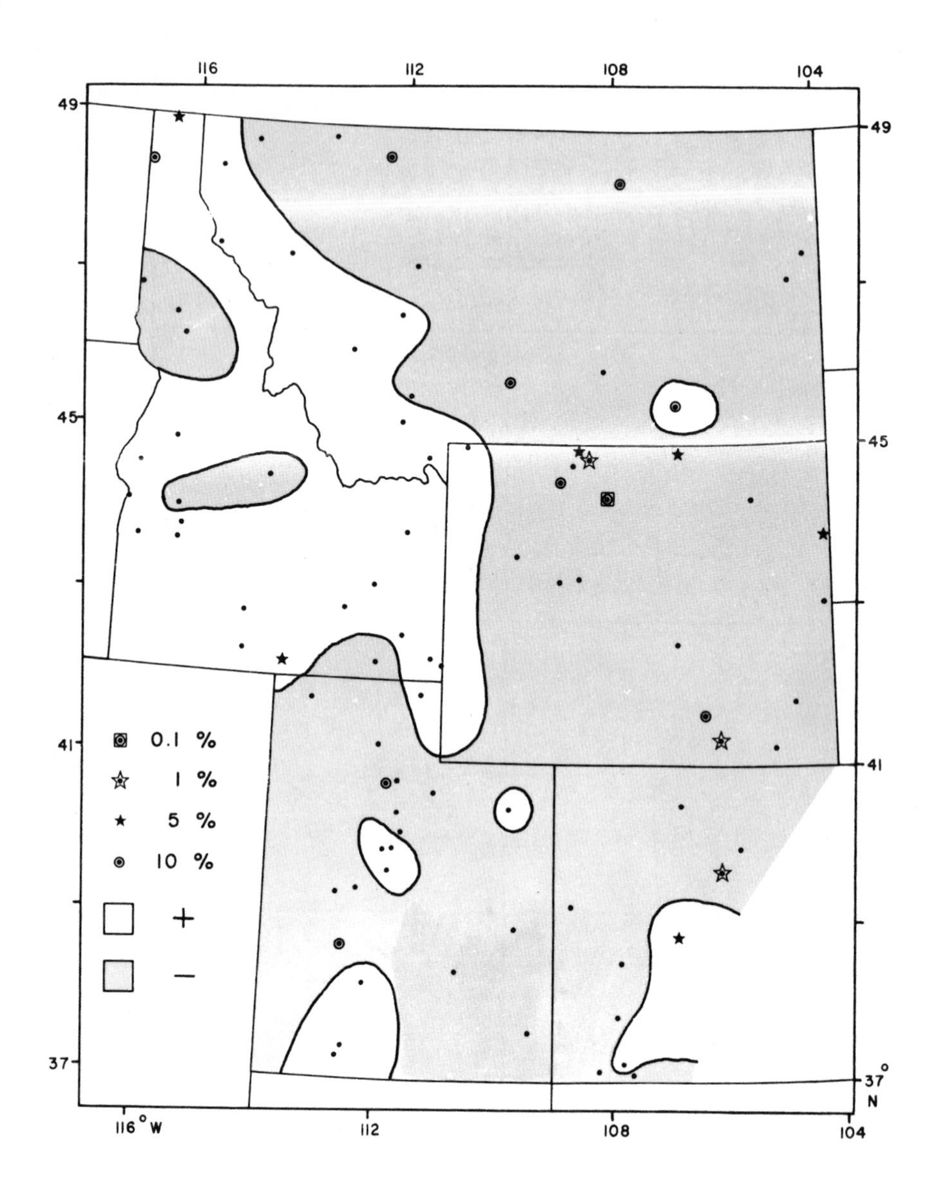

Figure 91: Tau values for winter precipitation, 1920-1970.

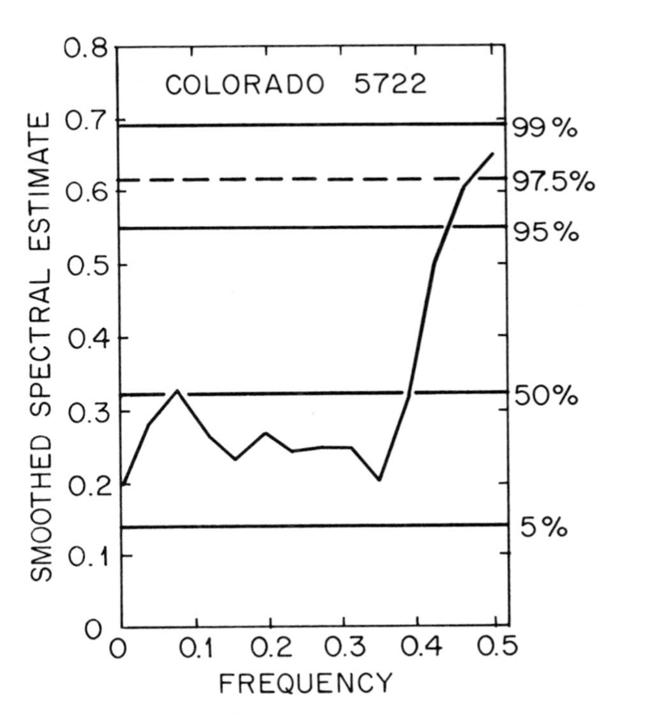

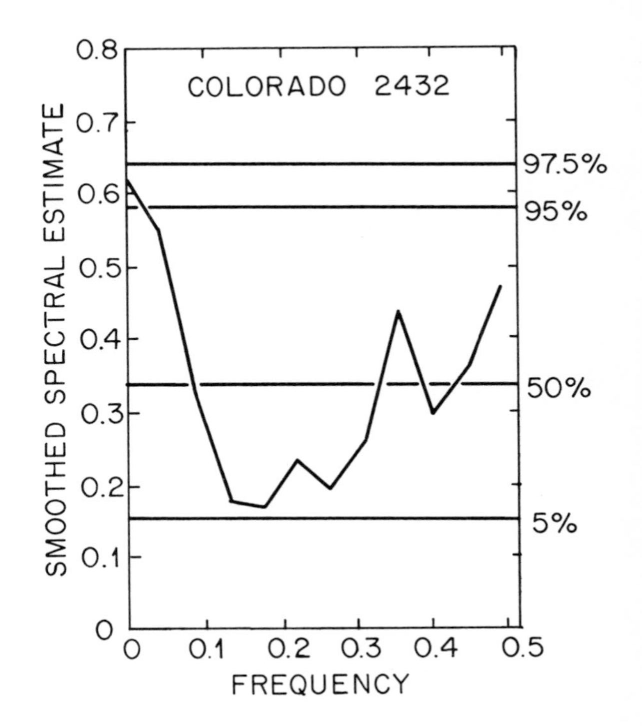

Figure 92: Power spectrum of summer precipitation at Bench Mark station Montrose (No. 2), Colorado, (1885-1970) and of winter precipitation at Durango, Colorado, (1894-1970). White noise null continuum and confidence limits are demonstrated in both cases. At Montrose, statistically significant peaks at frequencies of 0.46 and 0.5 (2-2.2 year periodicities) are shown. At Durango, only a long-term trend is indicated as statistically significant.

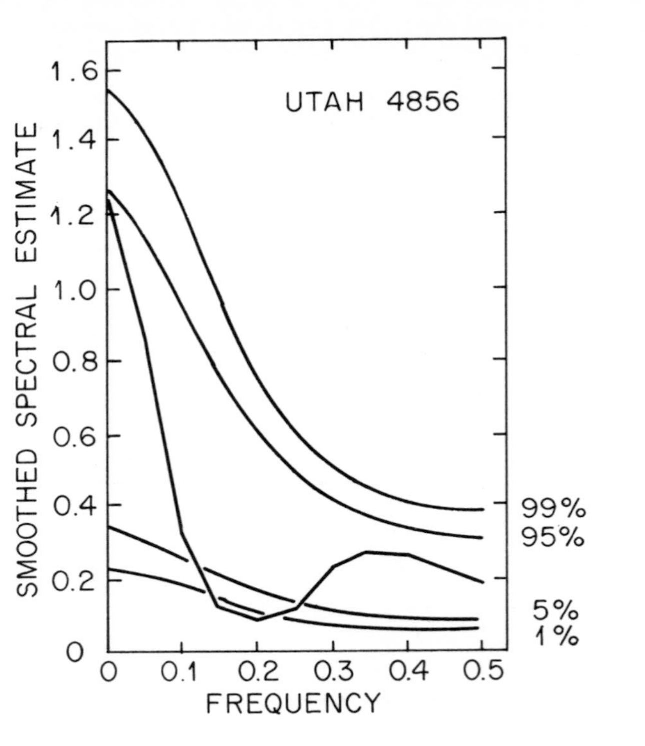

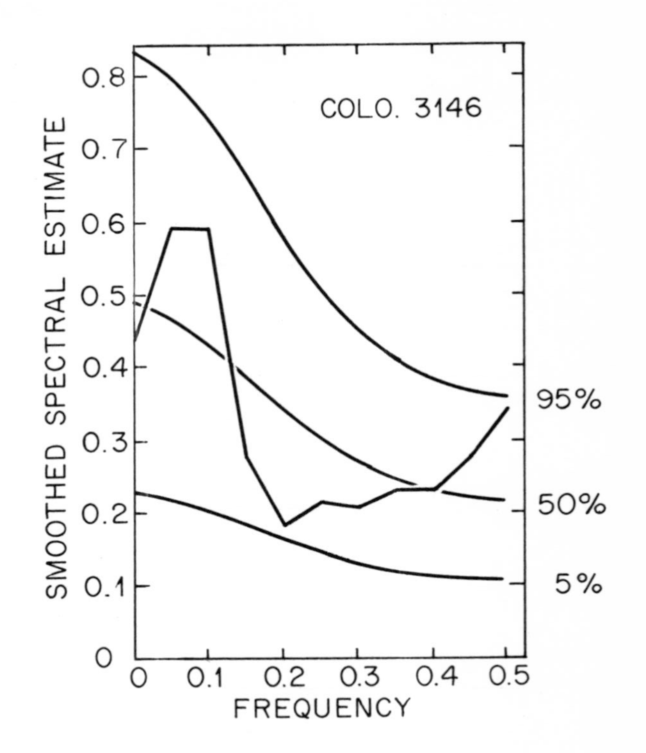

Figure 93: Power spectra of fall precipitation at Laketown, Utah (1900-1970), and Fruita, Colorado, 1902-1970) with red noise null continuum and confidence limits demonstrated. No statistically significant spectral peaks-

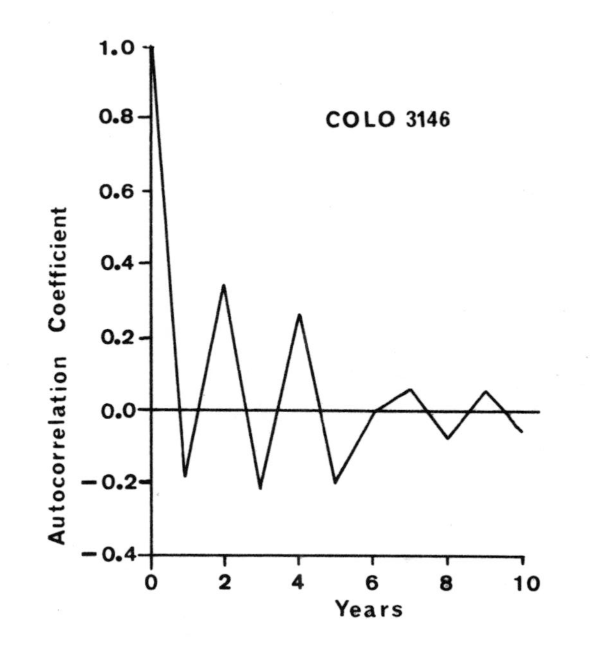

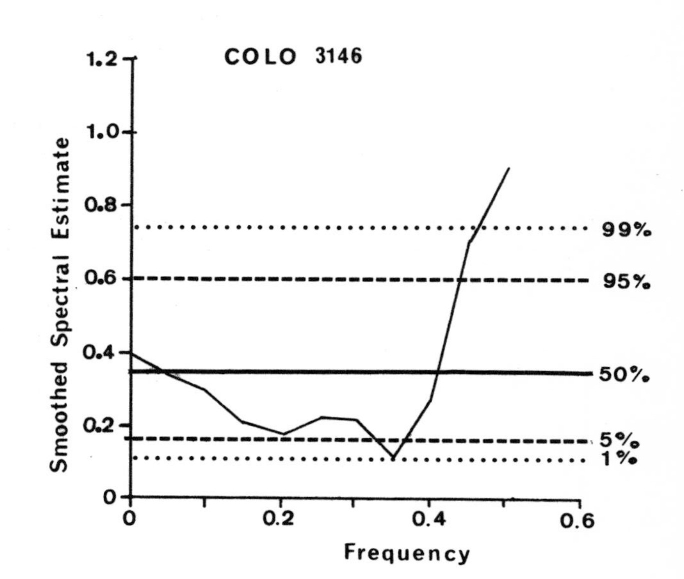

Figure 94: Correlogram and power spectrum of late summer precipitation at Fruita, Colorado, (1902-1970) demonstrating marked high frequency variation at 2-2.2 years.

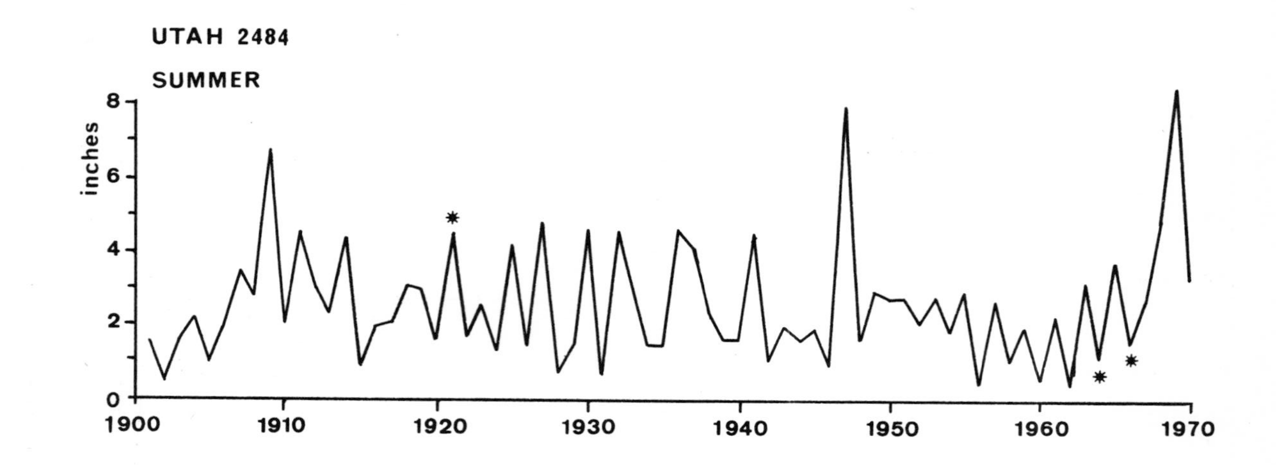

Figure 95: Summer precipitation record at Emery, Utah, (1901-1970) illustrating the approximately 2-year periodicity.

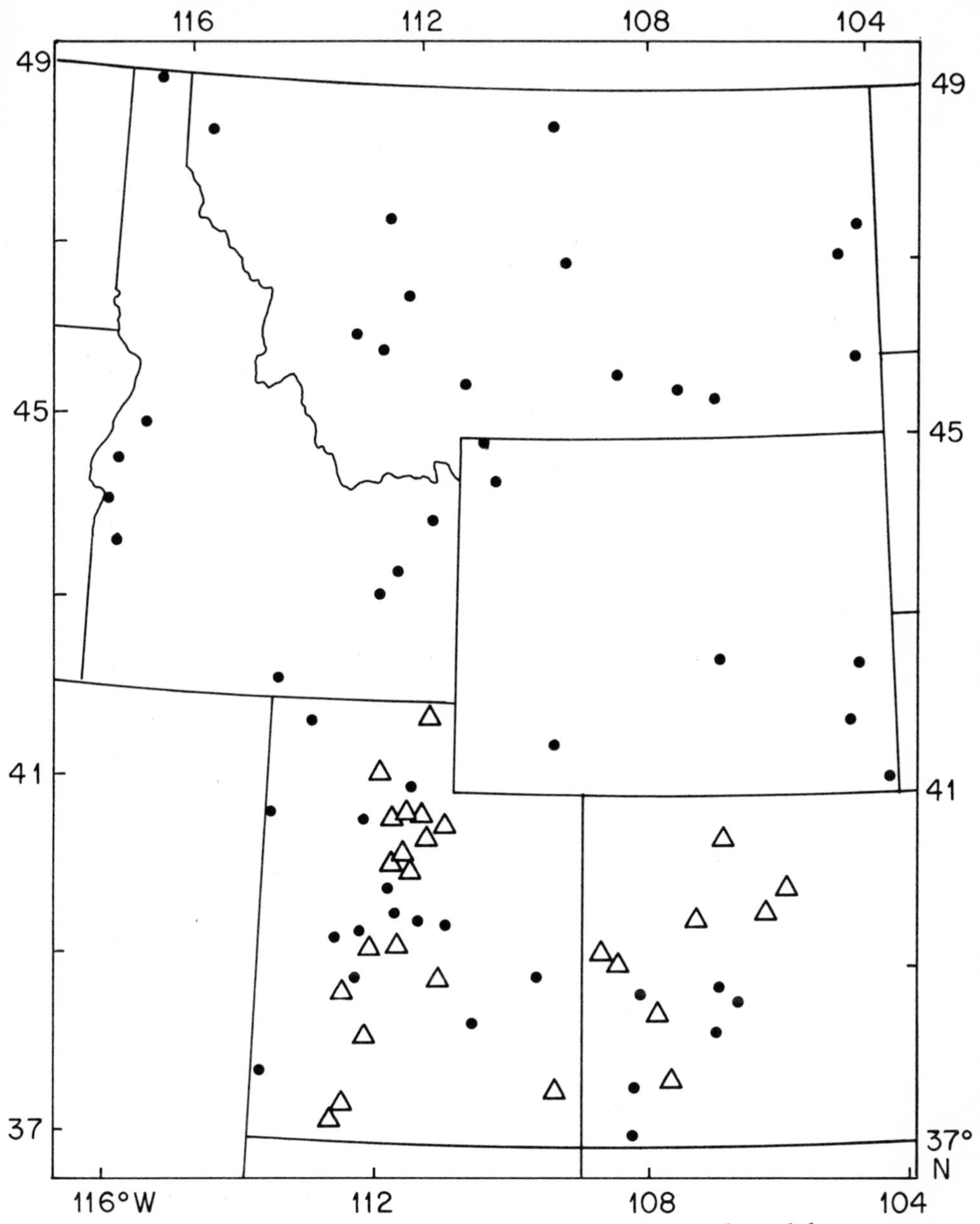

Figure 96. Summer precipitation records analyzed by power spectrum analysis. Those showing a statistically significant high frequency periodicity of 2-2.2 years are indicated by triangles.

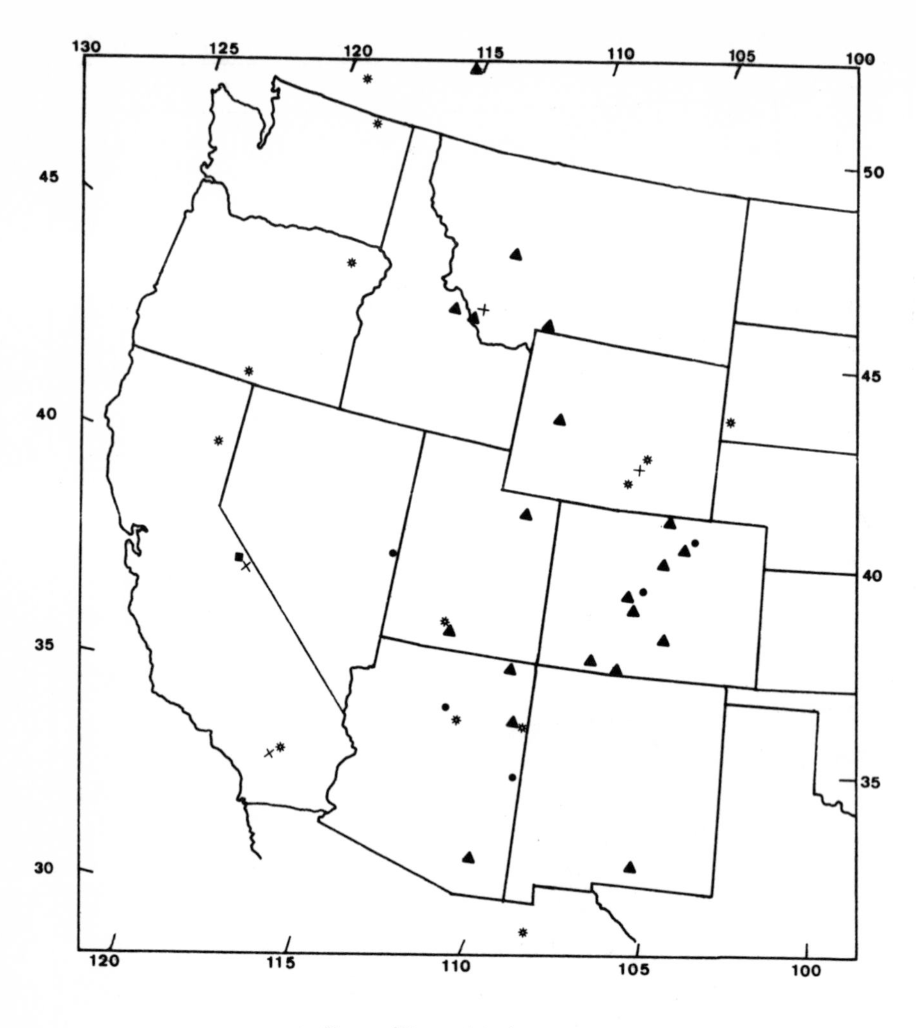

Figure 97. Dendrochronologies used in the study (data from Stokes *et al.*, 1973).

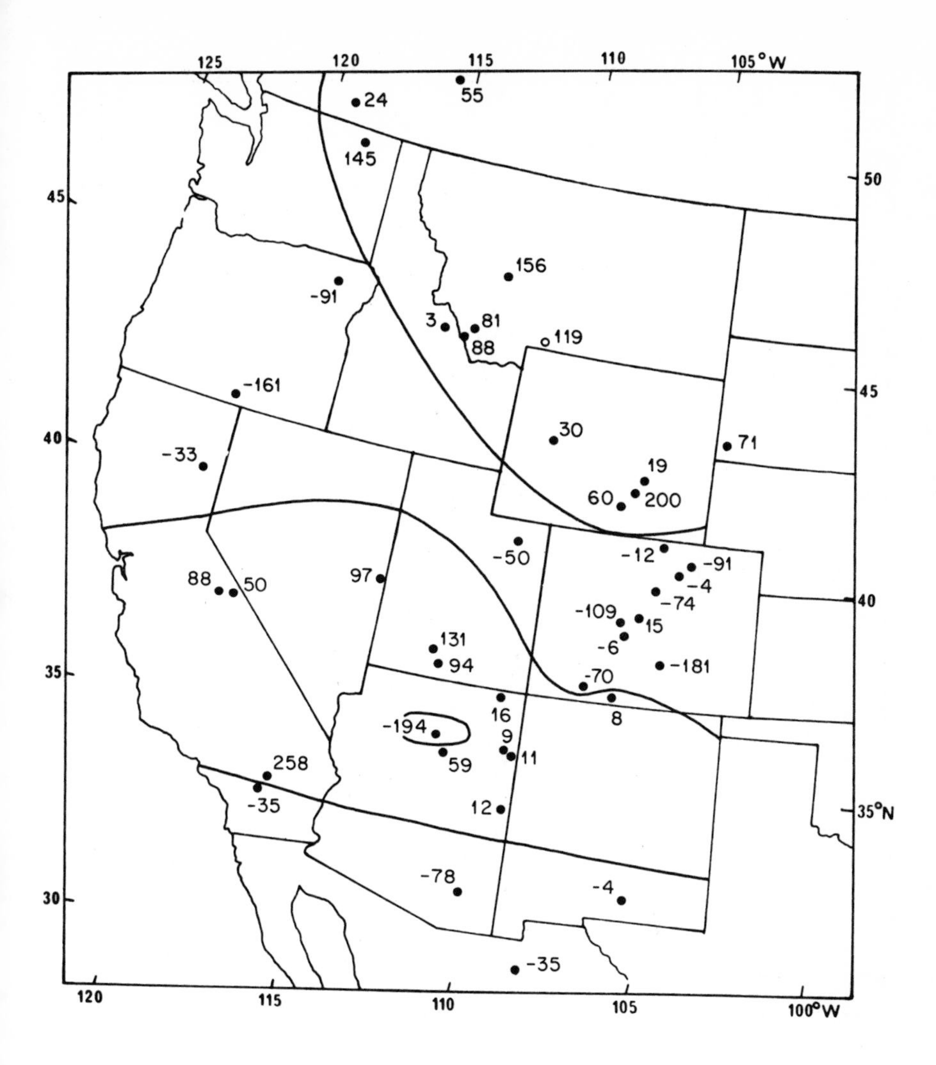

Figure 98: Deviations of mean tree-ring indices (1872-76) from the 1951-60 averages (expressed in standard deviation units of the 1951-60 period). Values in hundredths.

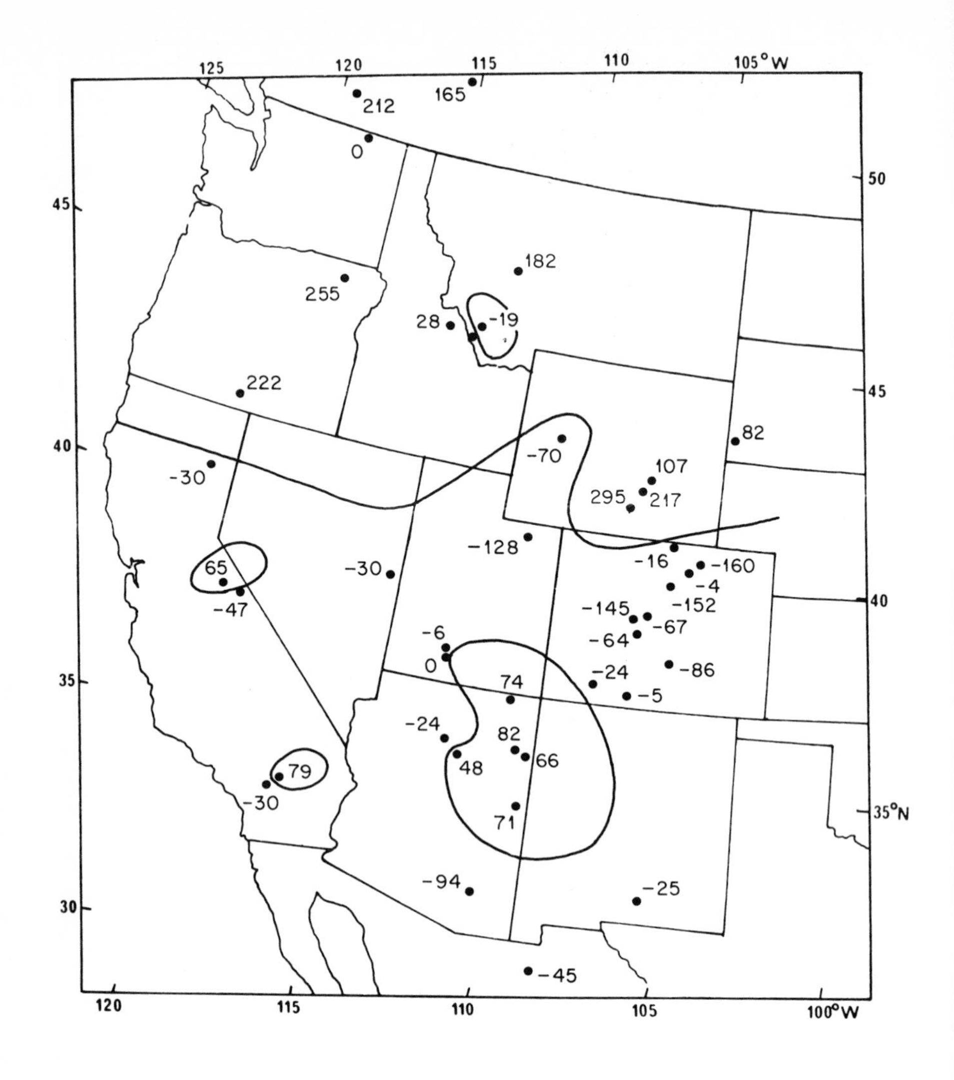

Figure 99: Deviations of mean tree-ring indices (1877-81) from the 1951-60 averages (expressed in standard deviation units of the 1951-60 period). Values in hundredths.

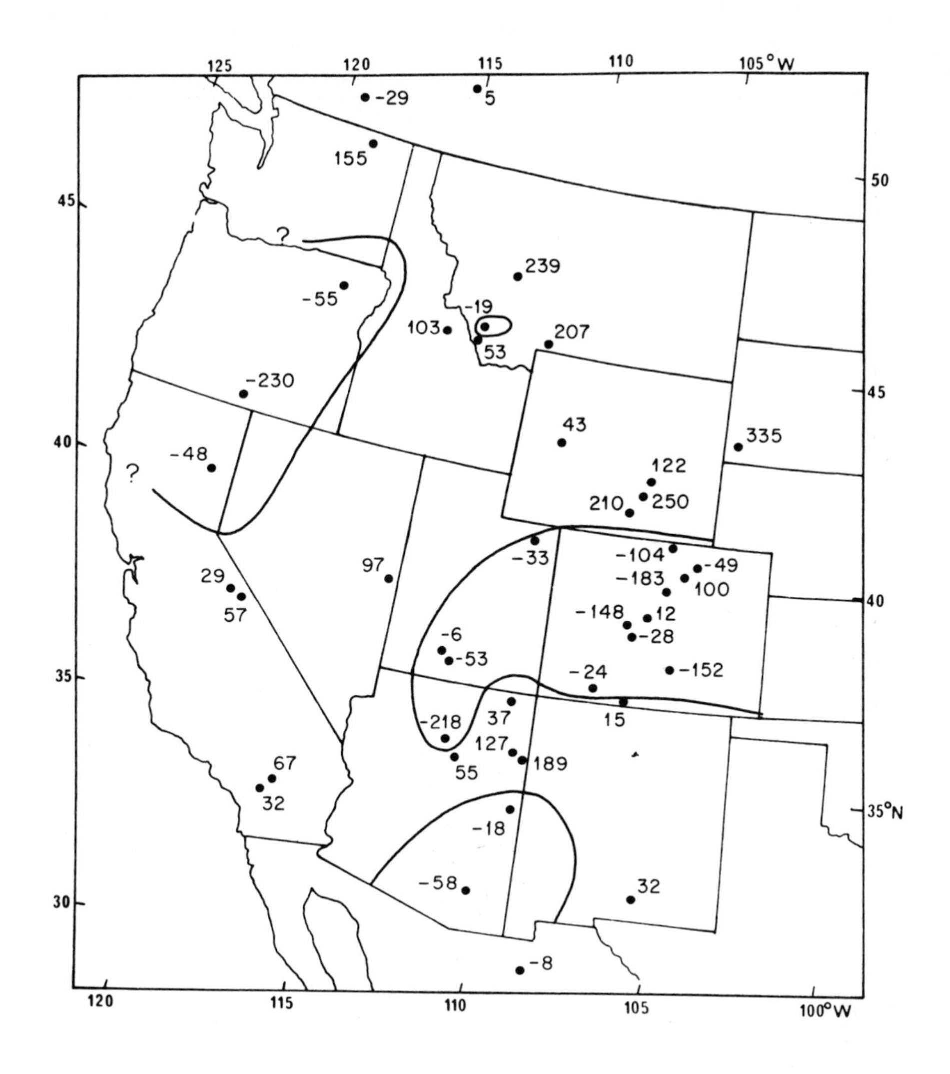

Figure 100: Deviations of mean tree-ring indices (1887-91) from the 1951-60 averages (expressed in standard deviation units of the 1951-60 period). Values in hundredths.

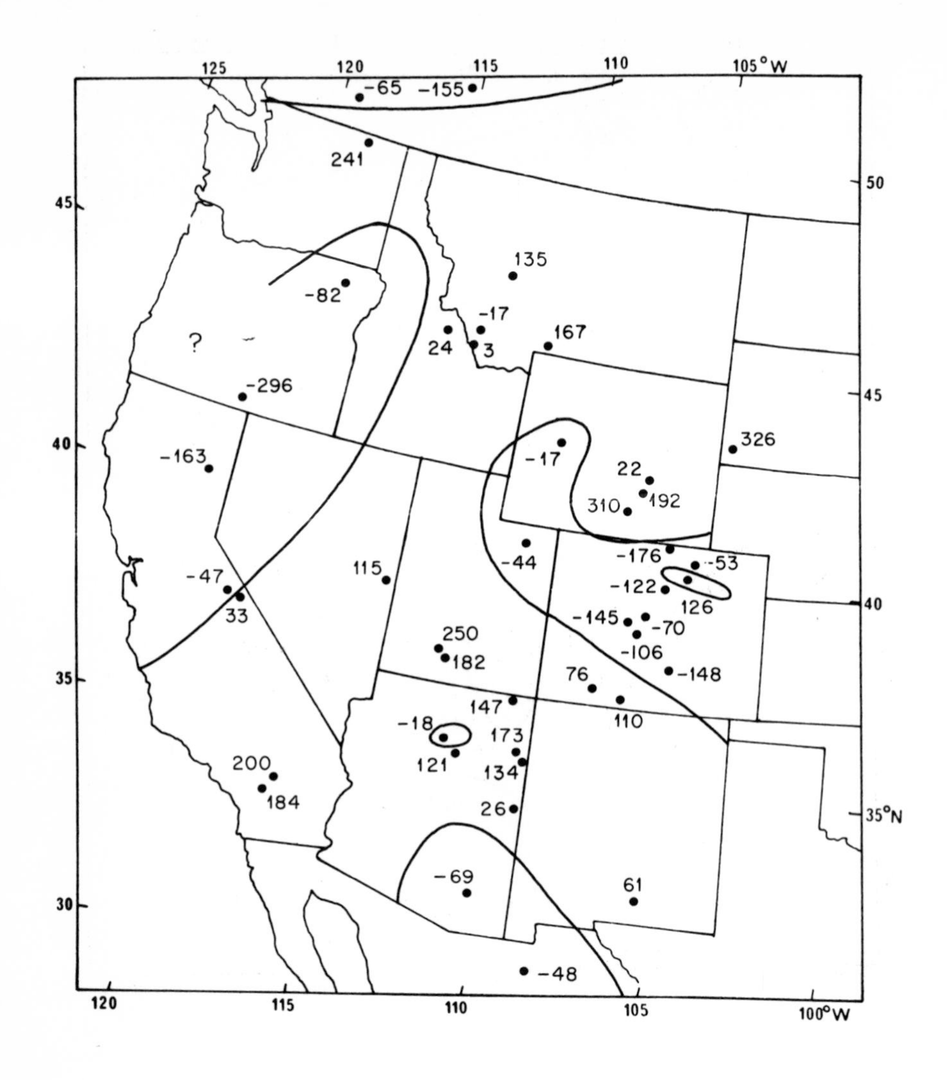

Figure 101: Deviations of mean tree-ring indices (1887-91) from the 1951-60 averages (expressed in standard deviation units of the 1951-60 period). Values in hundredths.

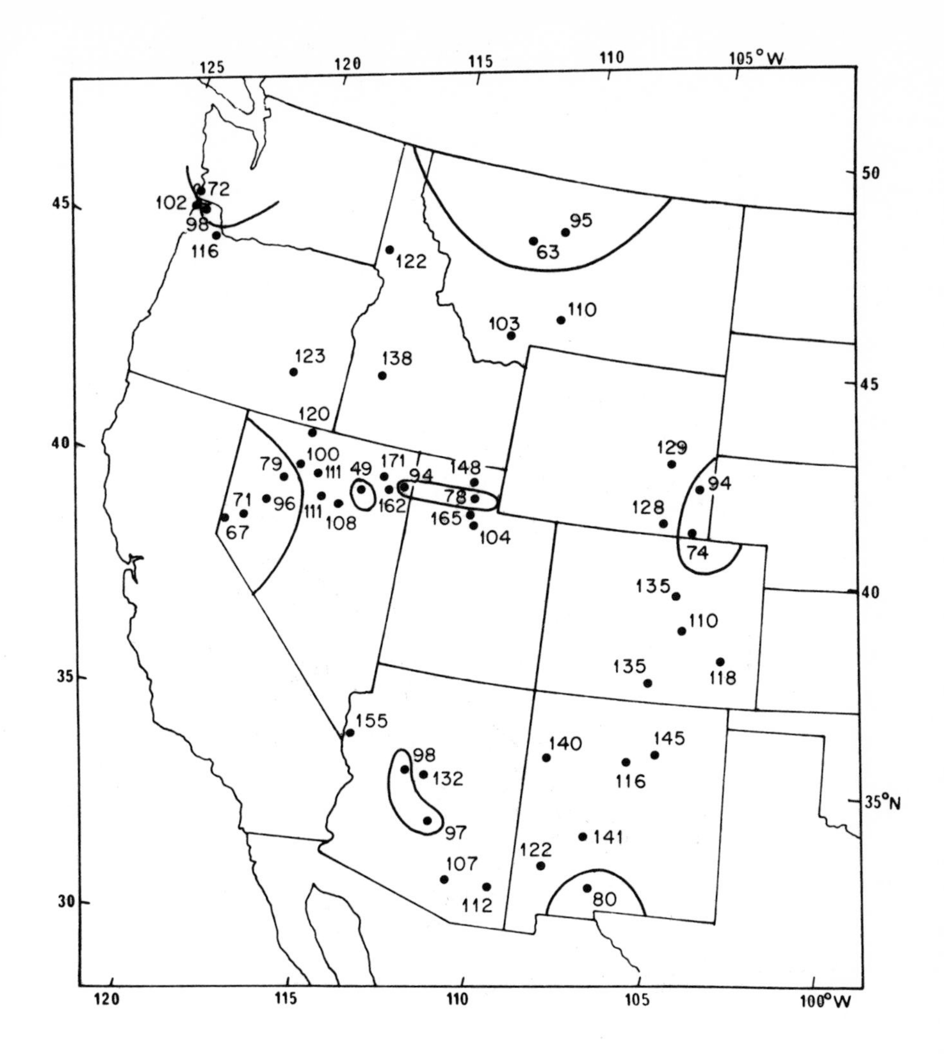

Figure 102: Average precipitation September 1971-August 1881 (5 years) as a percentage of 1951-60 averages (expressed in standard deviation units of the 1951-60 period). Values in hundredths.

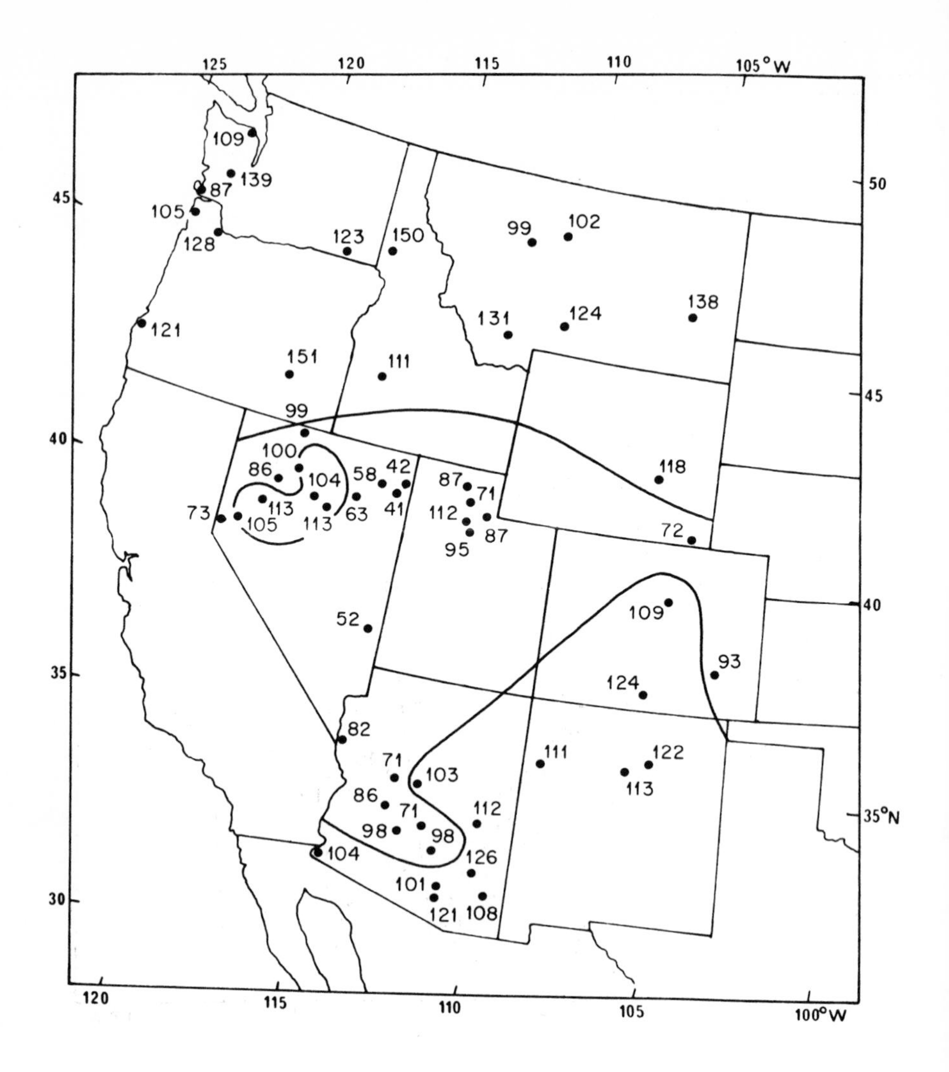

Figure 103. Average precipitation September 1876-August 1881 (5 years) as a percentage of 1951-60 averages.

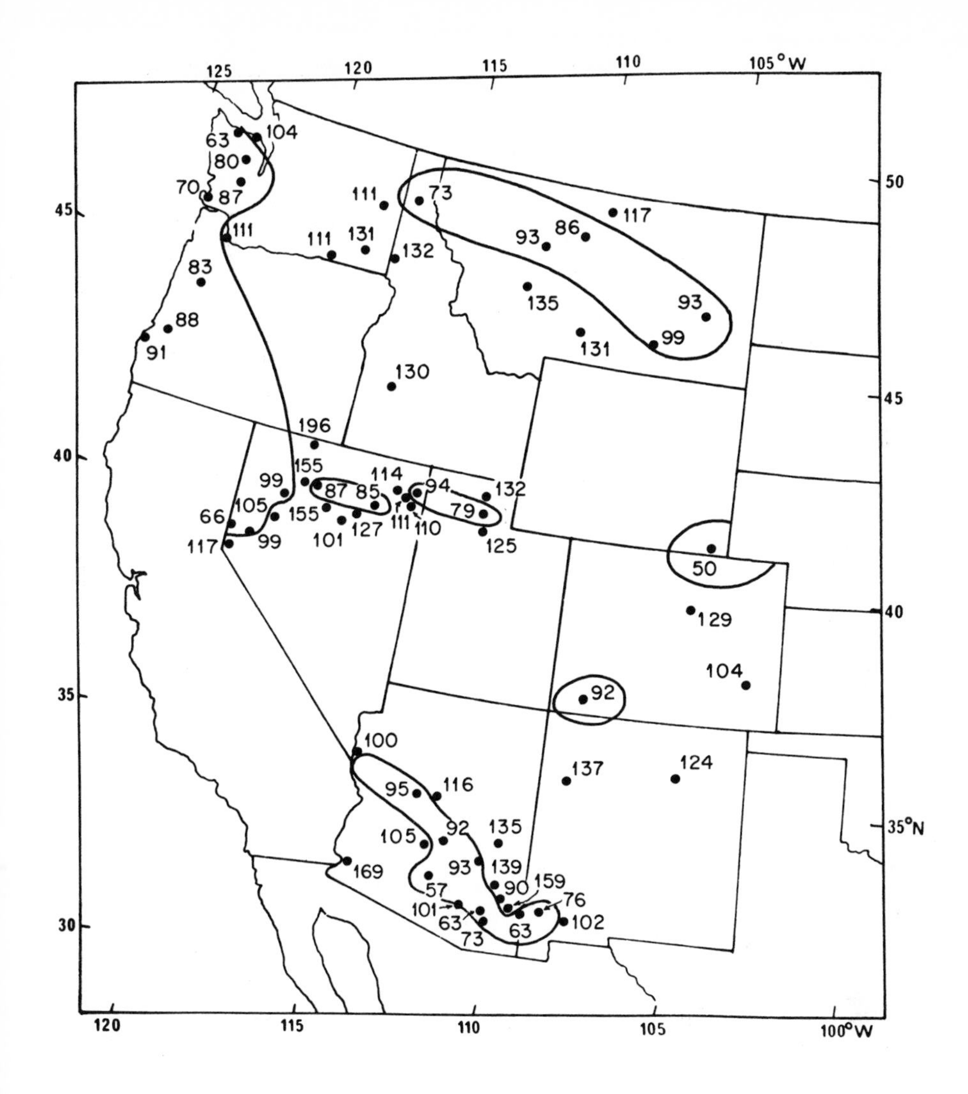

Figure 104. Average precipitation September 1881-August 1886 (5 years) as a percentage of 1951-60 averages.

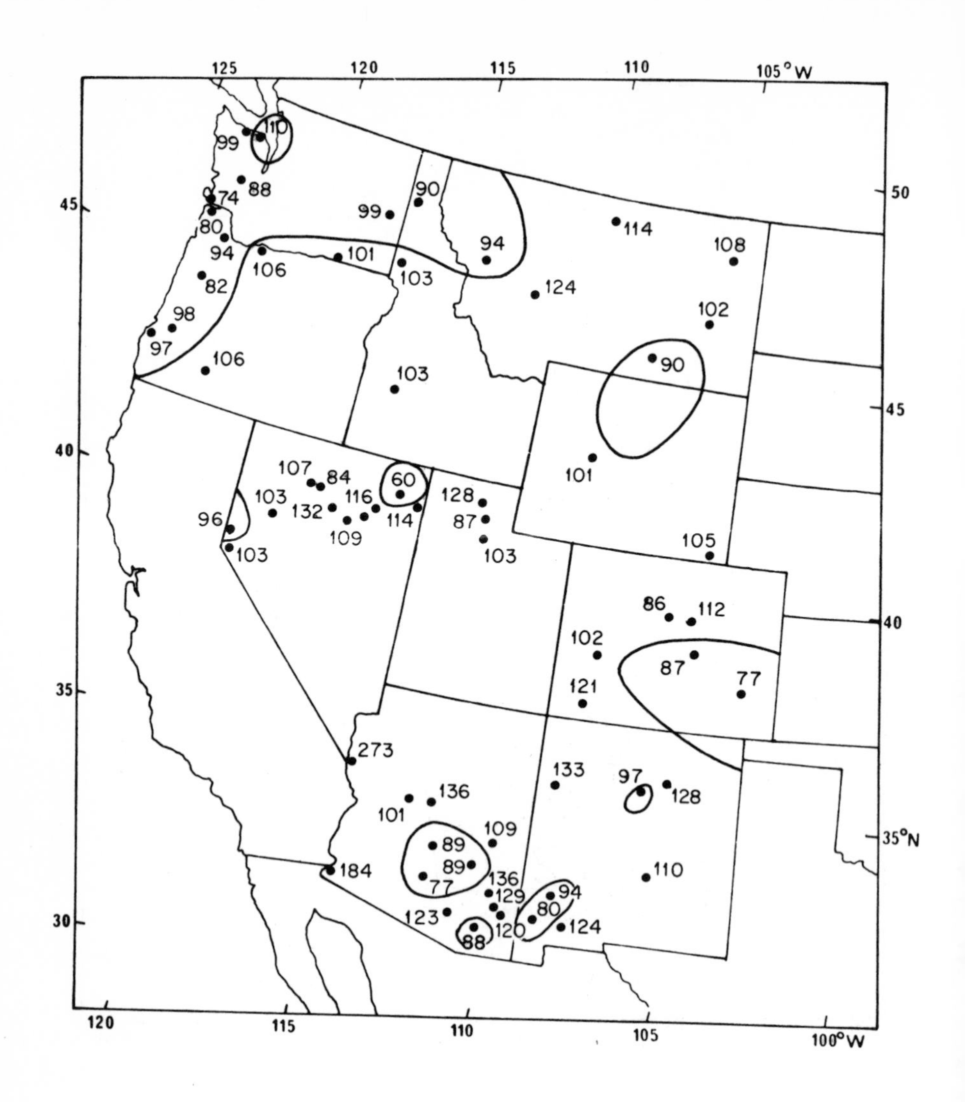

Figure 105: Average precipitation September 1886-August 1891 (5 years) as a percentage of 1951-60 averages.

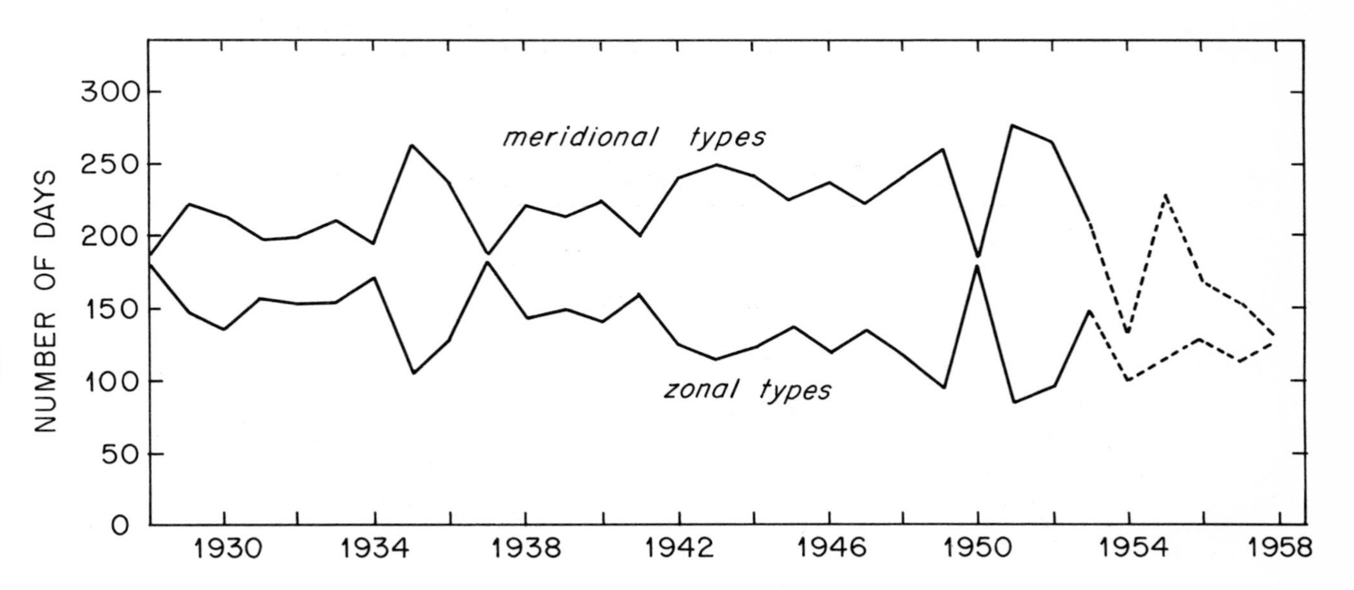

Figure 106: Synoptic weather types of North America: frequency of zonal and meriodional types. Dashed line indicates the minimum estimates due to absence of data.

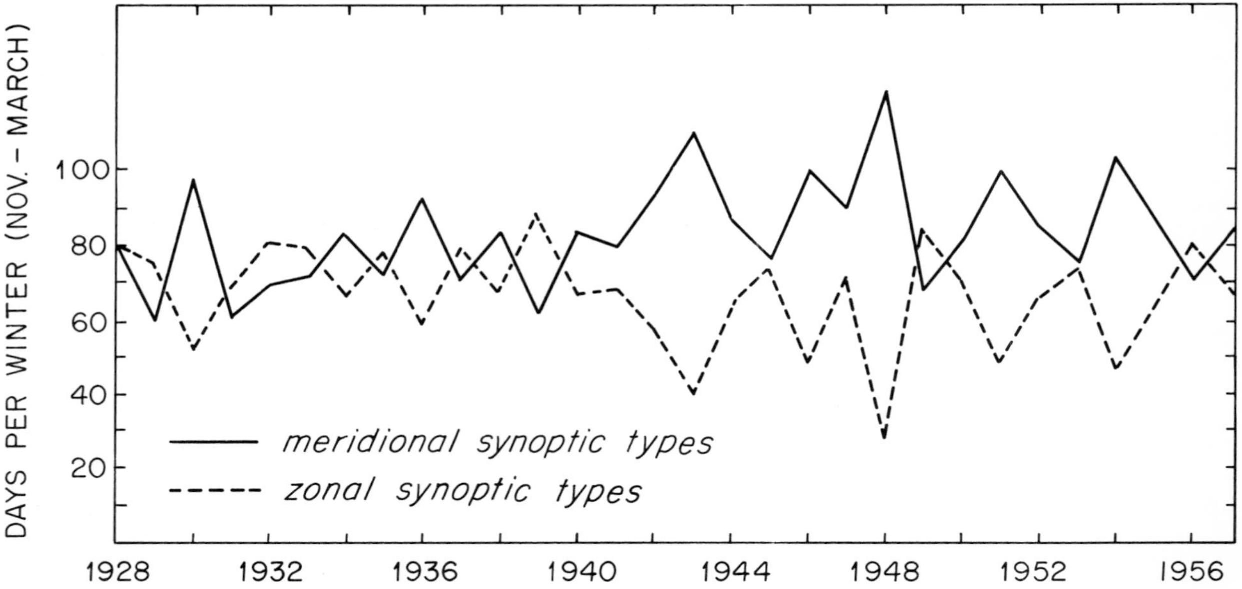

Figure 107: Synoptic weather types of North America: winter frequencies of zonal types (dashed line) and meriodional types (solid line). Years indicate beginning of winter seasons.

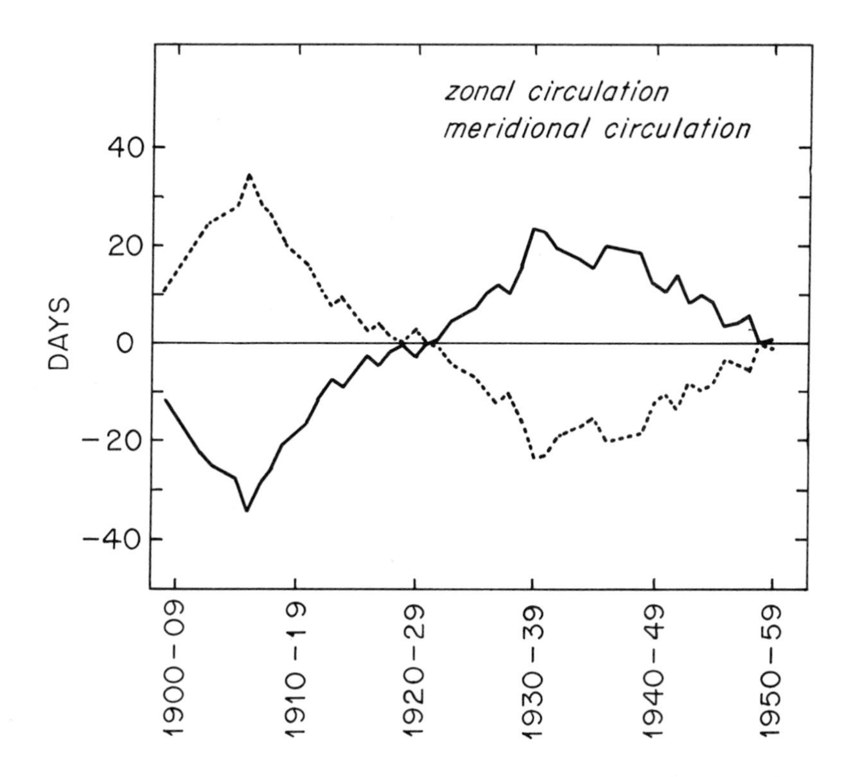

Figure 108: Course of meridional and zonal circulation in the 20th century (circulation epochs) (from Dzerdzeevski, 1966).

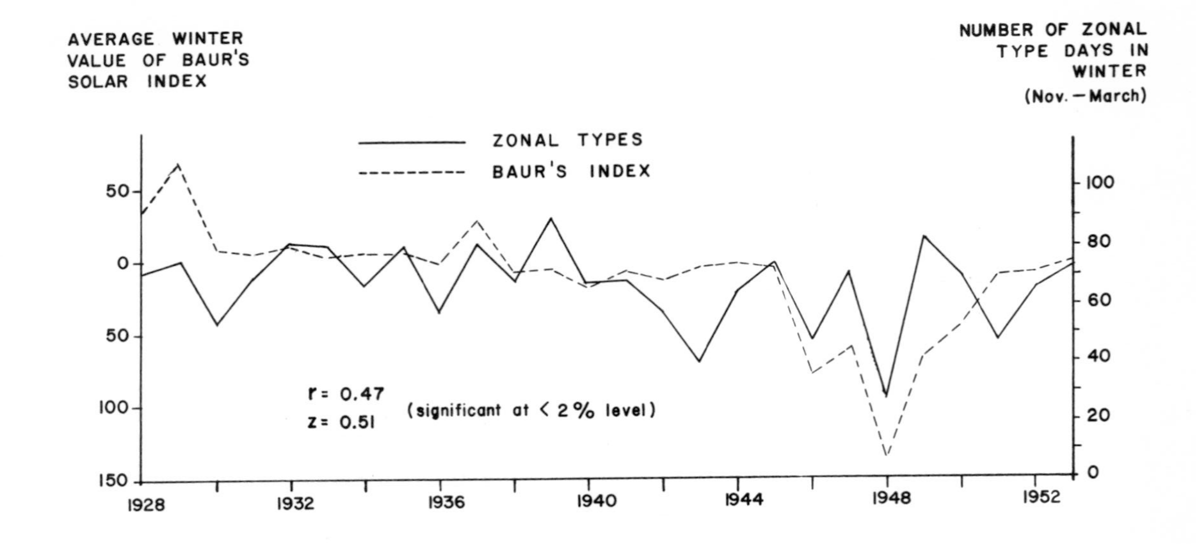

Figure 109. Frequency of zonal circulation types in winter months in relation to winter values of Baur's solar index.

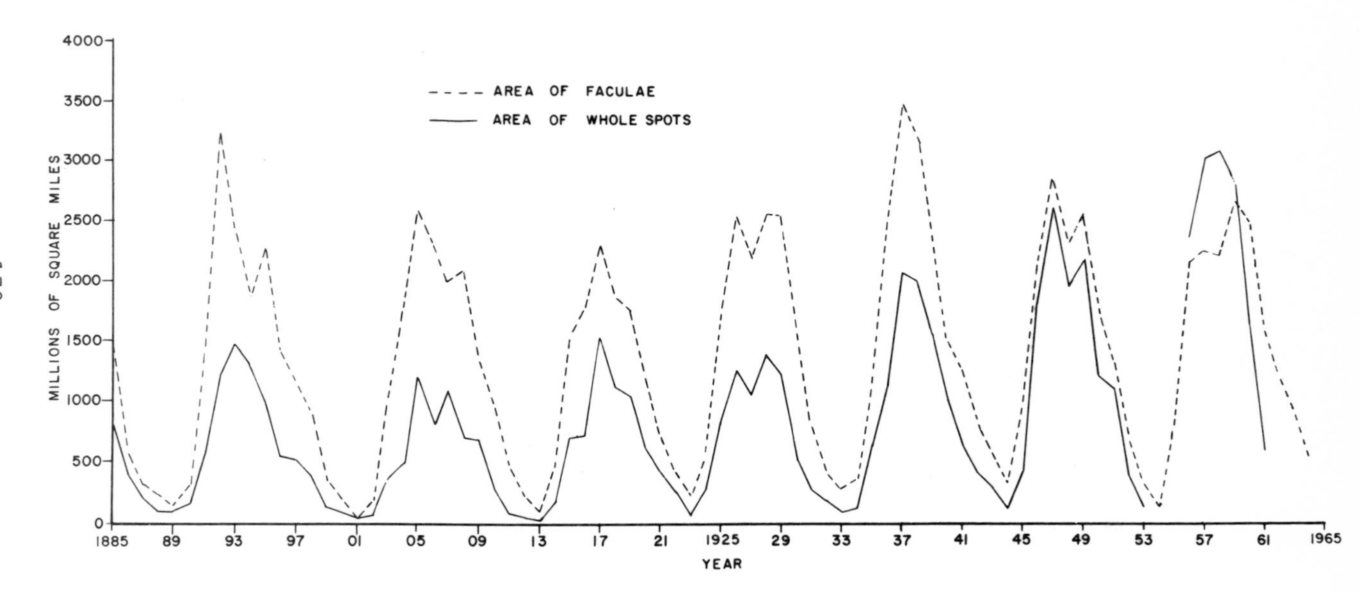

Figure 110: Secular changes in the area of solar faculae and whole spots (annual averages).

APPENDIX A

ANNOTATED BIBLIOGRAPHY OF SELECTED EARLY SOURCES AND INVENTORIES OF CLIMATOLOGICAL DATA FOR THE WESTERN UNITED STATES

APPENDIX A

Annotated Bibliography of Selected Early Sources and Inventories of Climatological Data for the Western United States[1]

Anon: Temperature and precipitation data from post hospitals and voluntary observers 1880-1890 vol. 1; Alabama - Kansas; vol. 2; Kentucky - New Mexico; vol. 3; New York - Wyoming. National Archives Record Group 27; entry 110.
- Hand-written summaries of many stations; not microfilmed.

Bigelow, F. H. 1908: The normals of the daily temperature and the daily precipitation in the United States. U.S. Department of Agriculture, Weather Bureau Bulletin R. 186 pp.
- Series of daily and longer interval "normals" of temperature and precipitation; years on which averages are based given.

Bigelow, F. H. 1912: Summary of the climatological data for the United States, by sections. U.S. Department of Agriculture, Weather Bureau Bulletin W.
- Data from 2,000 stations up to 1909; monthly precipitation data and long-term averages.
- See also Bulletin W (3rd edition), 1932-36 climatological data from the establishment of stations to 1930. Contains mostly precipitation data; mean temperature data given for only 1 or 2 representative stations per state.

Bishop, W. D., and Henry, J. 1861: Results of meteorological observations made under the direction of the U.S. Patent Office and the Smithsonian Institution from 1854-1859. Report of the Commissioner of Patents, 1st

---

[1]Although the search for data concentrated on states west of 104°W excluding California, many of the references cited include data for that state. However, other sources specifically related to California are available. The most notable is McAdie, A. G. 1903: Climatology of California. U.S. Weather Bureau Bulletin L.

Session, 36th Congress, House Executive Document No. 55, 1219 pp.
- Month by month tabulations of temperature and precipitations data but only 22 U.S. stations W of 104°W.

Blodget, L. 1857: Climatology of the U.S. and of the temperate latitudes of the N. American continent. J. B. Lippincott and Company, Philadelphia, 536 pp.
- Mostly average temperature and precipitation data for different time periods though some month by month tabulations for the selected stations are given.

Coolidge, R. M. 1861: Army meteorological register for five years 1854-59. Appendix to "Statistical Report on Sickness and Mortality in the Army of U.S." 36th Congress, 1st Session, Senate Executive Document No. 55, 1219 pp.

Darter, L. J. 1942: List of climatological records in the National Archives. Special List no. 1, National Archives, Washington, D.C. 160 pp.
- Extremely comprehensive inventory of climatic data in the Archives, much of which is now on microfilm.

Dunwoody, H. H. C. 1882: Tables of rainfall and temperature compared with crop production. Professional Papers of the Signal Service No. 10, 15 pp. War Department, Washington, D.C.
- Tables of excesses or deficiencies of temperature and rainfall 1875-1882.

Greely, A.W. 1881: Isothermal lines of the U.S., 1871-1880. Professional Papers of the Signal Service No. 2, War Department, Washington, D.C.
- 12 monthly charts of mean temperature mainly for the period 1871-80 or 1874-80.

Greely, A. W. 1889: Climate of Oregon and Washington. Senate Executive Document No. 282. 50th Congress, 1st Session.
- Month by month tabulations of temperature and precipitation data.

Greely, A. W. 1891a: Report on the climatology of the arid regions of the U.S. with reference to irrigation. House Executive Document No. 287, 51st Congress, 2nd Session.
- Excellent source of month by month temperature and precipitation data for Colorado, Utah, Arizona, New Mexico, Nevada and California.

Greely, A. W. 1891b: Index of meteorological observations in the U.S. from the earliest records to January 1890. Signal Office, War Department, Washington, D.C., 297 pp.
- Inventory of stations operating for >1 year. Includes 3939 stations, only 1745 stations of which were operative in 1890.

Greely, A. W. and Glassford, W. A. 1888: Rainfall of the Pacific slope and the Western states and territories. 50th Congress, 1st Session, Senate Executive Document No. 91, 101 pp.
- Average maximum, minimum and monthly precipitation totals for periods of different length (years given).

Harrington, M. W. 1894: Rainfall and snow of the U.S. compiled to the end of 1891, with annual, seasonal, monthly and other charts. U.S. Department of Agriculture, Weather Bureau Bulletin C, Washington, D.C.
- Annual precipitation totals, tables of abstracted precipitation data for different time periods (years given).

Henry, A. J. 1897: Rainfall of the U.S. (with annual, seasonal and other charts). Department of Agriculture, Weather Bureau Bulletin D, Washington, 58 pp.
- Monthly means for different time periods (years generally given).

Lawson, T. 1855: Meteorological register for 12 years from 1843 to 1854, inc. compiled from observations made by the Officers of the Medical Department of the Army at the Military Posts of the U.S., War Department, Washington, 677 pp.
- Month by month tables of precipitation and temperature data; includes earliest military observations in western states.

Moore, W. L. 1911: Temperature departures, monthly and annual in the U.S.; January 1873 to June 1909. U.S. Department of Agriculture Weather Bureau Bulletin U.
- 474 charts of month by month departures of temperature from normals 1873-1905 (published in Bulletin S).

New Mexico State Engineer: Climatological summary of New Mexico; precipitation: 1849-1954. Technical Report No. 6. State Engineer's Office, Albuquerque, 407 pp.
- Month by month tabulations of precipitation.

New Mexico State Engineer, 1956: Climatological summary of New Mexico; temperature: 1850-1954; frost: 1850-1954; evaporation: 1912-1954. Technical Report No. 5, State Engineer's Office, Albuquerque.
- Month by month tabulations.

Schott, C. A. 1873: Tables and results of the precipitation in rain and snow in the U.S. and at some stations in adjacent parts of North America and in Central and South America. Smithsonian Institution Contributions to Knowledge XVII (S.I. Publication No. 222), Washington, D.C. 178 pp. Also, 2nd edition (1881): Smithsonian Institution Contributions to Knowledge XXIV (S.I. Publication No. 353), Washington, D.C., 209 pp.

Schott, C. A. 1876: Tables, distribution and variation of atmospheric temperature in the U.S. and some adjacent parts of America. Smithsonian Institution Contributions to Knowledge XXI (S.I. Publication No. 277), Washington, D.C., 360 pp.
- Monthly averages of temperature for different time periods (years stated) and annual averages by year.

Smithsonian Institution, 1864: Results of meteorological observations from 1854-1859; Volume II. Smithsonian Institution Publication No. 182, Washington, D.C., 546 pp.

Smithsonian Institution, 1869: Meteorological stations and observers of the Smithsonian Institution in North America and adjacent islands (from the year 1849 to the end of 1868). Smithsonian Institution Annual Report, 1868, Washington, D.C., 68-101.

Stockman, W. B. 1905a. Periodic variation of rainfall in the arid region. U.S. Department of Agriculture, Weather Bureau Bulletin N, Washington, D.C. 15 pp.
- Monthly and annual averages of precipitation (years not stated).

Stockman, W. B. 1905b: Temperature and relative humidity data. U.S. Department of Agriculture, Weather Bureau Bulletin O, Washington, D.C., 29 pp.
- Tables of monthly and annual mean maximum and minimum temperatures and extremes (years given).

Trimble, R. E. 1912: Colorado Climatology. Colorado Agriculture Experimental Station Bulletin. 182. Fort Collins, 56 pp.
- Tables of data for selected stations for different time periods.

Weather Bureau, 1942: Maps of seasonal precipitation. Percentage of normal by States, 1886-1938. U.S. Department of Commerce, Weather Bureau Publication No. 1353.
- Seasonal precipitation as a percentage of normal (1886-1938) by state divisions.

In addition, numerous reports of the Signal Service--Signal Service Notes, Professional Papers and Annual Reports--occasionally include meteorological data and/or charts. Similarly with early volumes of the Monthly Weather Review.

Daily Bulletin of the Signal Service with the synopses, probabilities and facts published monthly 1872-1877 provides daily observations from which weather charts were compiled.

Reports of the Chief of the Weather Bureau 1891-92 to 1935-35 superseded by the Meteorological Yearbook to 1940, provide detailed month by month data for selected stations.

Month by month tabulations of mean temperature data from the establishment of stations to the early part of the twentieth century (generally 1930 or 1940) were kept by State Climatologists but most of these data were discarded with the termination of the State Climatologist Offices in 1973. Only Montana (National Weather Service) appears to have kept the records. The only tabulations of early temperature data now available are the "Washington tabulations" at the National Climatic Center.

Most of the records listed in Darter's 1942 inventory (Record Groups 57 and 59) are available (by state) on microfilm from the National Archives. These are the monthly reports of observers and include daily temperatures, precipitation, wind, cloud and pressure where available.

In addition a few sources pertinent to a particular station or small number of stations were located. Some of these are listed below:

Anon, 1893: A salubrious climate. The Roseburg Review, Jan. 16th, 1893
- Tables and summary of data for Roseburg, Oregon, 1878-1890.

McPartin, T. A. 1877: History and climate of New Mexico. Annual Report of Smithsonian Institution (S.I. Publication No. 396), Washington, D.C., 323-336.
- Description of country en route to New Mexico; includes data for some Colorado stations and Santa Fe for mid-1870s.

Parshall, M. 1939: Analysis of a 50 year record of meteorological data taken at the Colorado Experiment Station, Fort Collins. Colorado Agricultural Experiment Station Bulletin 456, 15 pp.

Theissen, A. F. 1917: The weather and climate of Salt Lake City. Proceedings of the 2nd Pan American Science Congress, Section II, 205-225.
- Detailed tables of climatic data and charts for the period 1875-1914.

Wetmore, E. L. 1893: Tucson climate, 1875-1893. Arizona Daily and Weekly Citizen, Tucson, Arizona, September 1893.
- Summary tables and notes about U.S. Signal Service stations.

Additional references concerning individual stations may be found in Annual Reports of the Smithsonian Institution for the 1850s and 1860s.

APPENDIX B

SEASONAL PRECIPITATION FLUCTUATIONS BY STATE FOR PENTADS 1851-55 TO 1886-90

ARIZONA

Although observations began in this state in 1851 (at Fort Defiance) it is not until the late 1860s that pentad averages are available; five stations began operating in 1867, eight in 1868, and ten in 1869. Prior to 1867 only two stations were recording monthly precipitation and temperature--Fort Defiance (1852-60) and Fort Buchanan (1857-61). Consequently the early 1850s are climatologically unknown and the 1860s are only slightly better. Because of the sparse data coverage, this early period is dealt with separately here.

In all seasons, the period 1856-60 at Fort Defiance was drier than in the four years 1852-55. Spring precipitation 1856-1860 was 81% of the 1852-55 amounts while in summer, fall and winter the amounts were 71%, 41%, and 68%, respectively. Compared to 1851-60 averages, precipitation (1858-61) at Fort Buchanan (southeastern Arizona) was below average in all seasons.

As the nearest equivalent station operating in the 1950s is over 1,000 feet below the stated elevation of Fort Buchanan, it seems likely that precipitation in this area in the late 1850s was considerably below the 1950s average even allowing for the imperfect inter-station comparison. At Fort Defiance, precipitation in the 1850s was considerably above average in all seasons except spring when it was slightly below the average. Subsequent years are dealt with, by season, below:

Spring 1867-1890 (Figures 111, 112)

Most stations show only very slight variations from pentad to pentad and generalizations are difficult. The 1870s were slightly wetter than the 1880s but their relationship to the 1850s and 1860s is unclear. Compared to 1951-60 averages, the 1880s were close to or slightly below average and the 1870s were slightly above the average. Figure 113 shows spring precipitation in the 1870s and 1880s along a transect from southeastern to northwestern Arizona, from Fort Bowie to Fort Mojave.

Summer 1867-1890 (Figures 114-116)

Precipitation amounts in the southern part of the state are generally the inverse of those further north. At the "northern" stations precipitation generally decreased from 1871-75 to 1886-90 with the pentad 1886-90 being the driest whereas in the south and west the pentad 1886-90 is the wettest for 15-20 years at most stations. Comparisons with the 1951-60 averages reveal no general patterns--in each pentad about half of the stations are above and half below the average. However, pentad averages conceal the great year to year variability of the records which is characteristic of this area in summer months (see Figures 117 and 118 for records of Fort Bowie, "southern" Arizona and Whipple Barracks, "northern" Arizona).

Fall 1866-1890 (Figures 119-121)

At almost all stations, maximum fall precipitation occurred in the pentad 1886-90, and many stations show a gradual increase in precipitation from 1871-75, or earlier, to 1886-90 (e.g., Fort Bowie, Whipple Barracks, Fort Verde, Fort McDowell). At some stations the late 1860s were slightly wetter than the early 1870s. Compared to the 1951-60 average, precipitation was above average during the last pentad but generally below average from 1866-70 to 1876-80 or 1881-85. Representative fall records from Fort McDowell and Fort Bowie are shown in Figure 122.

Winter 1867/8 to 1889/90 (Figures 123-125)

Precipitation was generally at a maximum in the years 1881-85 (mainly due to an exceptionally wet winter in 1883), though at some northern and western stations the years 1886-90 were as wet or wetter (Figure 126). The early 1870s were also relatively moist but at most stations, precipitation amounts in the latter half of the 1870s were low. In general, winter precipitation in the 1880s was higher than in the 1870s (e.g., Fort McDowell, Figure 127, Fort Bowie), and several stations show a definite upward trend from the late 1860s to the late 1880s (e.g., Fort Whipple, Fort Verde, Figure 126).

Finally, compared to 1951-60 averages, winter precipitation throughout the period 1866-90 was well above average with few exceptions.

COLORADO

Earliest records in the state are from Fort Massachusetts (October, 1852). However, records are extremely fragmentary over the subsequent forty years and only five stations have continuous records of ten years or more within this period (Denver, Fort Garland, Pike's Peak, Fort Lewis and Fort

Lyon). Generalizations are therefore extremely difficult, particularly in view of the great variations in station elevation and topographic situation (cf. Pike's Peak at 14,134 feet and Fort Lyon at 4,000 feet).

Spring 1867-1890 (Figures 128, 129)

No common trends between stations are apparent. At Denver (and Pike's Peak), maximum precipitation 1871-90 occurred in the pentad 1881-85, whereas at Fort Lyon on the plains to the southeast, the pentad of maximum precipitation was 1871-75. The relationship of pentad averages to the 1951-60 averages is unclear; however, at most stations the latter half of the 1880s was below average.

Summer 1867-1890 (Figure 130)

Again generalizations are difficult but the 1880s were probably drier than the 1870s and the late 1860s. In particular, the 1880s were relatively dry at Pike's Peak and Fort Lyon. The period 1867-70 to 1881-85 (and at some stations to 1886-90) was slightly below the average at several western stations.

Fall 1866-1890 (Figure 131)

Precipitation amounts fell from 1871-75 to 1881-85 then rose slightly in the years 1886-90. The pentad 1881-85 was probably the driest for the entire period. Compared to the 1951-60 averages, mean precipitation for the decade 1866-75 was generally above average and below average in the fifteen years 1876-90.

Winter 1866-1889 (Figures 132, 133)

Precipitation in the 1880s was probably lower than in the 1870s though this is an extremely tenuous suggestion. Precipitation amounts were lower than the 1951-60 average in the late 1880s at most stations (the above average record for this period at Fort Lewis is related to very high precipitation receipts in the winter of 1889-90). No firm conclusions about the earlier periods can be reached.

IDAHO

Although some of the earliest meteorological measurements in the Rocky Mountains physiographic province were taken in Idaho, there are few long-term stations in the state. Only five stations kept records continuously for more than nine years prior to 1890, and there are few other short-term stations. Generalizations are hence rather difficult.

Spring 1866-1890 (Figure 134)

At Fort Lapwai precipitation was at a maximum in the latter half of the 1870s and declined through the 1880s. The pentad 1871-75 was probably the period of lowest spring precipitation. At Boise, southern Idaho, however, 1866-70 was the pentad of maximum precipitation and the subsequent twenty years were considerably drier. There is not enough agreement between stations to permit a realistic assessment of spring precipitation amounts compared to 1951-60 averages.

Summer 1871-1890 (Figure 135)

Precipitation amounts rose gradually from the early 1870s through the 1880s and the pentad 1886-90 was probably the period of maximum precipitation at most stations. Compared to 1951-60 averages, most of the period seems to have been below average.

Fall 1867-1890 (Figure 136)

Precipitation amounts were at a maximum in the years 1876-80 (Fort Lapwai/Lewiston) and 1881-85 (Boise). During the period 1867-75 in the north and 1871 to 1880 in the south, precipitation amounts were considerably less, and slightly below 1951-60 averages.

Winter 1866-1889 (Figure 137)

Precipitation was markedly above 1951-60 averages throughout the period with mean winter precipitation amounts at Boise for 1867-89 of 141% of the 1951-60 levels. At Fort Lapwai, 1866-80, the corresponding figure is 170%. Precipitation was at a maximum in the late 1860s and early 1870s and declined thereafter reaching a minimum in the final period 1886-89.

MONTANA

Earliest records for the state are from Fort Benton (October 1862) though observations were intermittent from then until November 1860. Only seven other stations have records for the 1860s;[1] none of these are earlier than September 1866 and none are more than 40 months in total length. Clearly then, very little can be said about the climate of Montana prior to the early 1870s.

---

[1]Benton City, Camp Cooke, Deer Lodge, Fort C. F. Smith, Fort Ellis, Fort Shaw, and Helena.

Spring 1871-1890 (Figures 138, 139)

Precipitation was at a maximum in the pentad 1876-80 and was considerably less in the subsequent decade. The early 1870s were also relatively dry. At eastern stations (Miles City, Poplar, and Fort Custer) precipitation in the latter half of the 1880s was slightly drier than the former half. This may not have been true at stations to the west (Helena, Fort Assiniboine). Precipitation was markedly above 1951-60 averages in the pentad 1876-80 but below average in other pentads.

Summer 1871-1890 (Figures 140, 141)

Maximum precipitation occurred in the pentad 1881-85 having risen from a low point in the years 1871-75. The latter half of the 1880s was slightly drier than the previous pentad. Compared to 1951-60 averages, precipitation at most stations was above average in the years 1876-85, and at Fort Shaw and Fort Benton the pentad 1871-75 was well below average.

Fall 1867-1890 (Figure 142)

Precipitation was at a maximum at most stations in the years 1881-85 though at eastern stations this period was relatively dry. At western stations precipitation amounts increased from the early 1870s to 1881-85 then fell to 1870s levels in the latter half of the 1880s. At eastern stations, however, the late 1880s were wetter than the early 1880s. Precipitation was generally above 1951-60 averages in the decade 1876-85 and below average during the period 1866-75.

Winter 1867-1890 (Figures 143, 144)

Like the spring and fall seasons, winter precipitation was at a maximum at most stations in the years 1876-80 or 1881-85. Precipitation was particularly heavy in the winter of 1880 at Fort Assiniboine, Fort Benton, and Fort Shaw (more than twice the period average at Forts Benton and Shaw) and in 1880 and 1881 at Fort Ellis to the south. The late 1860s and early 1870s were relatively dry and below 1951-60 averages at most stations and the subsequent decade was generally above average.

NEVADA

The earliest monthly data available for this state are from Fort Churchill (December 1860). However, there are not enough records for meaningful generalizations until approximately 1870 when the Pacific Railway Company established stations across the state. Most records are consequently restricted to a transect across northern Nevada. The only station for the early 1860s is Fort Churchill (western Nevada). Only one pentad average is available for this station and it indicates that precipitation was above the 1951-60 average for winter and spring (particularly winter) and below average for summer and fall. If this comparison is reasonable, it would indicate that the below normal rainfall of the 1870s and 1880s in summer and fall was also characteristic of the 1860s (see below). Similarly, the above average precipitation in winter was also characteristic of the 1870s and 1880s.

Spring 1866-1870 (Figures 145, 146)

Maximum pentad precipitation occurs in 1876-80 at most western stations (Reno, Wadsworth, Hot Springs, Brown's, Humboldt) and in 1881-85 at the majority of the others (Figure 147). Those stations with maxima in 1876-80 show decreases through 1886-90 whereas a few stations (Toano, Elko, Carlin) show an increase from the early 1870s to 1890 (Figure 147). Comparison with the 1951-60 average shows that precipitation was generally below average at most stations but the maximum pentad (1876-80 or 1881-85) exceeds the average in some cases.

Summer 1871-1890 (Figures 148, 149)

Interpentad variations are generally small with average precipitation at all stations less than one inch. At most stations precipitation was generally highest in the pentad 1881-85 but at some northcentral Nevada stations the maximum was in 1876-80 (Figure 150). When compared to average precipitation for 1951-60, almost all pentad means were well below average. As noted before, precipitation amounts in the early records were sometimes not recorded, particularly when light shower activity was involved. Such convective precipitation is common in the summer months in this area and thus the actual amounts recorded in the 1870s and 1880s may be artificially low. However, in view of the consistency of the pattern throughout the area and for the entire period it is probable that the low amounts are in fact real and that the summer period 1870-1890 in northern Nevada at least, was relatively dry.

Fall 1866-90 (Figures 151, 152)

At most stations the pentad 1881-85 was the period of maximum precipitation during the twenty-five years of record (Figure 153). A few stations show a maximum in the pentad 1886-90 (e.g., Wadsworth, Elko, Toano, see Figure 153) though, again, interpentad differences are small. In nearly all cases, the pentad 1881-85 precipitation is above the 1951-60 average; other pentads are generally below normal though at many stations the average for 1886-90 is also above the 1950s average.

Winter 1866-90 (Figures 154-157)

Although no one pentad stands out as the period of maximum precipitation, the early 1870s were relatively wet at most stations and so were the early 1880s (Figure 158). The period 1876-80 was the driest pentad at most stations.

Nearly all of the early records show precipitation throughout the period 1861-90 (if the Fort Churchill record is included) as above the 1951-60 average. The main exception to this is the late 1870s at some eastern stations (Toano, Tecoma, Wells). Although peaks in the winter precipitation record are not synchronous with those in the Arizona records, it is noticeable that at a number of stations recurrent peaks occur in 1870, 1874 or 1875, 1880 (less pronounced), 1885 and 1889 (see, for example, the Reno records, Figure 159).

NEW MEXICO

Although the earliest records in the mountain and desert west were kept in New Mexico, historical records of five years or more are sparse. Only three long-term records (>15 years) are available--all in northern New Mexico and hence generalizations are inevitably weighted towards these records. With this in mind, the following generalizations are made.

Spring 1852-1890 (Figure 160)

Precipitation amounts appear to have risen from the late 1850s to the 1880s (e.g., Fort Union, Figure 161, Santa

Fe[?], Fort Craig[?]). Compared to 1951-60 averages, the following generalizations can be made: 1852-55, above normal in southern New Mexico, below normal further north; 1855-70, below normal; 1871-75, above normal; 1876-80, indeterminate; 1880s, generally above normal except in southern New Mexico.

Summer 1851-1890 (Figures 162, 163)

Little coherence within the 40-year period is apparent on the basis of the few records available; long-term records show no consistency in the pentad of maximum or minimum precipitation, presumably reflecting the random nature of convective precipitation activity in these months. However, one pattern is clear--amounts were considerably higher than the 1951-60 average for the entire period at most stations. In particular, at several stations the late 1850s were well above the average (Fort Stanton, Fort Thorn, Fort Union, Figure 164). The only period of below average precipitation was the early 1870s in southern New Mexico.

Fall 1851-1890 (Figures 165, 166)

Precipitation in the 1850s appears to have been higher than in other decades though the evidence is scant (only two stations from northern New Mexico). However, the latter part of the 1850s at Fort Stanton and Fort Craig, south-central New Mexico (Figure 167) was also relatively wet. Little can be said about the other pentads; precipitation generally seems to have fallen from the early fifties to the late sixties and then risen slightly to the end of the 1880s. Compared to 1951-60 averages, precipitation throughout the period was above average, particularly in the 1850s.

Data for the 1860s and 1870s are inconclusive but precipitation in the 1880s was generally above the 1951-60 averages.

Winter 1851-1889 (Figures 168, 169)

Again, generalizations are difficult due to the few data available. However, it would appear that precipitation amounts in the 1870s were greater than in the 1880s but slightly less than in the 1850s (Figure 170). The 1860s are problematical but were probably drier than both the 1850s and the 1870s. Compared to 1951-60 averages, precipitation was generally high from 1866-70 to 1886-90. The early fifties were above normal but the late 1850s and early 1860s are indeterminate.

OREGON

Of all the states considered in this study, the earliest records are from Fort George which was probably close to the site of Astoria, Oregon today (Roden, 1966, see discussion above, p. 19). This station operated intermittently during the period 1821-24 and recorded temperatures at 6 am, noon, and 6 pm. However, records of precipitation have not been located and the earliest pentad averages for the state are for the latter part of the 1850s (Astoria, Fort Dalles, Fort Hoskins, and Fort Yamhill).

Spring 1856-1890 (Figures 171, 172)

Precipitation amounts rose from the earliest 1850s to the late 1870s or early 1880s at most stations (Figure 173). The late 1880s were dry--only slightly drier than the previous pentad at southern stations but more significantly drier at northern Oregon stations. The decade 1876-85 was the period of maximum spring precipitation, whereas the pentads 1856-60 and/or 1886-90 were probably the driest. Compared to 1951-60 averages, precipitation was probably above average for the period 1866-85 and close to or below average during other pentads.

Summer 1856-1890 (Figures 174, 175)

Precipitation was probably at a maximum in the 1860s when amounts were higher than in the pentads 1856-60 and 1871-75 and approximately similar to amounts in the decade 1876-85 (Figure 176). In the final pentad precipitation amounts fell slightly at northern stations but rose at a number of southern Oregon stations (Roseburg, Ashland, Fort Klamath). Precipitation levels were slightly above 1951-60 averages in the 1860s, and at some southern stations, in the late 1880s. For most of the period, however, precipitation was probably slightly below average.

Fall 1856-1890 (Figures 177-179)

Precipitation amounts were very high in the years 1858-60 at northwestern Oregon stations and well above 1951-60 averages (e.g., Fort Hoskins, Fort Yamhill, Fort Dalles and to some extent Astoria, see Figure 180). In the 1860s and early 1870s, however, fall precipitation was considerably less and equally well below 1951-60 averages. The decade 1876-85 was relatively wet but in the late 1880s precipitation fell markedly throughout the state. In short, the period was generally well below 1951-60 averages apart from the latter half of the 1850s, the early 1880s, and perhaps the late 1870s at some stations.

Winter 1851-1889 (Figures 181-183)

At the only station for which there are winter precipitation data throughout the 1850s (Fort Dalles), precipitation was twice as heavy in the latter half of the 1850s compared to the former half. Precipitation was also heavy at Fort Hoskins in the pentad 1856-60 compared to the 1951-60 average, though this was not the case at Astoria. At Fort Dalles and Fort Yamhill precipitation fell from 1856-60 to 1861-65 but rose at Astoria. In the subsequent three pentads precipitation increased quite regularly, reaching maxima at all stations in the pentad 1876-80 (Figure 184). The early 1880s were then quite dry but precipitation amounts increased at most stations in the latter half of the 1880s. At most stations, precipitation was above 1951-60 averages in the pentad 1856-60 and the years 1866-80 but well below average in the early 1850s(?), early 1860s and early 1880s. The latter half of the 1880s was slightly below average.

UTAH

Earliest monthly climatic records for this state began in 1850 but no pentad averages are available prior to the late 1860s. Many of the long-term records are from the Pacific Railway Company's stations and form an eastern extension of the Nevada station network mentioned above. In fact, all long-term nineteenth century records for the state are from the area north of 40°30'. Hence in both Utah and Nevada there is a vast area of the Great Basin for which we have almost no climatic data prior to the 1890s. With this in mind, the following generalizations are made.

Spring 1866-1890 (Figure 185)

At stations close to the Wasatch Front, precipitation amounts fell from the early 1870s to 1886-90 (e.g., Salt Lake City, Ogden, Douglas (?), Corrine (?), Figures 186, 187). Stations to the west do not show such a trend and precipitation there was at a maximum in the pentads 1876-80 or 1881-85 (Figure 187). Comparisons with the 1951-60 averages suggest that the period 1866-1875 was above average; the pentad 1886-90 is consistently below average but the number of stations on which to base a conclusion is small.

Summer 1866-1870 (Figure 188)

The first half of the 1880s was relatively moist at a number of stations but few other features are noticeable. Generally the first five years of the 1870s and 1880s were wetter than the second five years (Figure 189). Precipitation was below the 1951-60 average in the latter half of the 1870s and 1880s and above the average in the early 1880s, and perhaps the early 1870s.

Fall 1866-1890 (Figure 190)

Precipitation amounts increased from the early 1870s to 1881-85 then fell slightly in the pentad 1886-90. The average for 1881-85 is the period of maximum fall precipitation at several stations (Figure 191). Precipitation was at or below the 1951-60 average prior to 1876-80 and then above average for about the next fifteen years.

Winter 1866-1890 (Figures 182, 193)

Maximum precipitation occcurred in the decade 1866-75 though the 1880s were also relatively wet (Figures 194-196). Lowest precipitation amounts were received in the pentad 1876-80, when totals were generally less than the 1951-60 average. During the decades before and after this low point, however, precipitation was above the average.

WASHINGTON

Observations for the state began at Fort Steilacoom in 1849 but records at this station in the early fifties are noted as being doubtful or incorrect (Greely, 1889). However, there were two other long-term stations operating in the state during the 1850s--Fort Vancouver and Fort Walla Walla. Thus, for most seasons information is available from several stations for each pentad during the period 1851-90.

Spring 1852-1890 (Figures 197, 198)

Precipitation fell from the early 1850s to the latter part of the decade and then rose from the early 1860s to the late 1870s. Precipitation amounts in the 1880s were similar to or below the late 1870s levels (Figure 199). In particular, the latter half of the 1880s was relatively dry. The decade 1876-85 (or 1871-80) was the wettest at most stations whereas the decade 1856-65 was probably the driest spring period. Compared to 1951-60 averages, the late 1870s and early 1880s were above average (particularly the early 1880s), whereas the late 1880s were probably close to average. The 1850s were probably also above average but the 1860s are indeterminate.

Summer 1857-1890 (Figures 200, 201)

The summer record is characterized by alternating pentads of relatively high and low precipitation amounts. Thus the late 1850s were relatively wet, the early 1860s dry, late 1860s wet, etc. However, like the spring record there is some evidence of an underlying upward trend from the early 1860s to the late 1870s at least. The late 1870s were probably the wettest at some stations but at other stations the late 1880s were wetter. The first half of the 1860s was probably the driest pentad of the period. Compared to 1951-60 averages, the overall levels at most stations were probably about the same with the "high" pentads slightly above average and the "low" pentads slightly below average.

Fall 1852-1890 (Figures 202, 203)

The latter half of the 1850s was relatively wet compared to the pentads both before and after (Figure 204). Precipitation levels rose during the 1860s to (slightly?) higher averages during the latter half of the 1870s and early 1880s. The late 1880s, however, were well below the previous pentad and/or decade averages at most stations (Figures 202, 203). Precipitation amounts were above average in the years 1856-60 and 1876-85 and probably below average for the rest of the period, though there are very few data for the years 1866-75.

Winter 1851-1889 (Figures 205-207)

Precipitation amounts are extremely heavy at coastal stations in the state but decrease markedly to the east (e.g., Fort Walla Walla) and in the lee of coastal mountain ranges (e.g., Port Townsend). The coastal stations are characterized by large interpentad variations as shown by a comparison of the late 1870s (wet) and the early 1880s (dry) at Cape Disappointment (Fort Canby) and Olympia.

Precipitation fell from relatively high levels in the 1850s to low amounts in the pentad 1861-65 (Figure 208). On the coast the subsequent three pentads were increasingly wet though individual wet winters (e.g., 1866) were superimposed on this upward trend. The 1880s were then drier, particularly the first half of the decade, which was exceptionally dry at several stations. However, this pattern does not appear to be characteristic of eastern stations (Spokane Falls, Walla Walla). At these stations, precipitation decreased slightly from the late 1870s through the 1880s. Precipitation was above 1951-60 averages in the 1850s and below average in the first half of the 1860s. The next decade is indeterminate but is followed by a pentad of well above average precipitation and a second pentad of well below average precipitation. The late 1880s are also indeterminate, being above average at some stations and below average at others.

WYOMING

Some of the earliest records in the Rocky Mountain region are from Fort Laramie, Wyoming, for September 1849. However, only four other stations operated in the state during the 1850s (Camp Walbach, Deer Creek Agency, Fort Bridger, and Gilbert's Trading Post) and none of these has records of more than 30 months. Similarly, in the 1860s only seven stations were operated in the state[1] and only one of these (Fort Bridger) has a continuous record of more than five years. Because of this, generalizations about the climate of Wyoming in the 1850s and 1860s (Wahl and Lawson, 1971) are extremely suspect.

Spring (Figure 209)

Maximum precipitation 1866-1870 was probably reached in either 1871-75 or 1881-85. The latter part of the 1880s was relatively dry over a wide area. Compared to 1950s averages the late 1880s were well below average whereas the early 1870s and early 1880s were relatively wet.

Summer (Figure 210)

Again the pentad 1871-75 stands out as relatively wet--in fact, the only pentad clearly above 1951-60 averages. All other pentads were quite dry.

Fall (Figure 211)

Precipitation 1866-90 was heaviest in 1871-75 through the 1870s and 1880s were both probably above 1950s averages.

Winter (Figure 212)

As in the other seasons, the wettest pentad from 1866-90 was clearly 1871-75 though at Cheyenne this was not the case. Precipitation was well above 1950s averages in 1871-75 but other pentads can not clearly be placed in relation to this decade.

---

[1]Fort Laramie (very broken record), Fort D. A. Russell(begins December 1869), Fort Fetterman (begins November 1868), Fort Fred Steele (begins January 1869), Fort Halleck (1862-66), Fort Phil Kearney (1867-68) and Fort Bridger (1863 +).

Table 33

Arizona: 19th Century Stations and Their 1950s Analogs

19th Century Station				20th Century Equivalent				
Station	Lat.	Long.	Elevation (feet)	Index No.	Station	Lat.	Long.	Elevation (feet)
Fort Defiance	35°43'	109°10'	6500	9410	Window Rock	35°41'	109°03'	6750
Ft. Buchanan	31°41'	110°55'	5330	7593	?Santa Rita Exp. Range	31°46'	110°51'	4300
Ft. Mohave	35°02'	114°36'	604	2439	?Davis Dam No.2	35°12'	114°34'	657
Camp Beal's Spring	35°20'	114°05'	2500	4639	?Kingman	35°11'	114°03'	3333
Camp Date Creek	34°18'	112°40'	3726	9166	?Walnut Grove	34°18'	112°33'	3764
Camp Hualpai	35°10'	113°50'	5322		?			
Camp Willow Grove	35°10'	113°50'	4170		?			
Prescott (Whipple Barracks)	34°33'	112°28'	5389	6796	Prescott	34°33'	112°27'	5419
Fort Verde	34°32'	111°47'	3160	5635	Montezuma Castle	34°37'	111°50'	3180
Camp McPherson	34°45'	112°18'	3726	4453	Jerome?	34°45'	112°07'	5245
				1654	Chino Valley?	34°46'	112°27'	4673
Fort Apache	33°48'	109°57'	5050	9271	White River	33°50'	109°58'	5280
Camp Colorado	34°10'	114°15'	?	6250	Parker	34°10'	114°17'	425
Stanwix	32°57'	113°21'	567	2430	Dateland?	32°48'	113°32'	452
Yuma	32°44'	114°36'	141	9662	Yuma	32°44'	114°37'	138
Fort McDowell	33°38'	111°38'	1250	0632	Bartlett Dam?	33°49'	111°38'	1650
				8214	Stewart Mtn?	33°34'	111°32'	1422

Table 33 (Continued)

Arizona: 19th Century Stations and Their 1950s Analogs

19th Century Station				20th Century Equivalent				
Station	Lat.	Long.	Elevation (feet)	Index No.	Station	Lat.	Long.	Elevation (feet)
Phoenix	33°28'	112°00'	1068	6486	Phoenix P.O.	33°28'	112°04'	1083
				6481	Phoenix WB AP	33°26'	112°02'	1108
Burkes	32°58'	113°16'	?		?			
Wickenburg	33°56'	112°42'	1400	9287	Wickenburg	33°58'	112°44'	2070
Camp Reno	34°45'	112°18'	3726	4453	Jerome	34°45'	112°07'	5245
				1654	Chino Valley	34°46'	112°27'	4673
Breckenridge (Old Camp Grant)	32°48'	110°36'	3800	7530	San Manuel?	32°37'	110°38'	3360
Maricopa	33°05'	112°00'	1190		?			
Florence	33°03'	111°20'	1553	3027	Florence	33°02'	111°23'	1500
Camp Goodwin	-3°05'	110°00'	2650		?			
Fort Grant	32°36'	109°53'	4860	3110	Fort Grant	32°37'	109°57'	4851/4880
Fort Lowell	32°16'	110°47'	2400	7355	Sabino Canyon?	32°18'	110°49'	2540
				8800	Tucson Magnet?	32°15'	110°50'	2526
Camp Crittenden	31°34'	110°39'	2000		?			
Tucson	32°14'	110°54'	2404	8815	Tucson, U of A	32°14'	110°57'	2423
Camp Walker	31°40'	110°50'	3000	7593	Santa Rite Exp. Range?	31°46'	110°51'	4300
Fort Bowie	32°12'	109°20'	4781	2648	Dos Cabezas?	32°10'	109°36'	5100
Benson	32°00'	110°22'	3580	0680	Benson	31°58'	110°18'	3585
Casa Grande	32°55'	111°45'	1398	1306	Casa Grande	32°53'	111°45'	1405

Table 34

Colorado: 19th Century Stations and Their 1950s Analogs

	19th Century Station				20th Century Equivalent			
Station	Lat.	Long.	Elevation (feet)	Index No.	Station	Lat.	Long.	Elevation (feet)
Colorado Springs	38°51'	104°47'	6032	1778	Colorado Springs	38°49'	104°42'	6173
Denver	39°45'	105°00'	6281	2225	Denver WB City	39°45'	105°00'	5221
Fort Garland	37°25'	105°23'	7937	0776	Blanca	37°26'	105°31'	7500
Fort Lewis	37°15'	107°57'	8500	3016	Fort Lewis	37°14'	108°03'	7610
Pike's Peak	38°50'	105°20'	14,134	4750	?Lake Moraine	38°49'	105°01'	10,265
Fort Collins	40°35'	105°02'	5000	3005	Fort Collins	40°35'	105°05'	5025
Fort Sedgwick	41°00'	102°27'	3660	4413	?Julesburg	41°00'	102°15'	3469
Hot Sulpher Springs	40°05'	106°10'	7600	4129	Hot Sulpher Springs	40°03'	106°09'	7800
Fort Morgan	40°18	103°42'	4500	3038	Fort Morgan	40°15'	103°48'	4321
Georgetown	39°43'	105°41'	8594	3261	Georgetown	39°42'	105°43'	8570
Golden City	39°44'	105°20'	5993	2557	?Edgewater	39°46'	105°03'	5450
				2795	Evergreen	39°37'	105°21'	7300
Hutchinson	39°32'	105°16'	?		?			
Dudley	39°30'	106°00'	10,500	0909	Breckenridge	39°29'	106°02'	9579
Fountain	38°40'	104°40'	5420	3063	Fountain	38°41'	104°42'	5546
Kit Carson	38°43'	102°47'	4289	4603	Kit Carson	38°46'	102°47'	4284
Canon City	38°30'	105°00'	4700	6410	Penrose	38°27'	105°04'	5501
Pueblo	38°18'	104°12'	4300		?			
Montrose	38°30'	107°56'	5795	5722	Montrose No. 2	38°29'	107°53'	5830
				5717	Montrose No. 1	38°29'	107°53'	5830
Fort Lyon	38°06'	103°30'	4000	1539	Cheraw	38°06'	103°30'	4072
Silverton	37°46'	107°46'	9400	7656	Silverton	37°48'	107°40'	9415
Summit	37°28'	106°35'	11,300		?			

Table 34 (Continued)

Colorado: 19th Century Stations and Their 1950s Analogs

19th Century Station				20th Century Equivalent				
Station	Lat.	Long.	Elevation (feet)	Index No.	Station	Lat.	Long.	Elevation (feet)
Fort Massachusetts	37°30'	105°33'	8365		Blanca			
Hermosa	37°24'	107°50'	6700		?			
Monte Vista	37°35'	106°05'	7665	5706	Monte Vista	37°35'	106°09'	7665
Trinidad	37°11'	104°28'	6070	8429	Trinidad	37°10'	104°29'	6030
Ft. Reynolds	38°15'	104°12'	4300	3079	?Fowler	38°08'	104°02'	4350
S. Pueblo	38°16'	104°37'	4732	6743	Pueblo City Resv.	37°17'	104°39'	4687
Husted	39°00'	104°49'	6540		?			
Las Animas	38°04'	103°12'	3899	4834	Los Animas IN	38°05'	103°13'	3892
T.S. Ranche	39°00'	108°15'			?			
Ranche (Nr. Como)	39°18'	105°35'	9500		?			

Table 35

Idaho: 19th Century Stations and Their 1950s Analogs

19th Century Station				20th Century Equivalent				
Station	Lat.	Long.	Elevation (feet)	Index No.	Station	Lat.	Long.	Elevation (feet)
Boise	43°37'	116°08'	2750	1022	Boise WB AP	43°34'	116°13'	2841
Fort Sherman	47°37'	116°50'	2175	1956	Couer D'Alene Ranger Station?	47°41'	116°45'	2158
Lewiston	46°23'	116°58'	757	5230	Lewiston	46°25'	117°02'	733
Fort Lapwai	46°23'	116°48'	891	5230	Lewiston?	45°25'	117°02'	733

Table 36

Montana: 19th Century Stations and Their 1950s Analogs

19th Century Station				20th Century Equivalent				
Station	Lat.	Long.	Elevation (feet)	Index No.	Station	Lat.	Long.	Elevation (feet)
Fort Missoula	46°55'	114°00'	3225	5740	Missoula 2ANW	46°53'	114°02'	3172
					Missoula WB AP	46°55'	114°05'	3200
Fort Shaw (Fairfield)	47°30'	111°49'	3983	2857	Fairfield	47°37'	111°59'	3983
Fort Benton	47°52'	110°43'	2630	3113	Fort Benton	47°49'	110°40'	2635
Fort Assiniboine	48°34'	109°45'	2505	3994	Havre WB City	48°34'	109°40'	2488
Fort Maginnis	47°11'	109°05'	4340		?			
Poplar	48°06'	105°12'	1994	6660	Poplar	48°07'	105°12'	2000
Fort Custer	45°37'	107°30'	3036	2112	Crow Agency	45°36'	107°27'	3030
Miles City	46°27'	105°53'	2378	5690	Miles City FAA AP	46°26'	105°52'	2629
Helena	46°34'	112°04'	4069	4055	Helena WB AP	46°36'	112°00'	3893
Fort Ellis	45°43'	111°05'	4900	1044	Bozeman Agricul. College	45°40'	111°03'	4856
Fort Logan	46°43'	111°12'	4750	3157	Fort Logan	46°41'	111°12'	4690
Virginia City	45°18'	111°56'	5822	8597	Virginia City	45°18'	111°56'	5835

Table 37

Nevada: 19th Century Stations and Their 1950s Analogs

19th Century Station				20th Century Equivalent				
Station	Lat.	Long.	Elevation (feet)	No.	Station	Lat.	Long.	Elevation (feet)
Beowawe	40°36'	116°32'	4695	0795	Beowawe	40°36'	116°29'	4695
Brown's	40°01'	118°41'	3929	4700	?Lovelock FAA AP	40°04'	118°33'	3900
Carlin	40°43'	116°07'	4897		?			
Carson City	39°08'	119°47'	4628	1485	Carson City	39°10'	119°46'	4675
Cedar Pass	41°08'	114°50'	?	8988	Wells?	41°07'	114°58'	5628
Elko	40°50'	115°46'	5065	2573	Elko WB AP	40°50'	115°47'	5075
Golconda	40°57'	117°34'	4392	3245	Golconda	40°57'	117°29'	4392
Halleck	40°56'	115°30'	5229		?			
Hot Springs	39°49'	119°02'	4072		?			
Humboldt	40°38'	118°14'	4236	7192	?Rye Patch Dam	40°28'	118°18'	4135
Iron Point	40°58'	117°21'	4375	3245	Golconda	40°57'	117°29'	4392
McDermit Camp	41°58'	117°45'	4700	4935	McDermitt	42°00'	117°43'	4427
Otego	41°09'	114°36'	?	6148	Pequop	41°04'	114°32'	6000
Palisade	40°39'	116°12'	4840	0795	Beowawe?	40°36'	116°29'	4695
Reno	39°33'	119°47'	4497	6779	Reno WB AP	39°30'	119°47'	4397
Tecoma	41°18'	114°07'	4812	5352	Montello	41°16'	114°12'	4877
Toano	41°07'	114°26'	5975	6148	Pequop	41°04'	114°32'	6000
Wadsworth	39°38'	119°19'	4077	2840	Femley	39°37'	119°15'	4160
Wells	41°07'	114°56'	5628	8988	Wells	41°07'	114°58'	5628
Winnemucca	40°59'	117°43'	4358	9171	Winnemucca WB AP	40°54'	117°48'	4299
Battle Mt.	40°38'	116°52'	5311	0691	Battle Mt FFA AP	40°37'	116°52'	4530
Fort Churchill	39°20'	119°05'	4284	4349	?Lahontan Dam	39°28'	119°04'	4158
Pioche	37°56'	114°26'	6110	6252	Pioche	37°56'	114°27'	6110

Table 38

New Mexico: 19th Century Stations and Their 1950s Analogs

19th Century Station				20th Century Equivalent				
Station	Lat.	Long.	Elevation (feet)	Index No.	Station	Lat.	Long.	Elevation (feet)
Albuquerque	35°05'	106°39'	5026	6079	Netherwood Park	35°06'	106°37'	5130
				0234	Albuquerque	35°03'	106°37'	5314
Cebolleta	35°20'	107°20'	6200		?			
Las Vegas	35°36'	105°12'	6418	4850	Las Vegas	35°36'	105°13'	6435
Santa Fe	35°41'	105°57'	7026	8072	Santa Fe	35°41'	105°57'	7013
Socorro	34°08'	106°55'	4565	8387	Socorro	34°04'	106°54'	4617
Fort Conrad	33°47'	106°48'	4576	1138	Bosque del Apache	33°46'	106°54'	4520
Fort Fillmore	32°13'	106°42'	3937	8535	State University	32°17'	106°45'	3881
Fort Union	35°54'	104°57'	6750	9330	Valmora	35°49'	104°56'	6330
Fort Webster	32°48'	108°04'	6350	3265	Fort Bayard	32°48'	108°09'	6152
Fort Thorn	32°40'	107°10'	4500	3855	Hatch	32°40'	107°09'	4042
Fort Craig	33°40'	107°01'	4619		?Bosque del Apache		see above	
Burgwin Camp	36°30'	105°40'	7900	8668?	?	36°25'	105°34'	6983
Fort Stanton	33°30'	105°26'	6154	3288	Fort Stanton	33°30'	105°31'	6230
Fort Bayard	32°47'	108°09'	6022	3265	Fort Bayard	32°48'	108°09'	6152
Fort Cummings	32°27'	107°40'	4750		?			
Deming	32°18'	107°48'	4327	2436	Deming	32°16'	107°45'	4331
Fort McRae	33°02'	107°05'	4500	2848	?Elephant Butte Dam	33°09'	107°11'	4576
Fort Selden	32°27'	106°55'	3937	4426	?Jornada Exp. Range	32°27'	106°44'	4265

Table 38 (Continued)

New Mexico: 19th Century Stations and Their 1950s Analogs

19th Century Station				20th Century Equivalent				
Station	Lat.	Long.	Elevation (feet)	Index No.	Station	Lat.	Long.	Elevation (feet)
Fort Wingate	35°28'	108°32'	6822	3305	Fort Wingate	35°28'	108°32'	7000
Fort Bascom	35°23'	103°27'	4000		?			
Los Pinos	34°50'	106°40'	5000	5147	Los Lunas	34°48'	106°44'	4885
Fort Fauntleroy	35°30'	108°40'	8000		?			
Fort Tulerosa	33°57'	108°15'	?		?			
La Mesilla	32°17'	106°48'	4124	8535	State University	32°17'	106°45'	3881
Silver City	32°46'	108°14'	5796	8324	Silver City	32°46'	108°17'	5895
Fort Sumner	34°19'	104°09'	4300?	3296	Fort Sumner 5S	34°22'	104°15'	4050
Lordsburgh	32°20'	108°40'	4247	5079	Lordsburgh	32°21'	108°42'	4245
Gallinas Spring	35°14'	104°51'	4800	2510	?Dilia 1SSE	35°11'	105°03'	5140
Lava	33°33'	106°59'	4703		?			
Springer	36°22'	104°33'	5766	8501	Springer	36°23'	104°36'	5857

Table 39

Oregon: 19th Century Stations and Their 1950s Analogs

19th Century Station				20th Century Equivalent				
Station	Lat.[1]	Long.[1]	Elevation (feet)	Index No.	Station	Lat.	Long.	Elevation (feet)
Albany	44°35'	122°50'	600	4603	Lacomb	44°35'	122°45'	665
Astoria	46°11'	123°48'/49?	50	0318	Astor Exp. Sta.	46°09'	123°49'	43
Block House	44°25'	123°30'	?	1877	Corvallis Water Bureau	44°30'	123°27'	510
Fort Dalles	45°33' 45°36'	120°50' 120°55' ?	350		?			
Eola	44°57'	122°54'	500		?			
Camp Harney	43°00'	119°00'	4600	1124	Buena Vista Sta.	43°04'	118°52'	4135
Fort Hoskins	45°06'? 02'	123°26'? 22'	?	9372	Willamina 2S	45°03'	123°30'	285
Fort Klamath	42°40'	121°50' 54'?	4200		?			
Oregon City	45°20'	122°18' 24'	200	6334	Oregon City	45°21'	122°36'	167
Portland	45°30' 24'?	122°36' 30'?	80	6761	Portland(WB City)	45°32'	122°40'	30
Port Orford	42°44'	124°29'	50	6779	Port Orford	42°44'	124°31'	292
Roseburgh	43°10'	123°20'	523	7326	Roseburgh(WB AP)	43°14'	123°22'	505
Salem	44°56'	122°45'	120	7500	Salem (WB AP)?	44°55'	123°01'	196
Fort Stevens	46°12'	123°57'	?	0324	Astoria	46°11'	123°50'	200
The Dalles	45°36'	121°12'	116	8407	The Dalles	45°36'	121°12'	102
Umatilla	45°55'	119°22'	290	8734	Umatilla	45°55'	119°21'	285

Table 39 (Continued)

Oregon: 19th Century Stations and Their 1950s Analogs

19th Century Station				20th Century Equivalent				
Station	Lat.[1]	Long.[1]	Elevation (feet)	Index No.	Station	Lat.	Long.	Elevation (feet)
Fort Umpqua	43°42'	124°10'	8	7082	Reedsport	43°42'	124°07'	55
Camp Warner	42°28'	119°42'	?		?			
	42°50'	120°00'						
Camp Watson	44°22'	119°48'	?		?			
	44°13'	119°45'						
Fort Yamhill	45°21'	123°15'	?	3705	Haskins Dam	45°19'	123°21'	721
		123°30'						
Bandon	43°05'	124°15'	?	0471	Bandon	43°07	124°25'	8
Klamath Agency	"5 ml NW of Chiloquin"		4200	1571	Chiloquin	42°35'	121°51'	4200
Hood River	"at Exp. Station"		300	4003	Hood River Exp. Station?	45°41'	121°31'	300
Linkville	42°15'	121°45'	4250?	4506	Klamath Falls	42°13'	121°47'	4098
	42°14'	120°40'	4169	4511	Klamath Falls AP	42°10'	121°44'	4084
La Grande	45°20'	118°07'	2600	4615	La Grande	45°20'	118°06'	2782
	45°22'	118°18'	?					

[1]The different locations of some stations are both from precipitation and temperature tables in Greely (1889).

Table 40

Utah: 19th Century Stations and Their 1950s Analogs

19th Century Station				20th Century Equivalent				
Station	Lat.	Long.	Elevation (feet)	Index No.	Station	Lat.	Long.	Elevation (feet)
Blue Creek	41°39'	112°28'	4379					
Corrine	41°30'	112°18'	4232	0506	Bear River Refuge	41°28'	112°16'	4208
Kelton	41°45'	113°08'	4222		?			
Promontory	41°35'	112°35'	?		?			
Terrace	41°30'	113°30'	4548		?			
Ogden	41°12'	111°57'	4340	6404	Ogden Pioneer Power House	41°15'	111°57'	4400
Coalville	40°56'	111°28'	5630	1588	Coalville	40°55'	111°24'	5550
				2385	Echo Dam	40°58'	111°26'	5500
St. Mary's	40°42'	111°00'	6200		?			
Wanship	40°49'	111°24'	6200	9165	Wanship Dam	40°48'	111°24'	5950
Deep Creek	40°32'	112°18'	?	8771	Tooele	40°32'	112°18'	4820
Camp Douglas	40°46'	111°50'	4800	8922	Univ. of Utah	40°46'	111°51'	4700
Salt Lake City	40°46'	111°54'	4354	7603	Salt Lake City	40°46'	111°54'	4260
Camp Floyd	40°16'	112°08'	4725	2696	Fairfield	40°16'	112°05'	4876
Fillmore	38°58'	112°18'	?	2828	Fillmore	38°58'	112°20'	5250
Harrisburg	37°16'	113°23'	2385		?			
St. George	37°09'	113°35'	?	7516	St. George PH	37°06'	113°34'	2700
Kanab	37°03'	112°32'	5500	4508	Kanab PH	37°03'	112°31'	5010
Mt. Carmel	37°12'	112°41'	5215	6534	Orderville	37°16'	112°38'	5460

Table 41

Washington: 19th Century Stations and Their 1950s Analogs

19th Century Station				20th Century Equivalent				
Station	Lat.	Long.	Elevation (feet)	Index No.	Station	Lat.	Long.	Elevation (feet)
Bainbridge Is.	47°32'	122°40'	50	0872	Bremerton (Navy Yard)	47°34'	122°40'	162
Bellingham Fort	48°45'	122°30'	80	0564	Bellingham 2N	48°47'	122°29'	112
Cape Disappointment (Fort Canby)	46°17'	124°02'	30	4752	Long Beach?	46°23'	124°02'	25
Fort Cascades	45°39'	121°50'	?	0897	Bonneville Dam?	45°38'	121°57'	85
Cathlamet(nr)	46°15'	123°12'	40	1207	Cathlamet 9NE?	46°19'	123°16'	476
Fort Colville	48°42'	118°02'	1962		?			
Dayton	46°19'	117°56'	1683	2030	Dayton	46°19'	117°59'	1620
Neah Bay	48°22'	124°37'	40	5801	Neah Bay 2E/1E	48°22'	124°36'	15
Olympia	47°03'	122°53'	36	6109	Olympia Priest Pt. Park	47°04'	122°53'	69
				6114	Olympia WB AP	46°58'	122°54'	190
San Juan Is.	48°28'	123°01'	150		?			
Camp Semiahmoo	49°01'	122°46'	11	0729	Blaine/Blaine 1E	48°59'	122°45'	40
Fort Simcoe	46°14'	120°40'	?		?			
Fort Steilacoom	47°10'	122°25'	300	6803	Puyallup 2W Exp. Station	47°12'	122°20'	50
Tatoosh Island Lighthouse	48°23'	124°44'	90	8332	Tatoosh Is. WB	48°23'	124°44'	101
Fort Townsend	48°05'	122°46'	135	6678	Port Townsend	48°07'	122°47'	71
Union Ridge	45°49'	123°42'	100		?			

Table 41 (Continued)

Washington: 19th Century Stations and Their 1950s Analogs

19th Century Station				20th Century Equivalent				
Station	Lat.	Long.	Elevation (feet)	Index No.	Station	Lat.	Long.	Elevation (feet)
Fort Vancouver	45°40'	122°30'	50	8773	Vancouver	45°38'	122°41'	100
				8778	Vancouver Port Dock	45°37'	122°40'	26
Walla Walla	46°05'	118°54'	930		?			
Fort Walla Walla	46°03'	118°20'	?	8931	Walla Walla WB City	46°02'	118°20'	949
Port Blakely	47°32'	122°40'	30	0872	Bremerton	47°34'	122°40'	162
Port Townsend	48°07'	122°45'	8	6678	Port Townsend	48°06'	122°46'	71
Seattle	47°32'	122°32'	20	7488	Seattle WB City?	47°36'	122°20'	14
Fort Spokane	47°30'	118°30'	1600	7938	Spokane	47°40'	117°25'	1875
Ediz Hook Light-House (Port Angeles)	48°08'	123°24'	12	6624	Port Angeles	48°07'	123°26'	99

Table 42

Wyoming: 19th Century Stations and Their 1950s Analogs

	19th Century Station				20th Century Equivalent			
Station	Lat.	Long.	Elevation (feet)	Index No.	Station	Lat.	Long.	Elevation (feet)
Cheyenne	41°06'	104°48'	6088	1675	Cheyenne WB AP	41°09'	104°49'	6139
Fort Fetteman	42°50	105°27'	4892	2690	Douglas FAA AP?	42°45'	105°22'	4853
Fort Sanders	41°20'	105°35'	7188	5410	Laramie	41°19'	105°35'	7200
Fort Laramie	42°13'	104°34'	4715	9604	Whalen Dam? Fort Laramie 11NNW	42°15'	104°38'	4294
Fort Washakie	43°02'	108°54'	5591	3570	Fort Washakie 2S	42°59'	108°52'	5583
Buffalo	44°22'	106°42'	4645	1160	Buffalo 5W?	44°22'	106°48'	5240
Fort Bridger	41°18'	110°32'	7010	3430	Fort Bridger AP?	41°24'	110°25'	7003

Note: The numbers 1, 2, 3, etc. beneath some pentads in the following figures indicate that some months' data were estimated. Generally the long-term monthly average for the month with missing data was substituted.

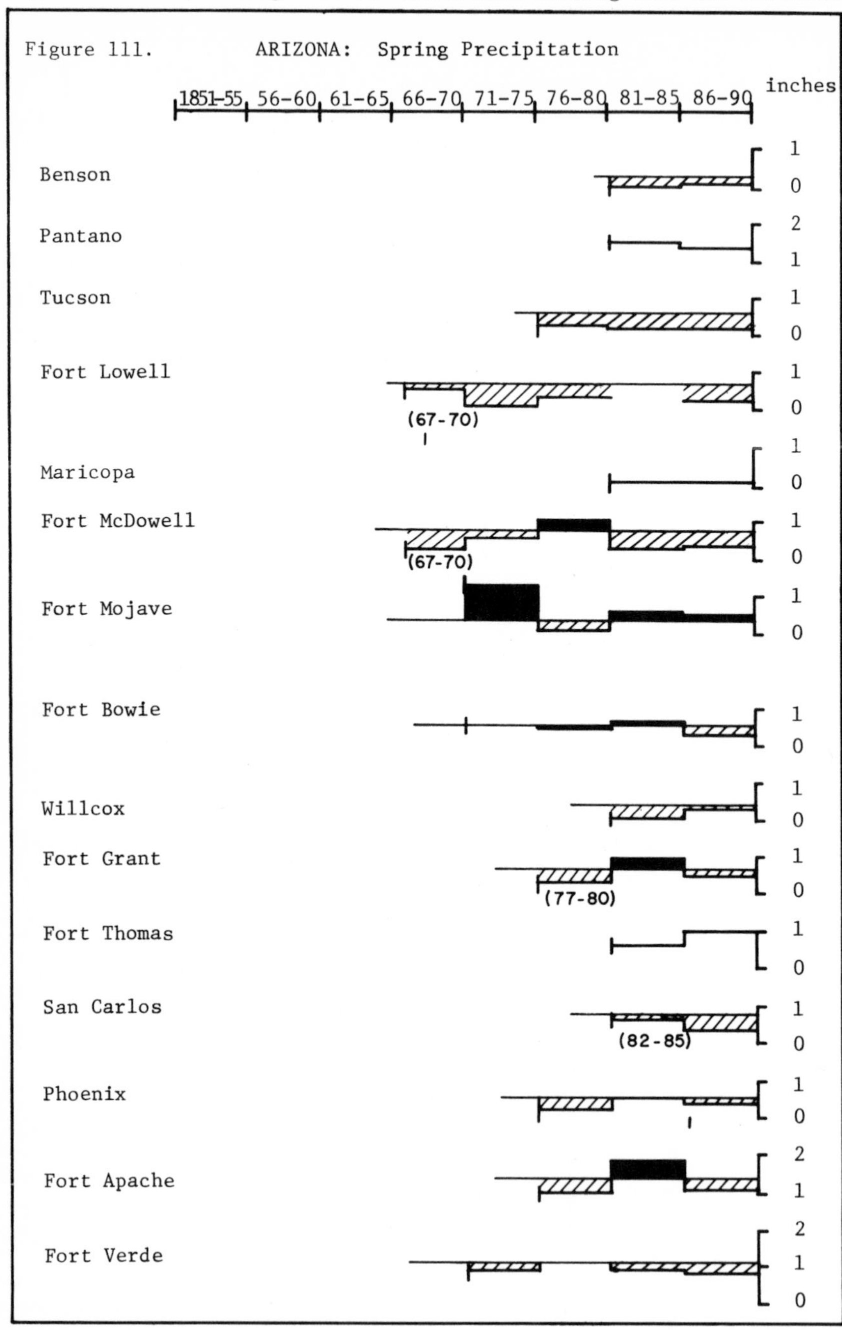

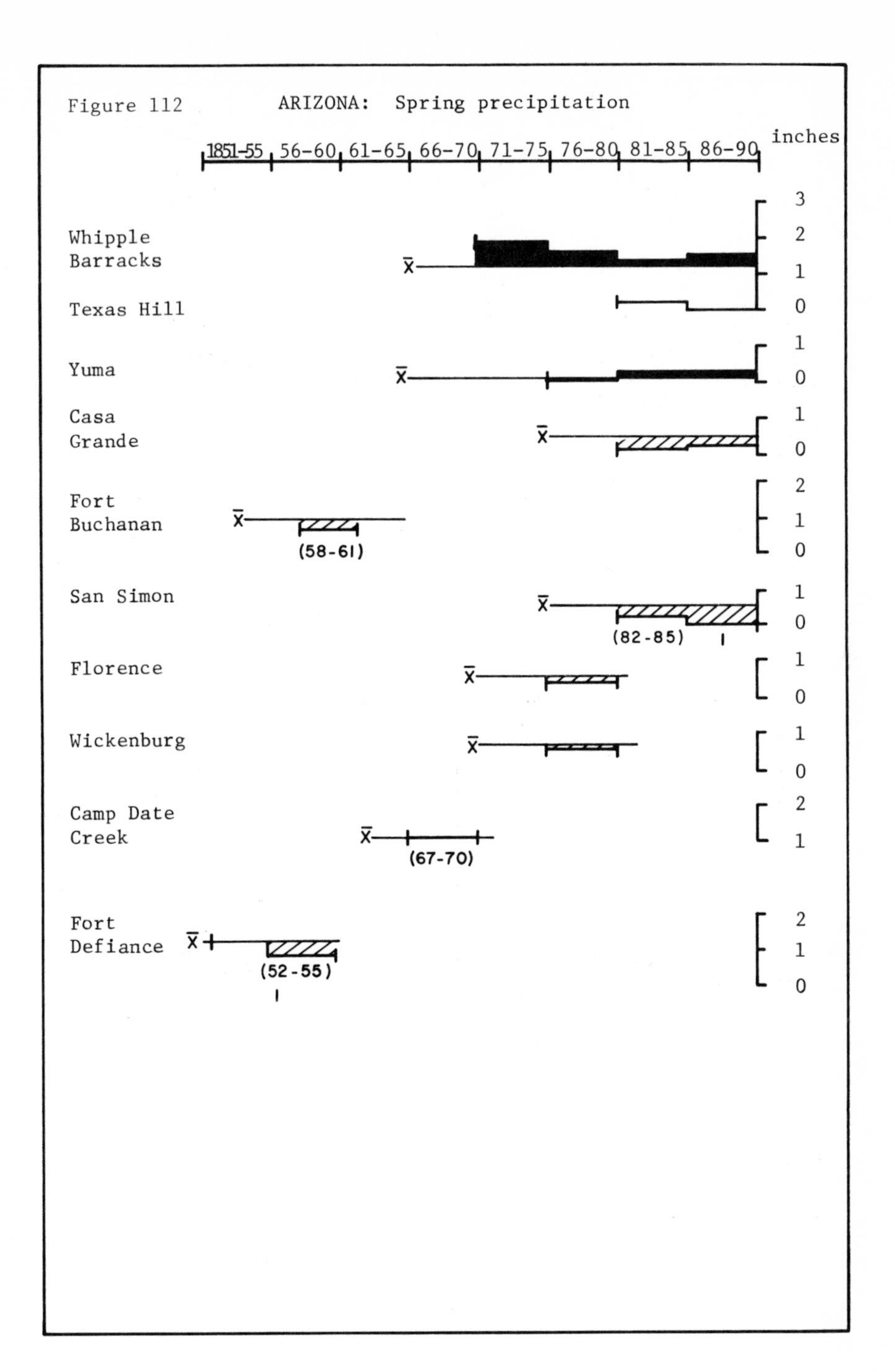
Figure 112
ARIZONA: Spring precipitation
1851-55
56-60
61-65
66-70
71-75
76-80
81-85
86-90
inches
Whipple Barracks
Texas Hill
Yuma
Casa Grande
Fort Buchanan
(58-61)
San Simon
(82-85)
Florence
Wickenburg
Camp Date Creek
(67-70)
Fort Defiance
(52-55)
3
2
1
0

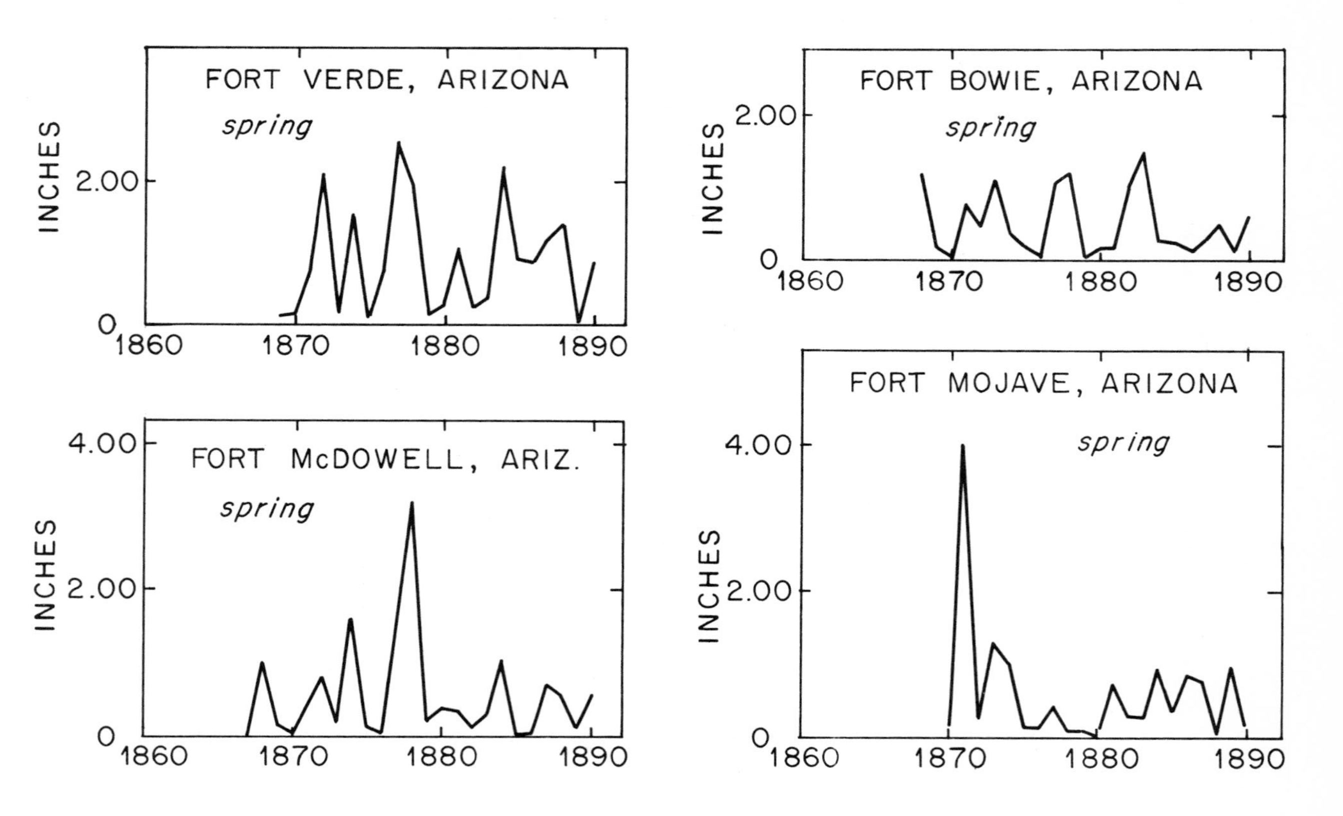

Figure 113: Arizona spring precipitation

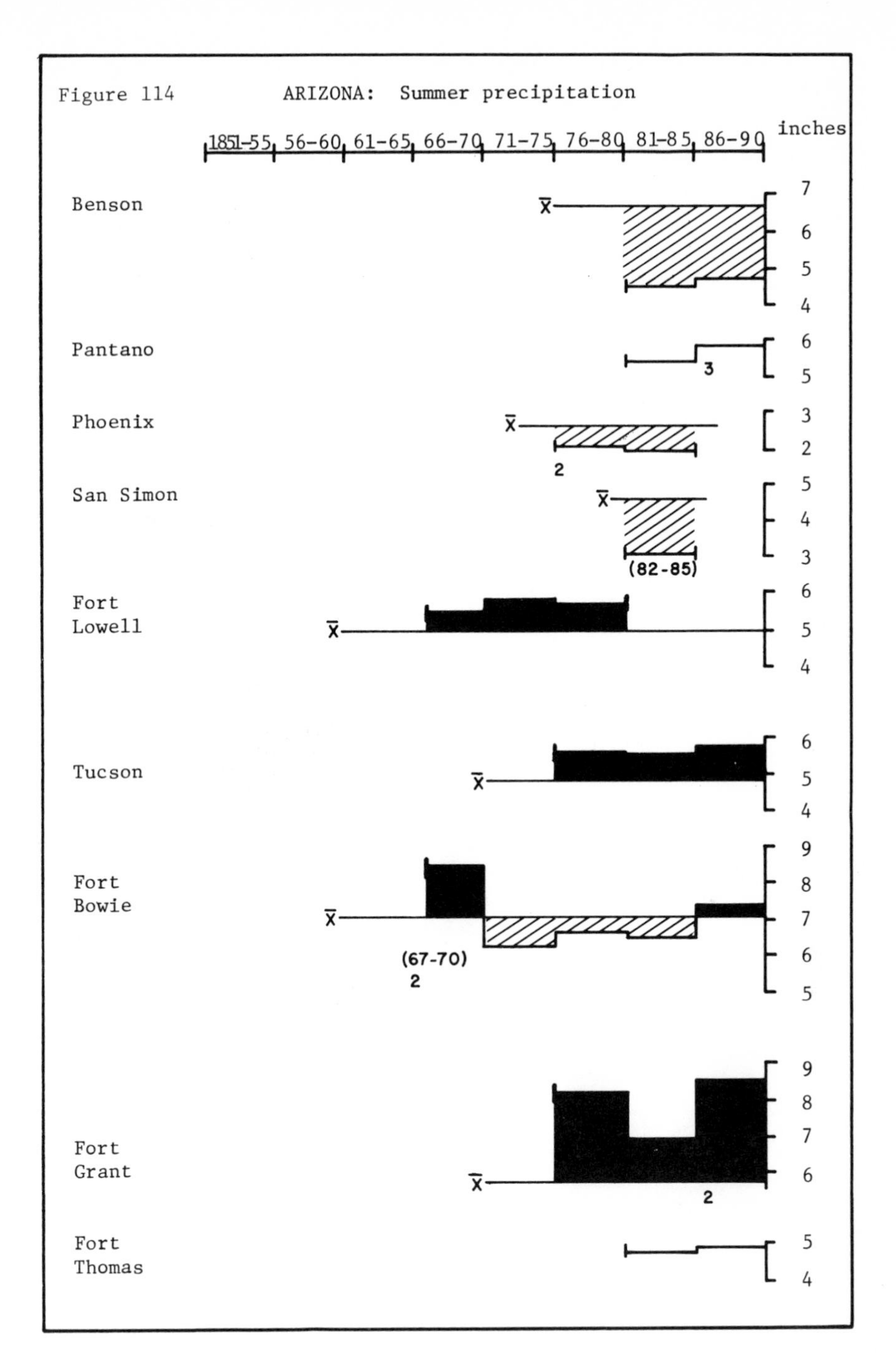
Figure 114
ARIZONA: Summer precipitation
1851-55
56-60
61-65
66-70
71-75
76-80
81-85
86-90
inches
Benson
Pantano
Phoenix
San Simon
Fort Lowell
Tucson
Fort Bowie
Fort Grant
Fort Thomas
(82-85)
(67-70)

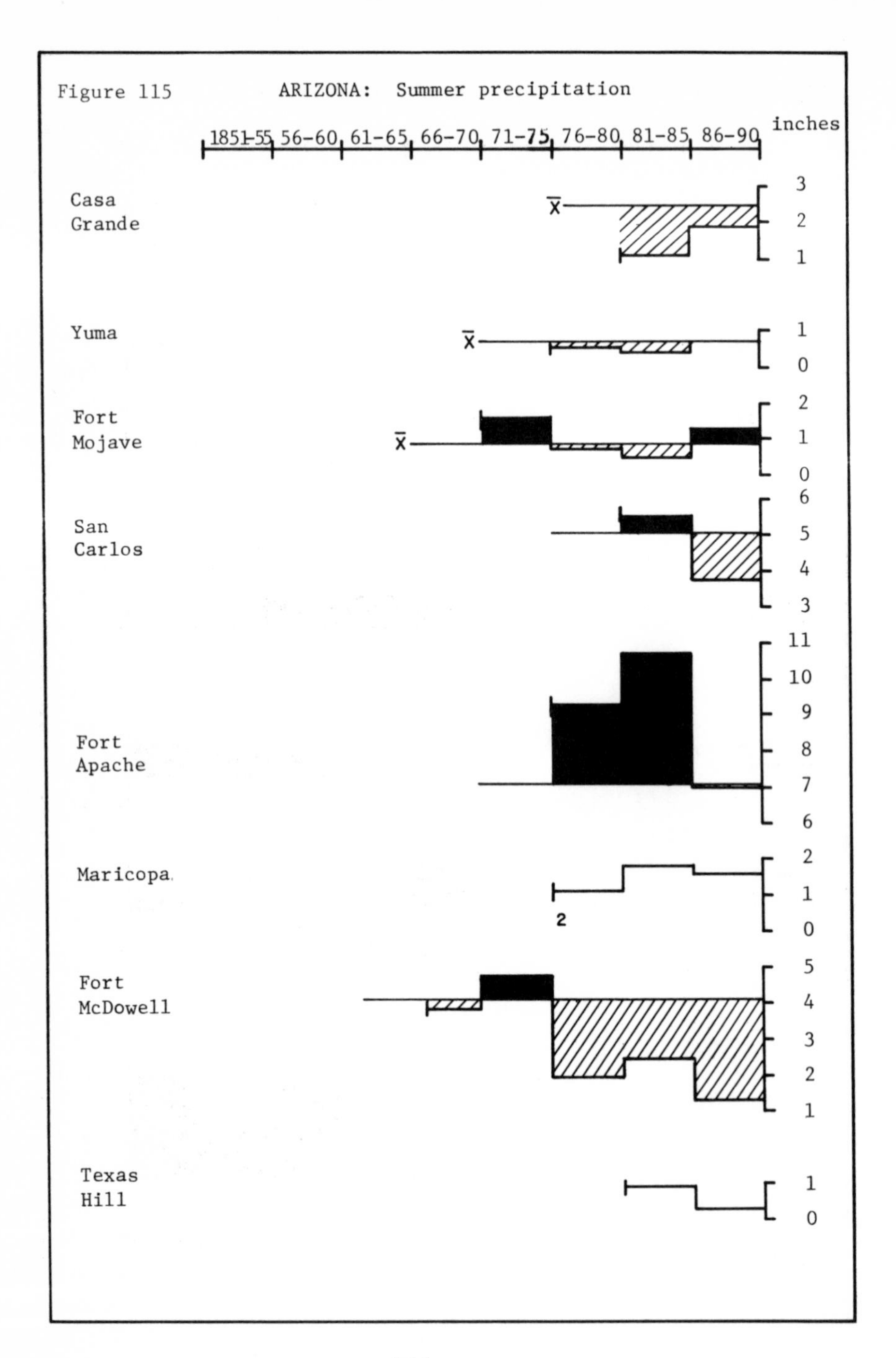
Figure 115
ARIZONA: Summer precipitation
1851-55
56-60
61-65
66-70
71-75
76-80
81-85
86-90
inches
Casa Grande
Yuma
Fort Mojave
San Carlos
Fort Apache
Maricopa
Fort McDowell
Texas Hill
X̄
2
3
2
1
1
0
2
1
0
6
5
4
3
11
10
9
8
7
6
2
1
0
5
4
3
2
1
1
0

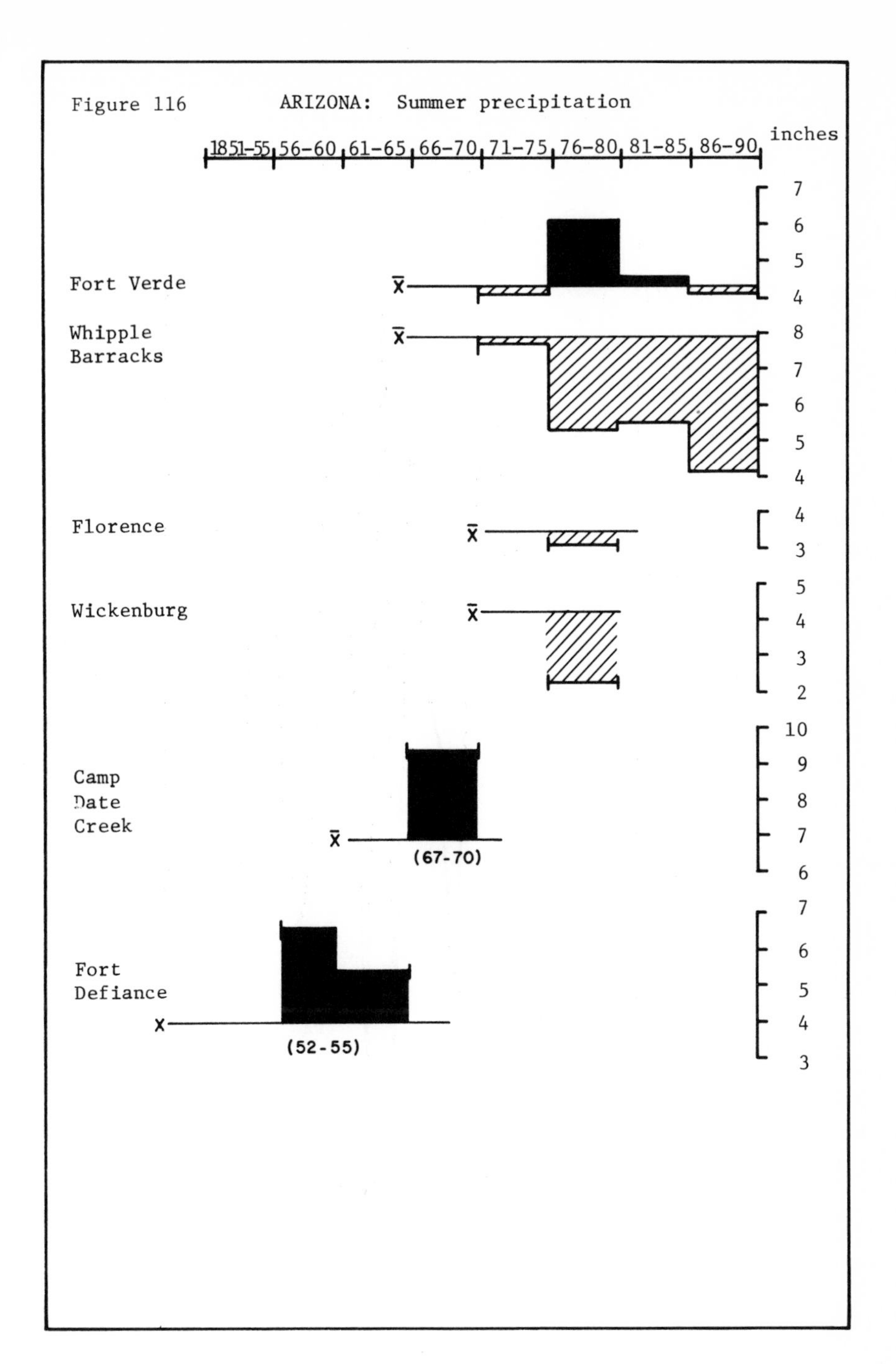

Figure 116
ARIZONA: Summer precipitation
1851-55
56-60
61-65
66-70
71-75
76-80
81-85
86-90
inches
Fort Verde
Whipple
Barracks
Florence
Wickenburg
Camp
Date
Creek
(67-70)
Fort
Defiance
(52-55)
7
6
5
4
8
7
6
5
4
4
3
5
4
3
2
10
9
8
7
6
7
6
5
4
3

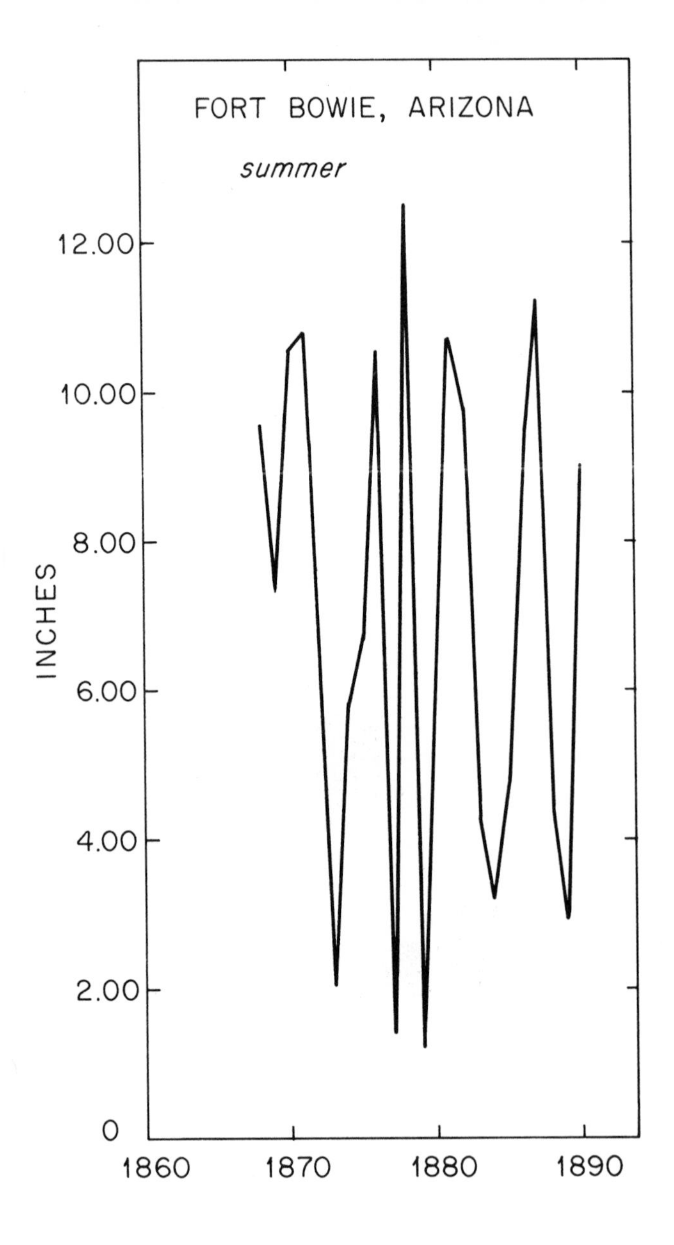

Figure 117: Arizona summer precipitation

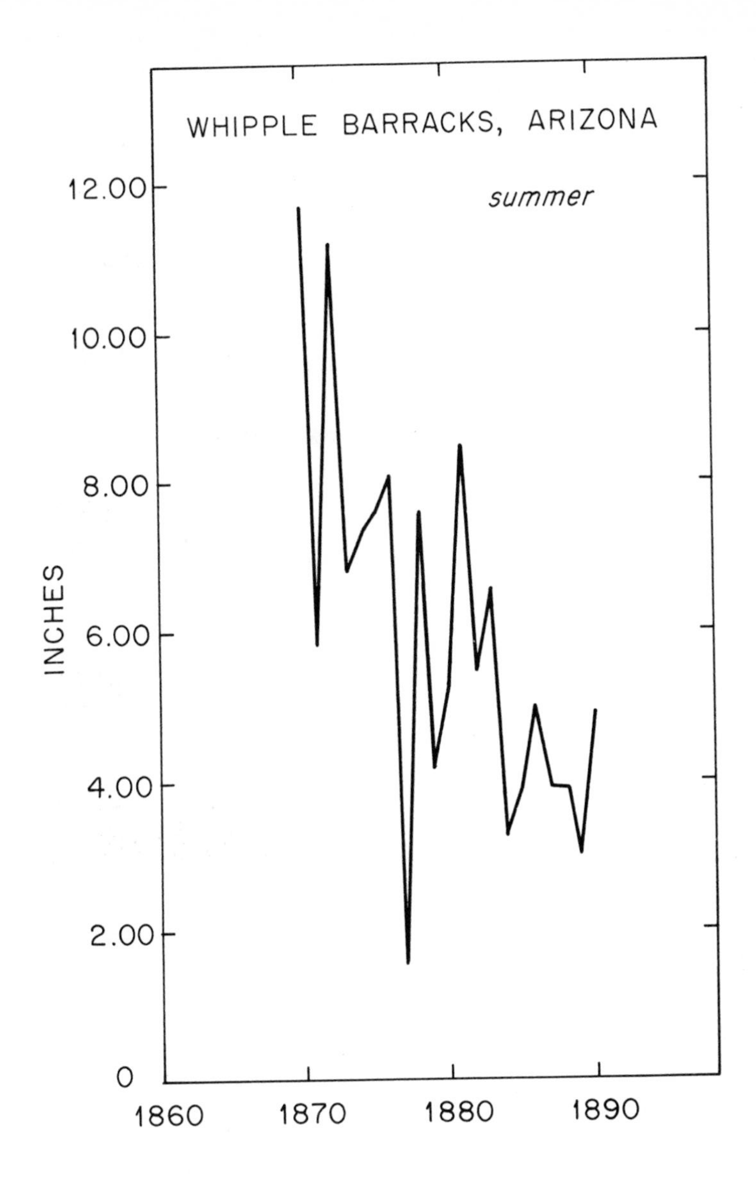

Figure 118: Arizona summer precipitation

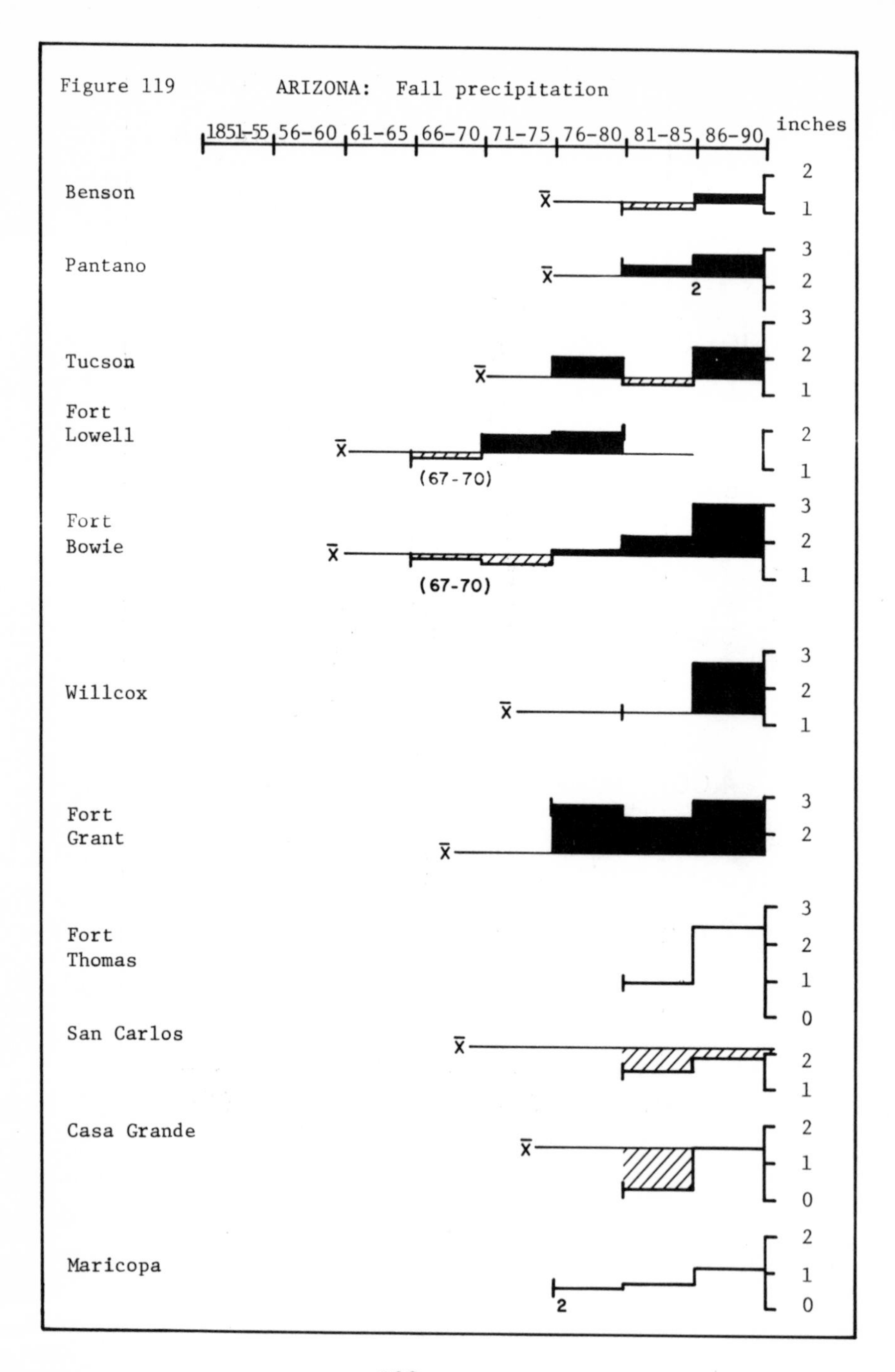

Figure 119
ARIZONA: Fall precipitation
1851-55
56-60
61-65
66-70
71-75
76-80
81-85
86-90
inches
Benson
Pantano
Tucson
Fort Lowell
(67-70)
Fort Bowie
(67-70)
Willcox
Fort Grant
Fort Thomas
San Carlos
Casa Grande
Maricopa

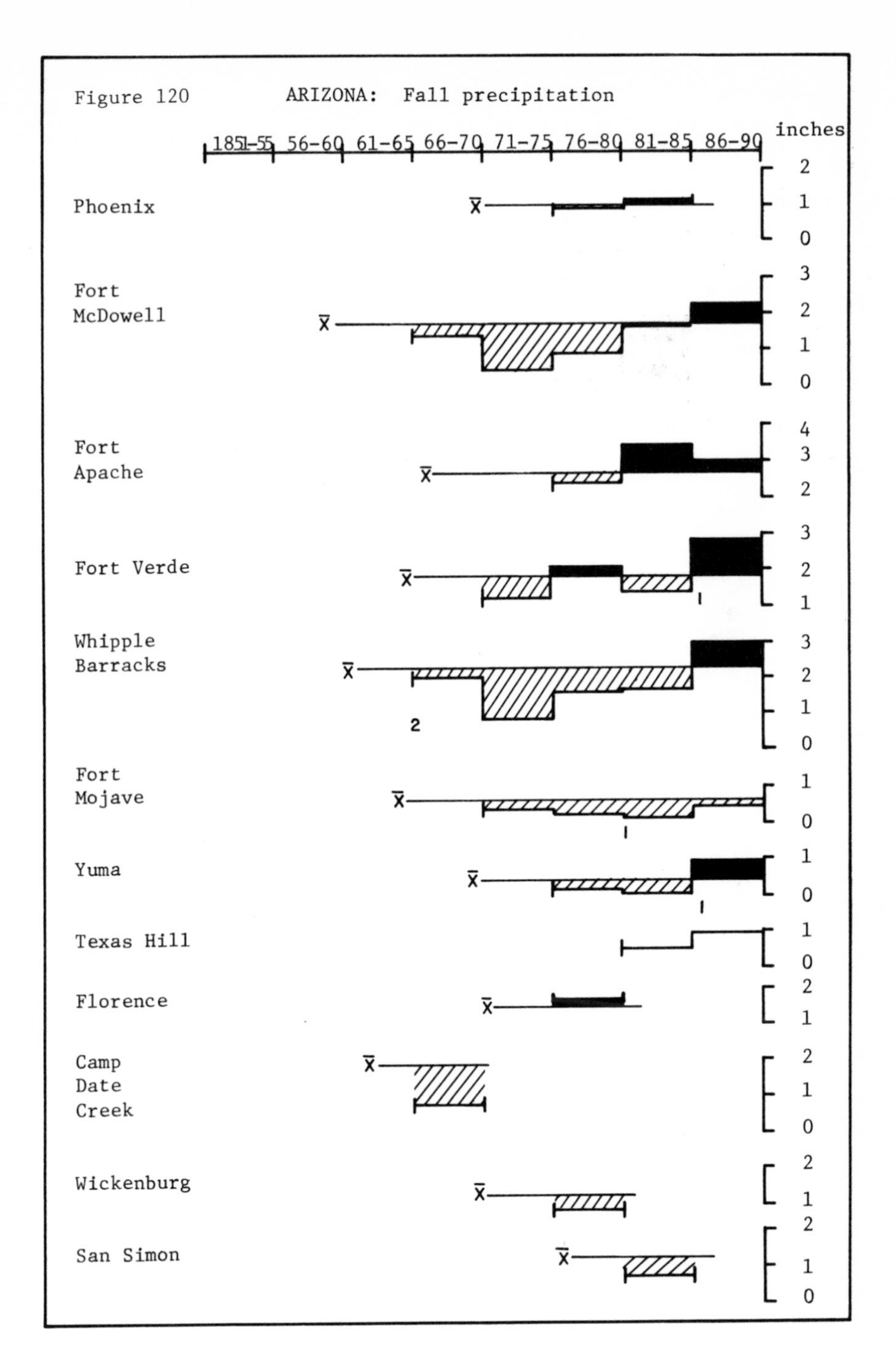
Figure 120
ARIZONA: Fall precipitation
1851-55
56-60
61-65
66-70
71-75
76-80
81-85
86-90
inches
Phoenix
Fort McDowell
Fort Apache
Fort Verde
Whipple Barracks
Fort Mojave
Yuma
Texas Hill
Florence
Camp Date Creek
Wickenburg
San Simon

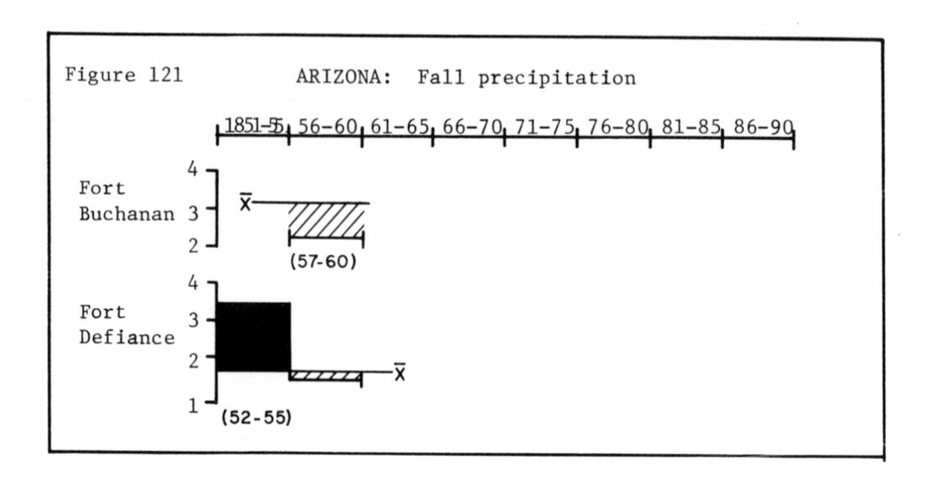
Figure 121
ARIZONA: Fall precipitation
1851-55
56-60
61-65
66-70
71-75
76-80
81-85
86-90
Fort
Buchanan
4
3
2
X̄
(57-60)
Fort
Defiance
4
3
2
1
X̄
(52-55)

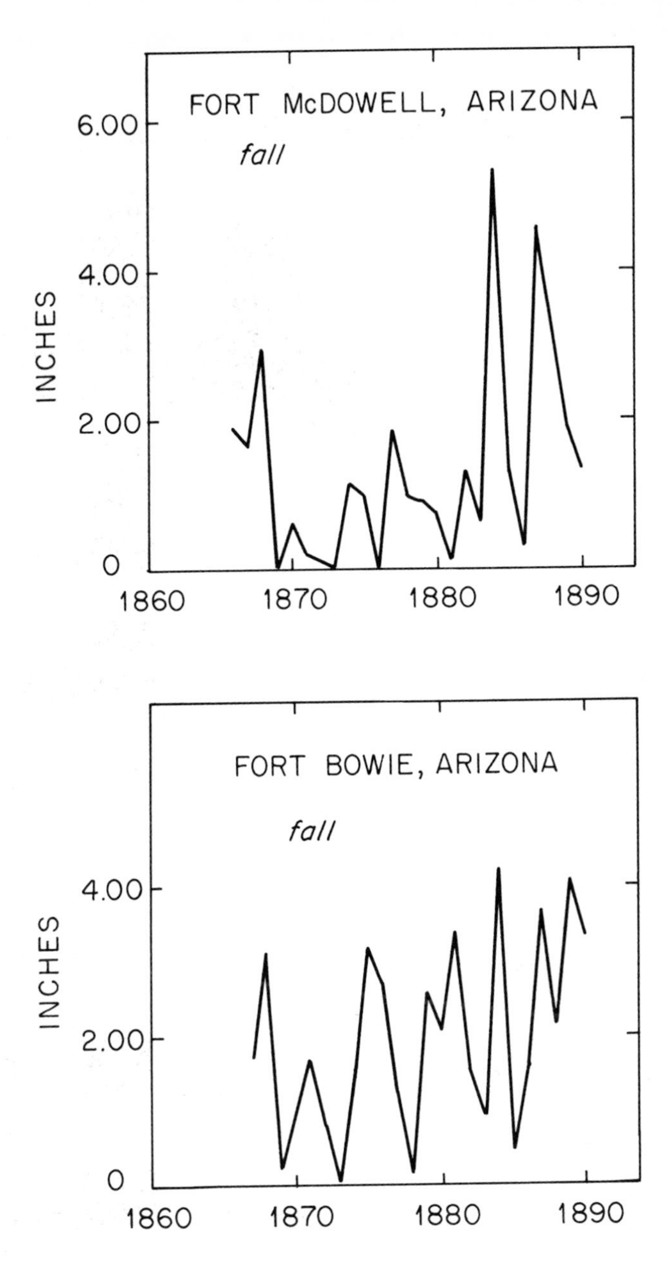

Figure 122: Arizona fall precipitation

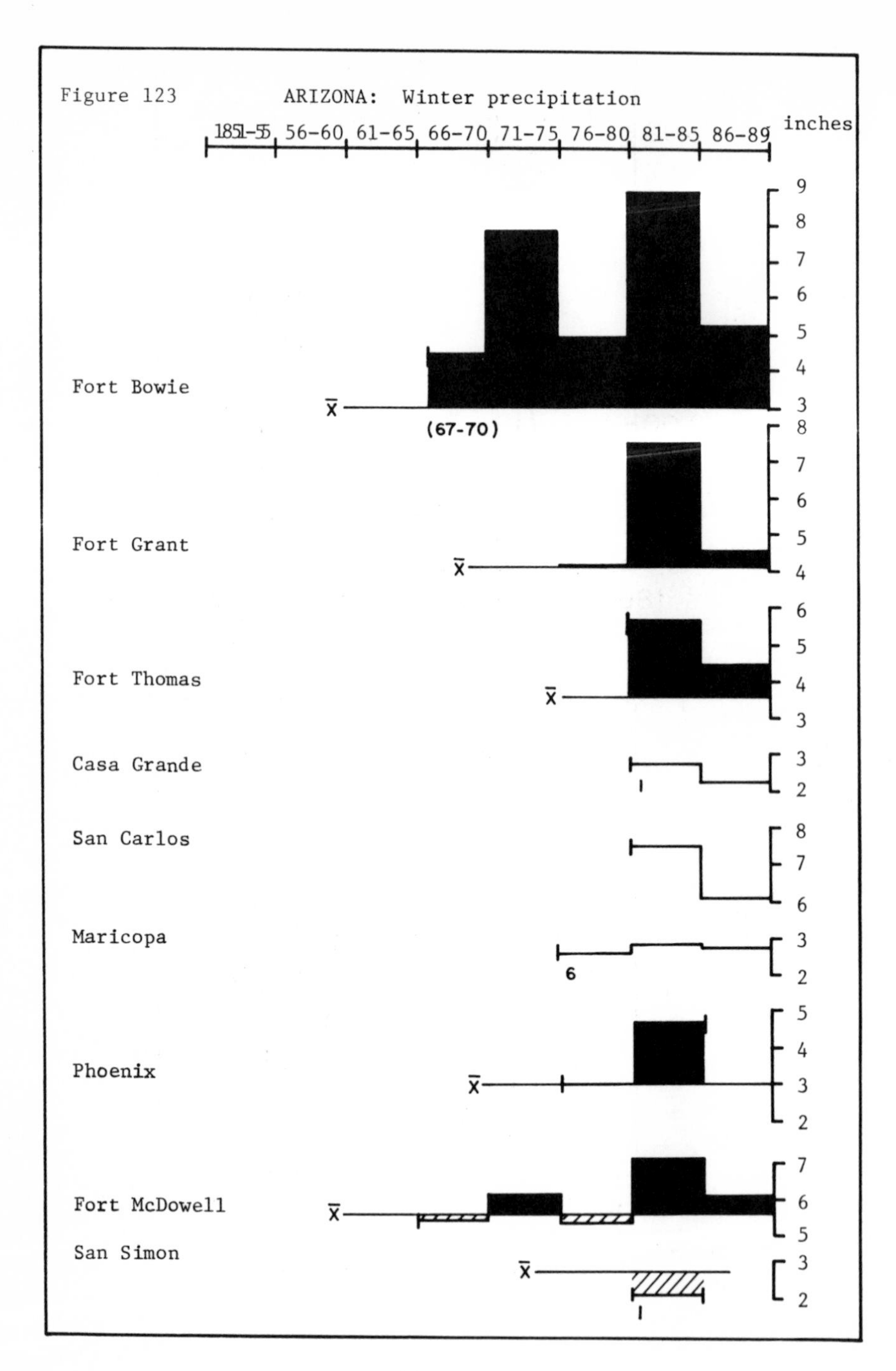

Figure 123
ARIZONA: Winter precipitation
1851-55
56-60
61-65
66-70
71-75
76-80
81-85
86-89
inches
Fort Bowie
(67-70)
Fort Grant
Fort Thomas
Casa Grande
San Carlos
Maricopa
6
Phoenix
Fort McDowell
San Simon
9
8
7
6
5
4
3
8
7
6
5
4
6
5
4
3
3
2
8
7
6
3
2
5
4
3
2
7
6
5
3
2

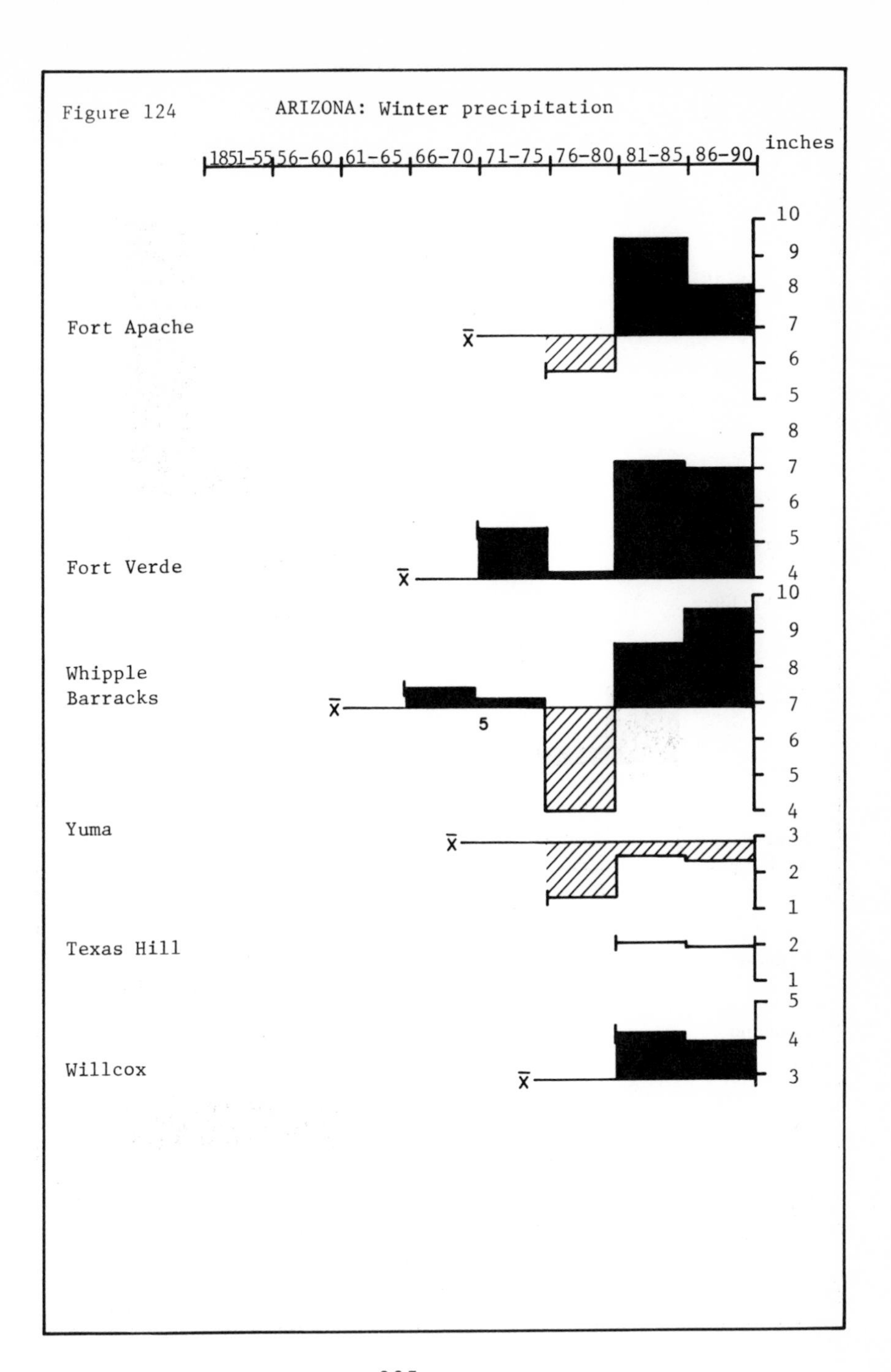
Figure 124
ARIZONA: Winter precipitation
1851-55
56-60
61-65
66-70
71-75
76-80
81-85
86-90
inches
Fort Apache
Fort Verde
Whipple
Barracks
Yuma
Texas Hill
Willcox
X̄
5
10
9
8
7
6
5
8
7
6
5
4
10
9
8
7
6
5
4
3
2
1
2
1
5
4
3

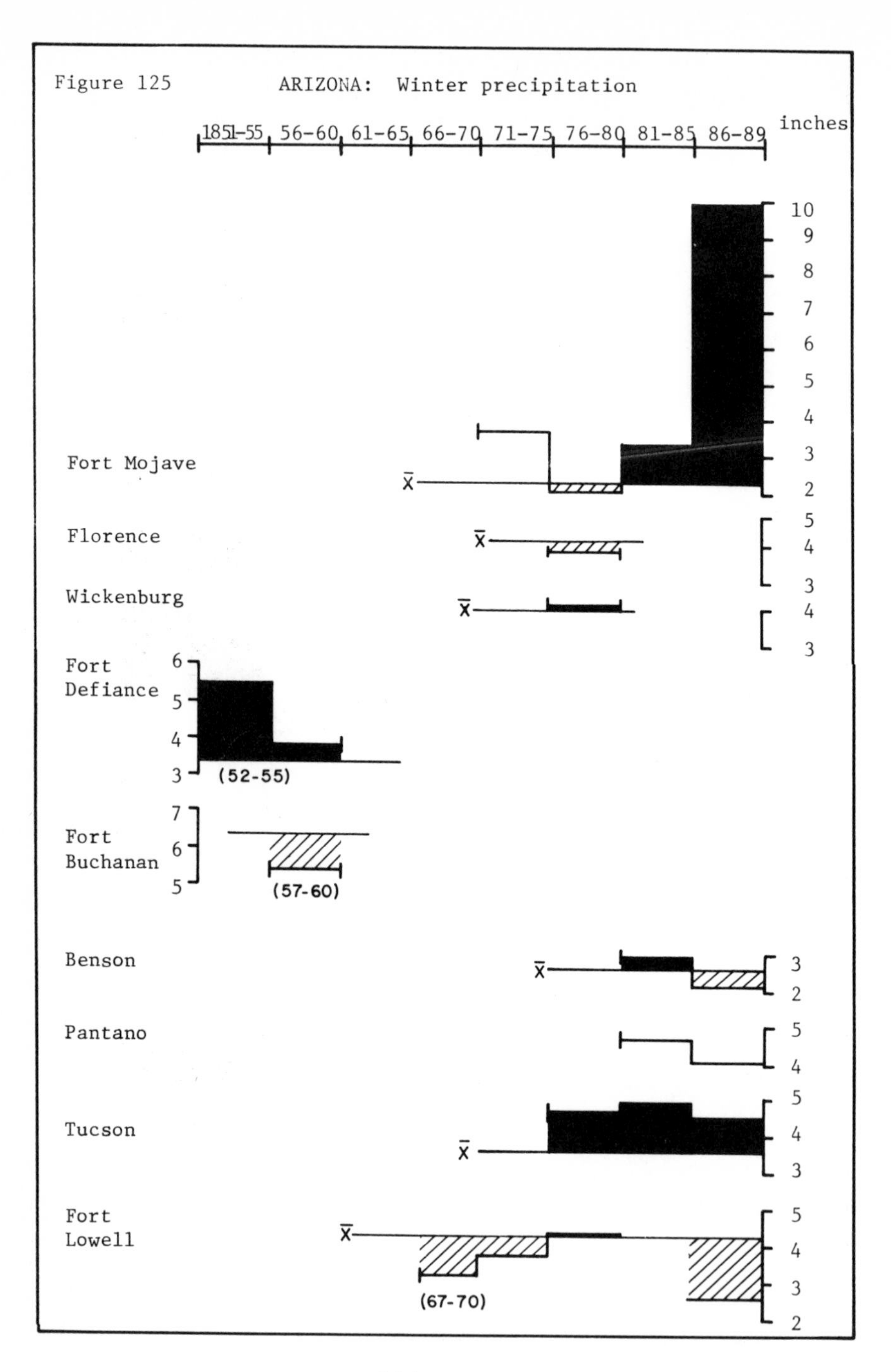

Figure 125
ARIZONA: Winter precipitation
1851-55
56-60
61-65
66-70
71-75
76-80
81-85
86-89
inches
Fort Mojave
Florence
Wickenburg
Fort Defiance
(52-55)
Fort Buchanan
(57-60)
Benson
Pantano
Tucson
Fort Lowell
(67-70)

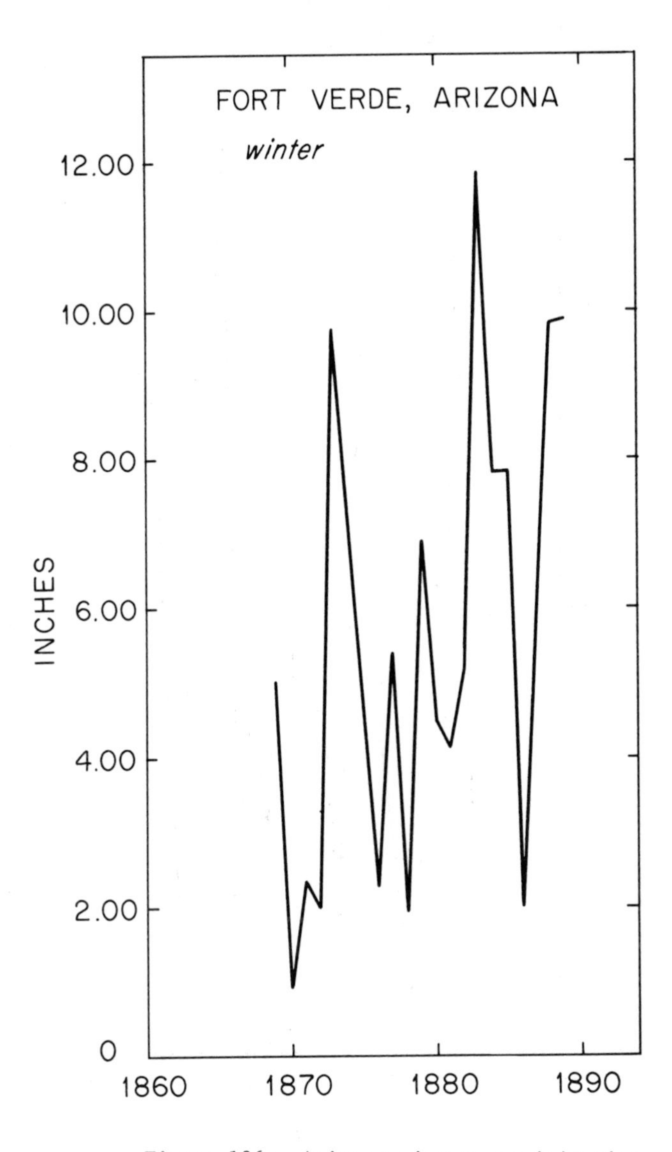

Figure 126: Arizona winter precipitation

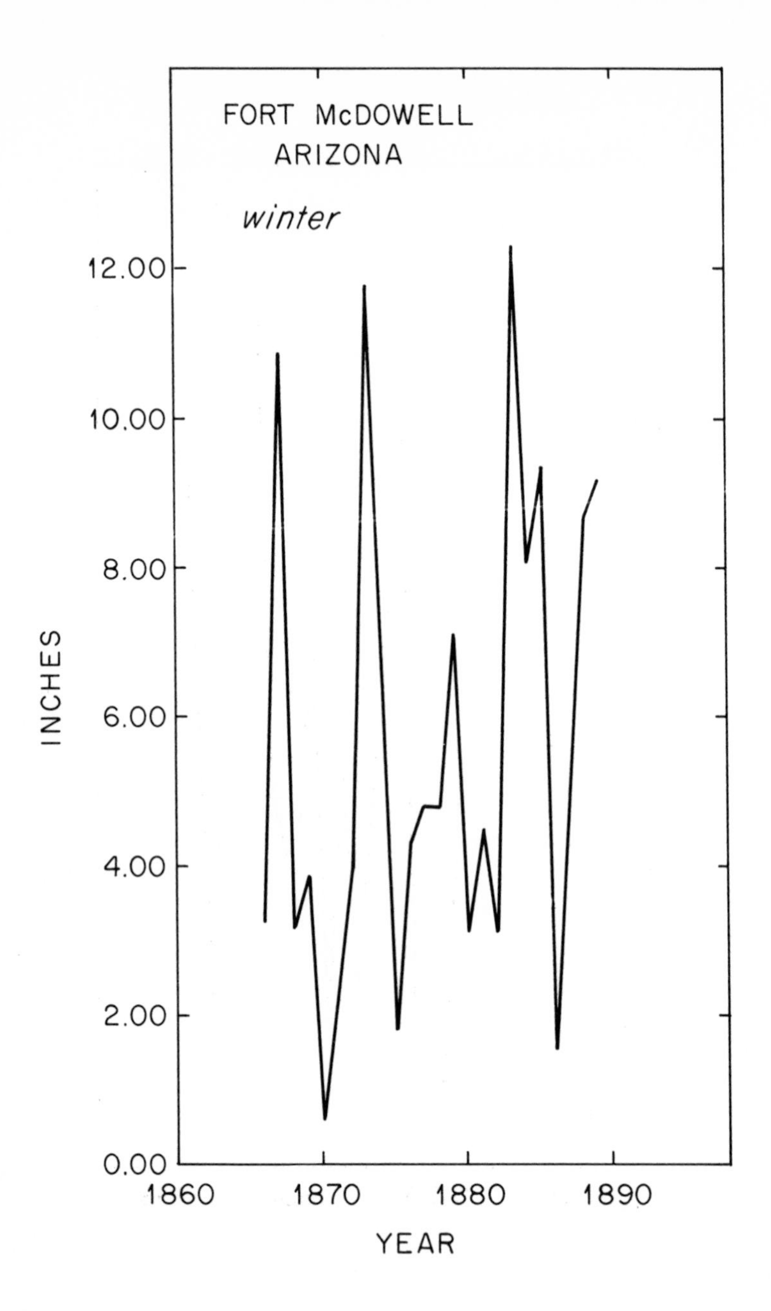

Figure 127: Arizona winter precipitation

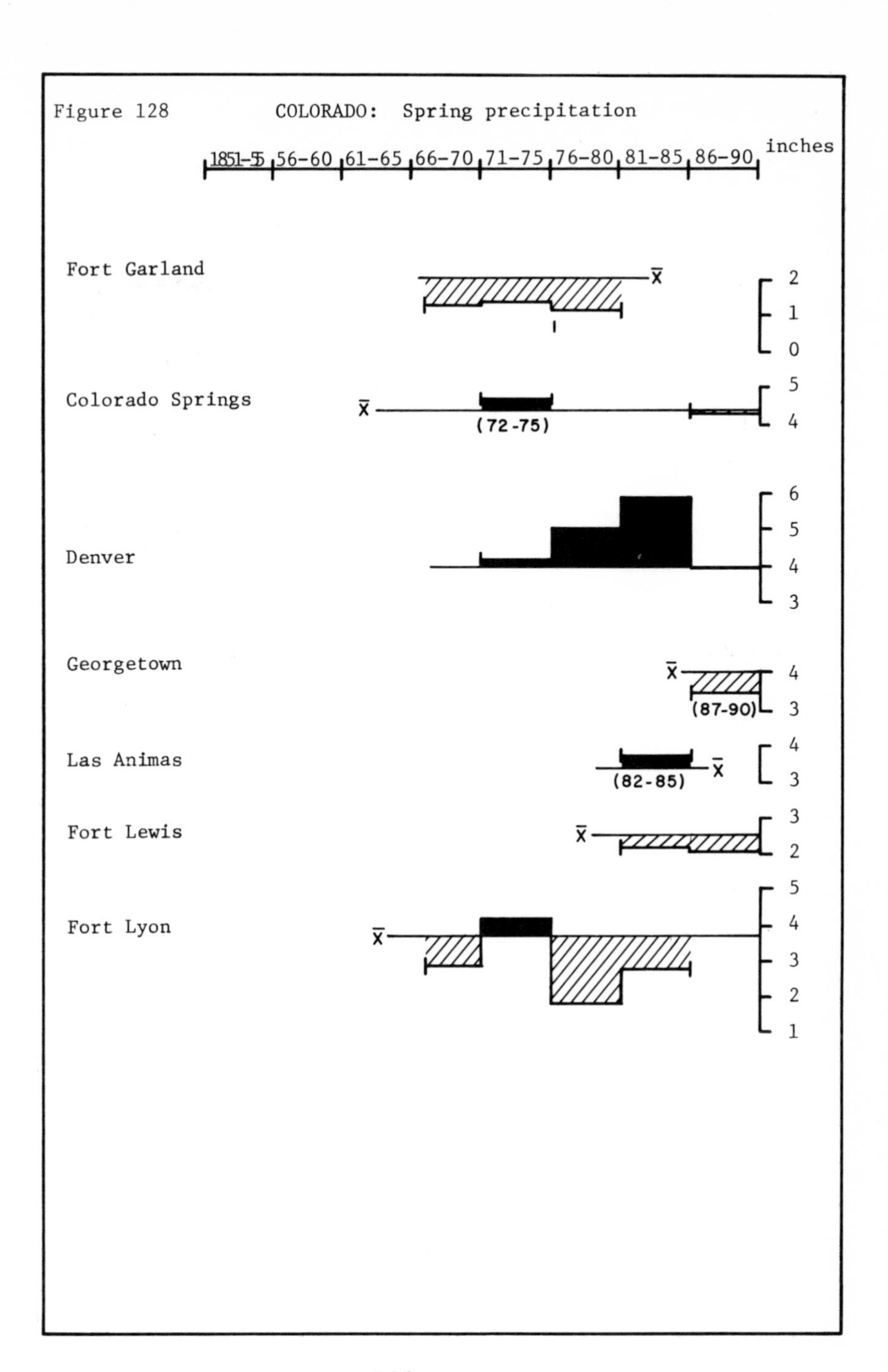
Figure 128 COLORADO: Spring precipitation
1851-55
56-60
61-65
66-70
71-75
76-80
81-85
86-90
inches
Fort Garland
Colorado Springs
(72-75)
Denver
Georgetown
(87-90)
Las Animas
(82-85)
Fort Lewis
Fort Lyon

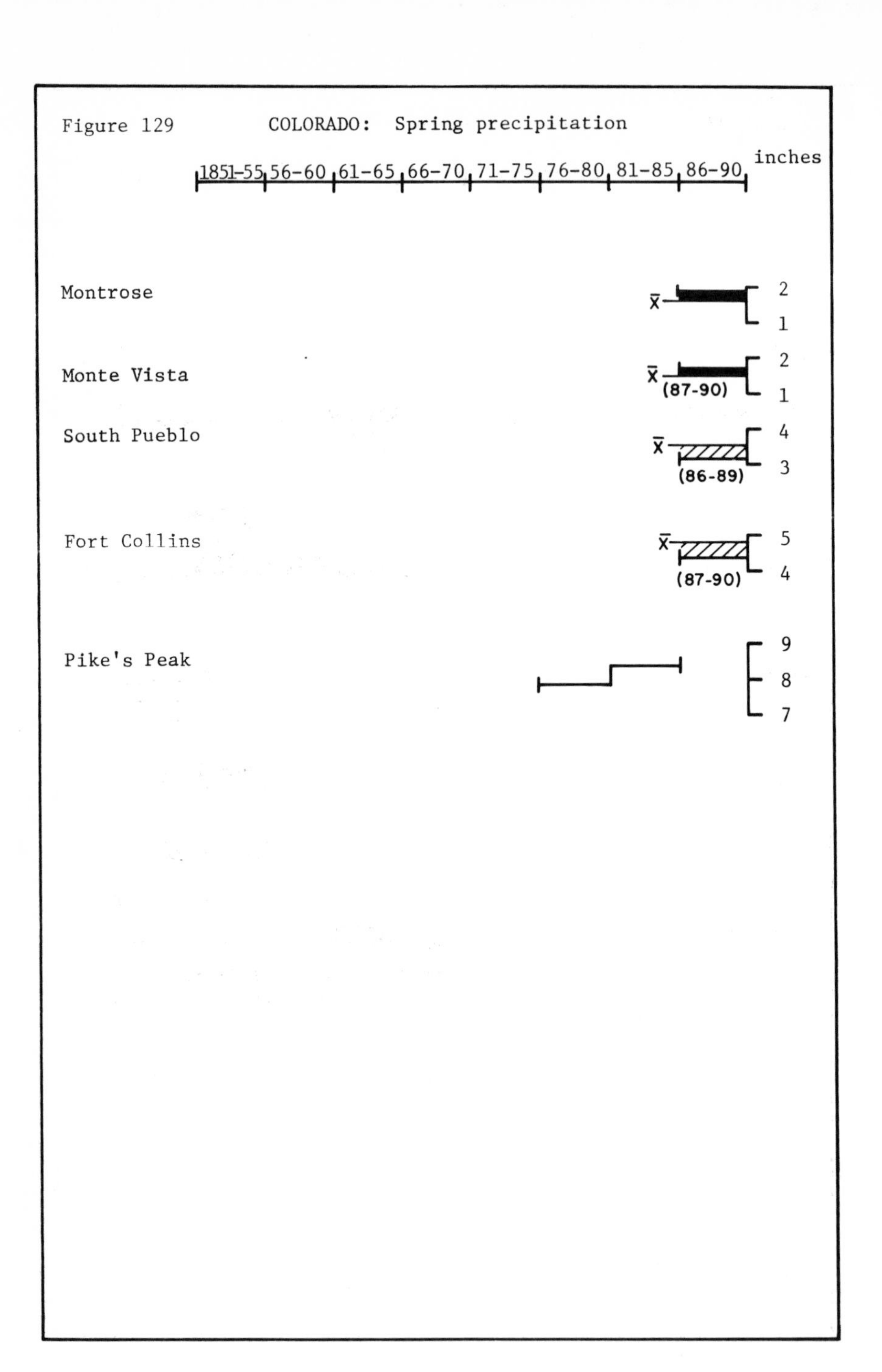

Figure 129
COLORADO: Spring precipitation
1851-55
56-60
61-65
66-70
71-75
76-80
81-85
86-90
inches
Montrose
x̄
2
1
Monte Vista
x̄
(87-90)
2
1
South Pueblo
x̄
(86-89)
4
3
Fort Collins
x̄
(87-90)
5
4
Pike's Peak
9
8
7

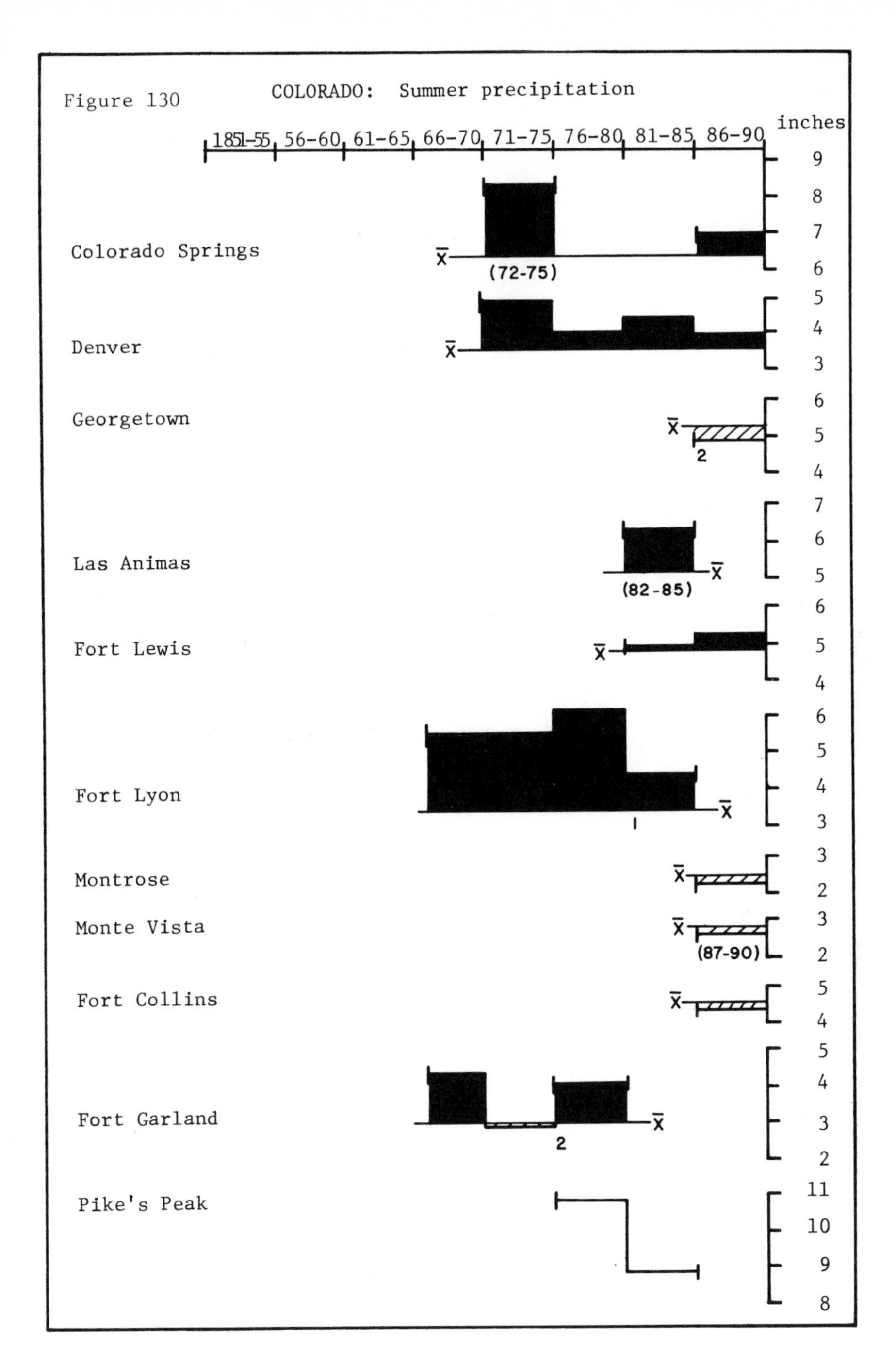

Figure 130
COLORADO: Summer precipitation
1851-55
56-60
61-65
66-70
71-75
76-80
81-85
86-90
inches
Colorado Springs
(72-75)
Denver
Georgetown
Las Animas
(82-85)
Fort Lewis
Fort Lyon
Montrose
Monte Vista
(87-90)
Fort Collins
Fort Garland
Pike's Peak

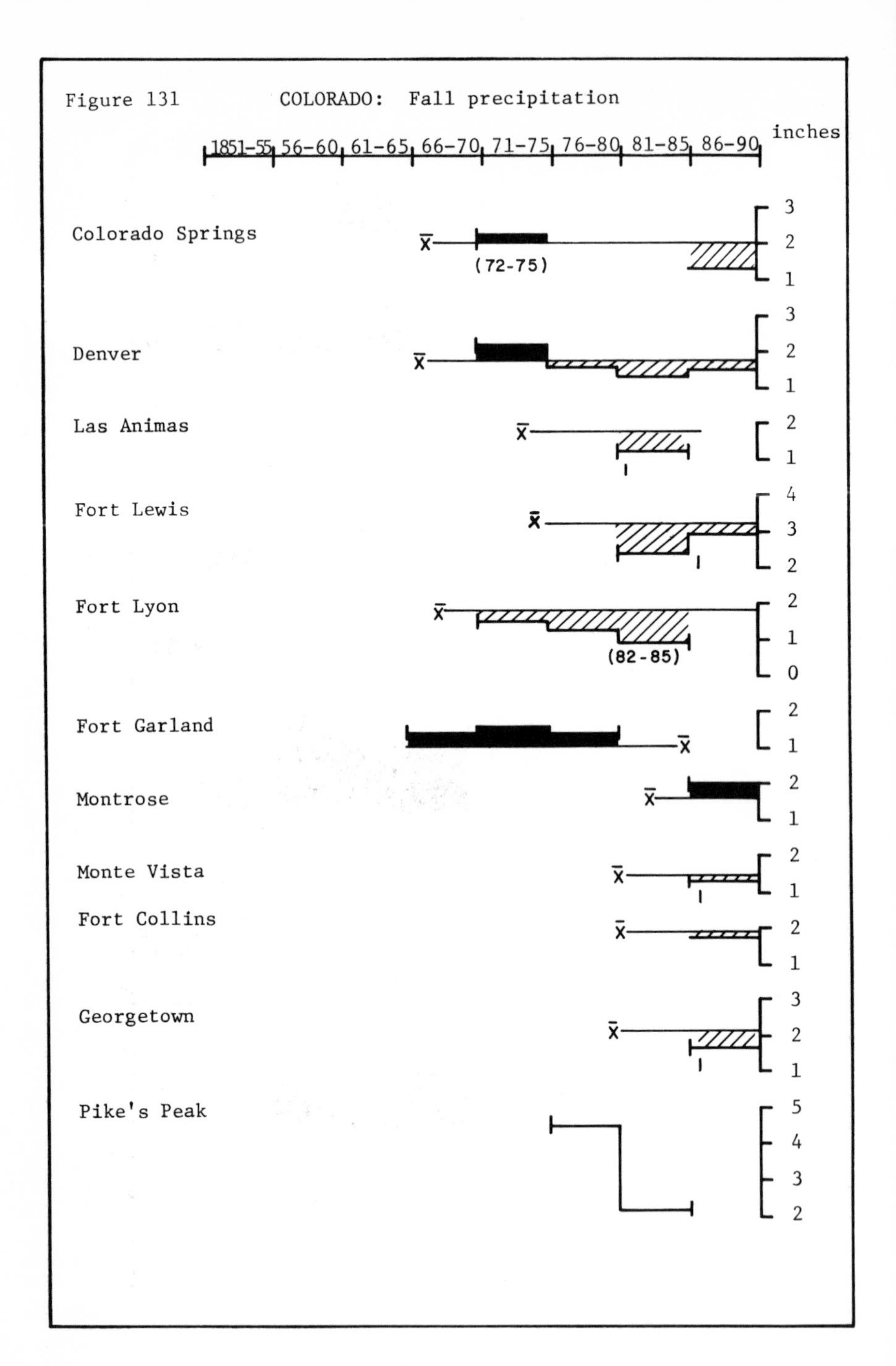
Figure 131
COLORADO: Fall precipitation
1851-55
56-60
61-65
66-70
71-75
76-80
81-85
86-90
inches
Colorado Springs
(72-75)
Denver
Las Animas
Fort Lewis
Fort Lyon
(82-85)
Fort Garland
Montrose
Monte Vista
Fort Collins
Georgetown
Pike's Peak

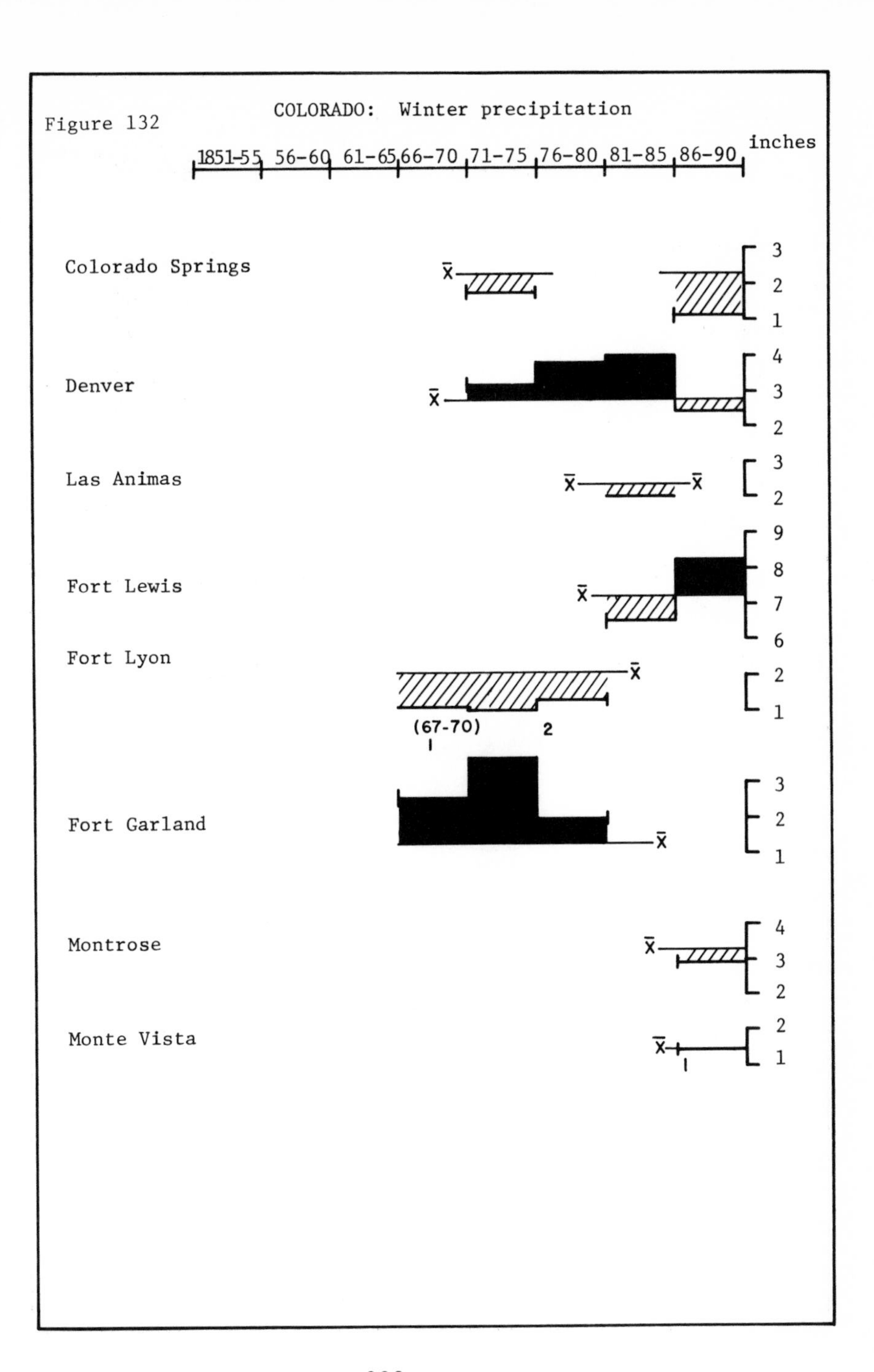
Figure 132
COLORADO: Winter precipitation
1851-55
56-60
61-65
66-70
71-75
76-80
81-85
86-90
inches
Colorado Springs
Denver
Las Animas
Fort Lewis
Fort Lyon
(67-70)
Fort Garland
Montrose
Monte Vista

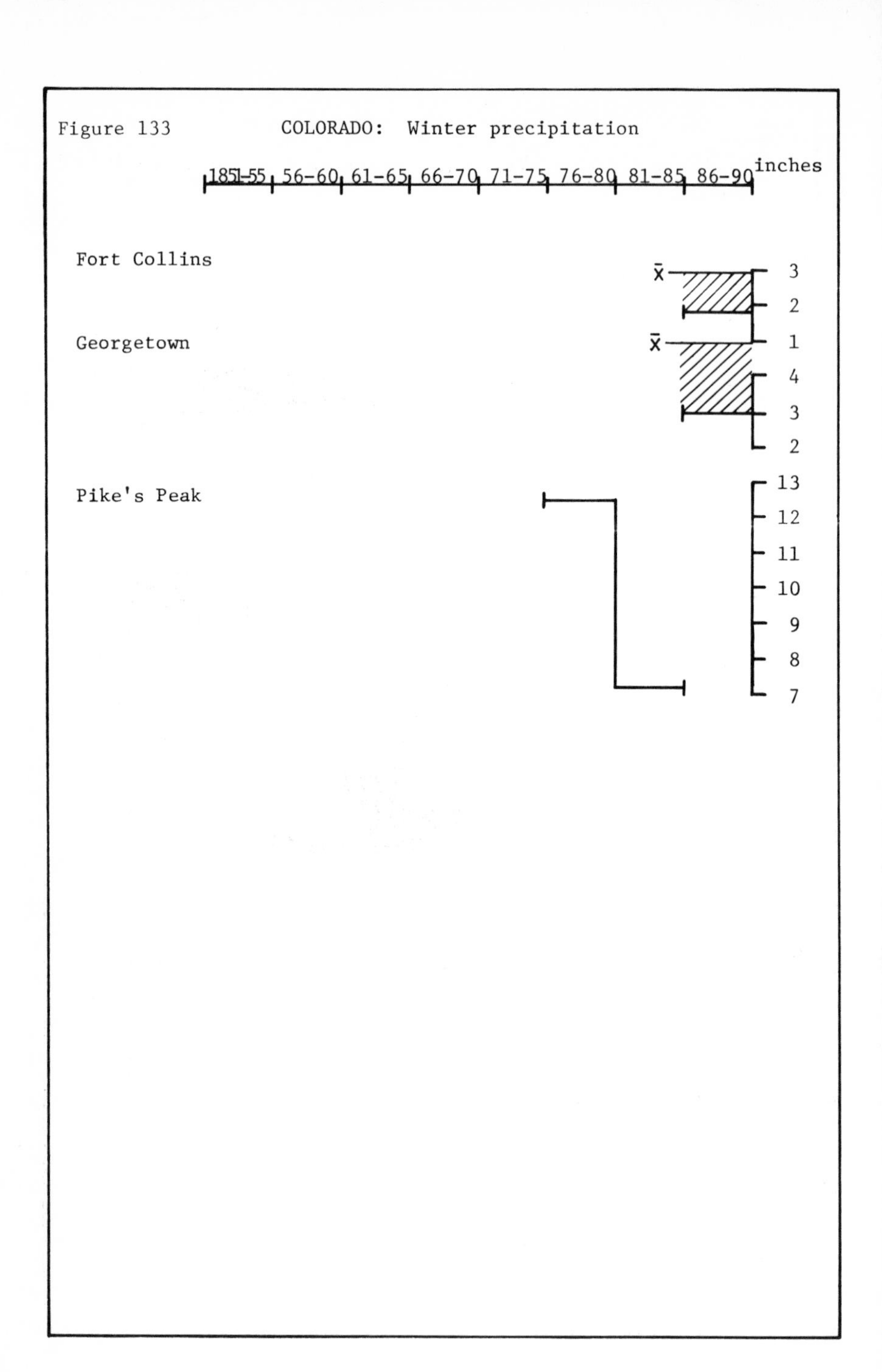

Figure 133
COLORADO: Winter precipitation
1851-55
56-60
61-65
66-70
71-75
76-80
81-85
86-90
inches
Fort Collins
Georgetown
Pike's Peak
$\bar{x}$
$\bar{x}$
3
2
1
4
3
2
13
12
11
10
9
8
7

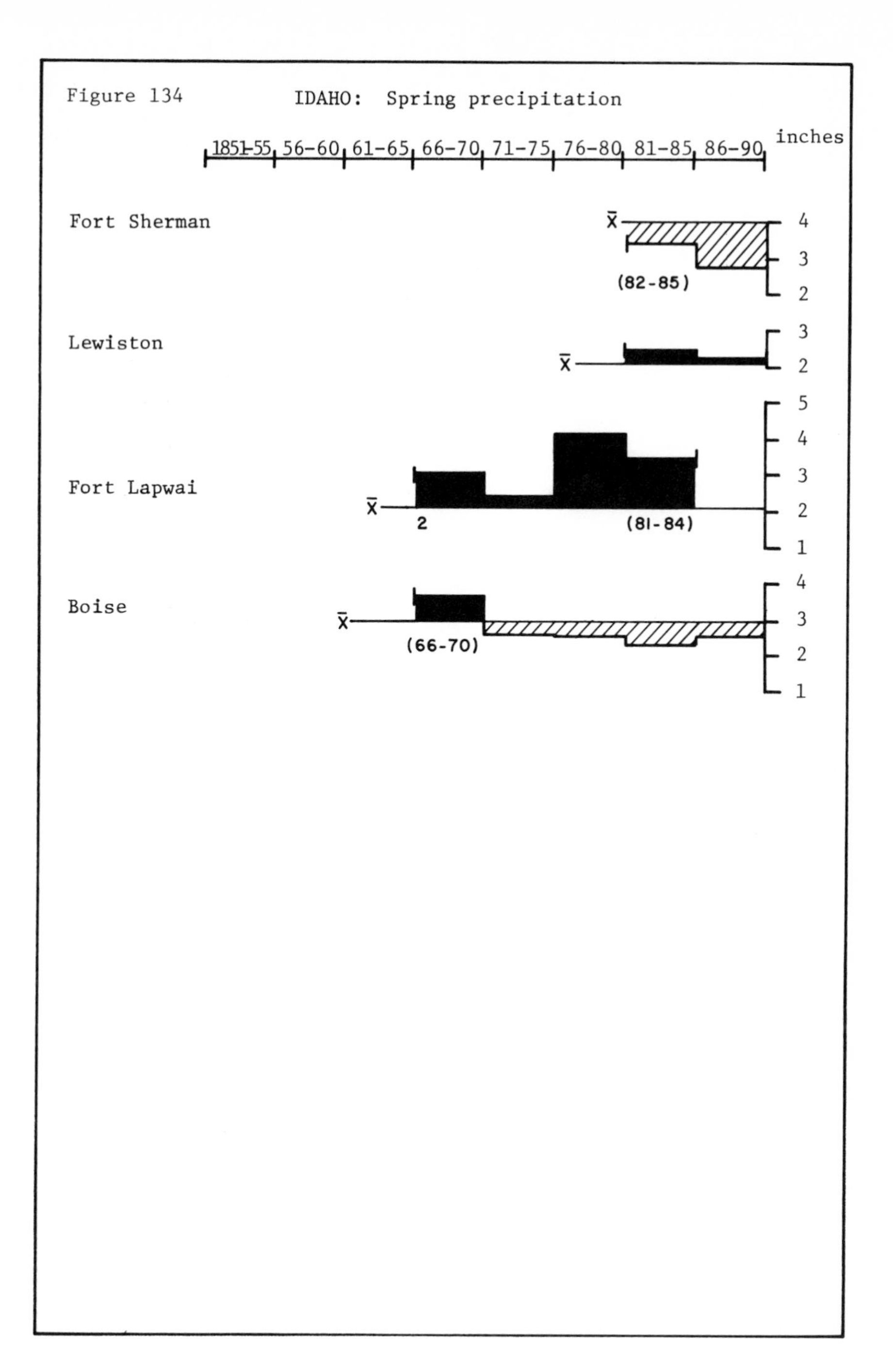
Figure 134
IDAHO: Spring precipitation
1851-55
56-60
61-65
66-70
71-75
76-80
81-85
86-90
inches
Fort Sherman
X̄
(82-85)
4
3
2
Lewiston
X̄
3
2
Fort Lapwai
X̄
2
(81-84)
5
4
3
2
1
Boise
X̄
(66-70)
4
3
2
1

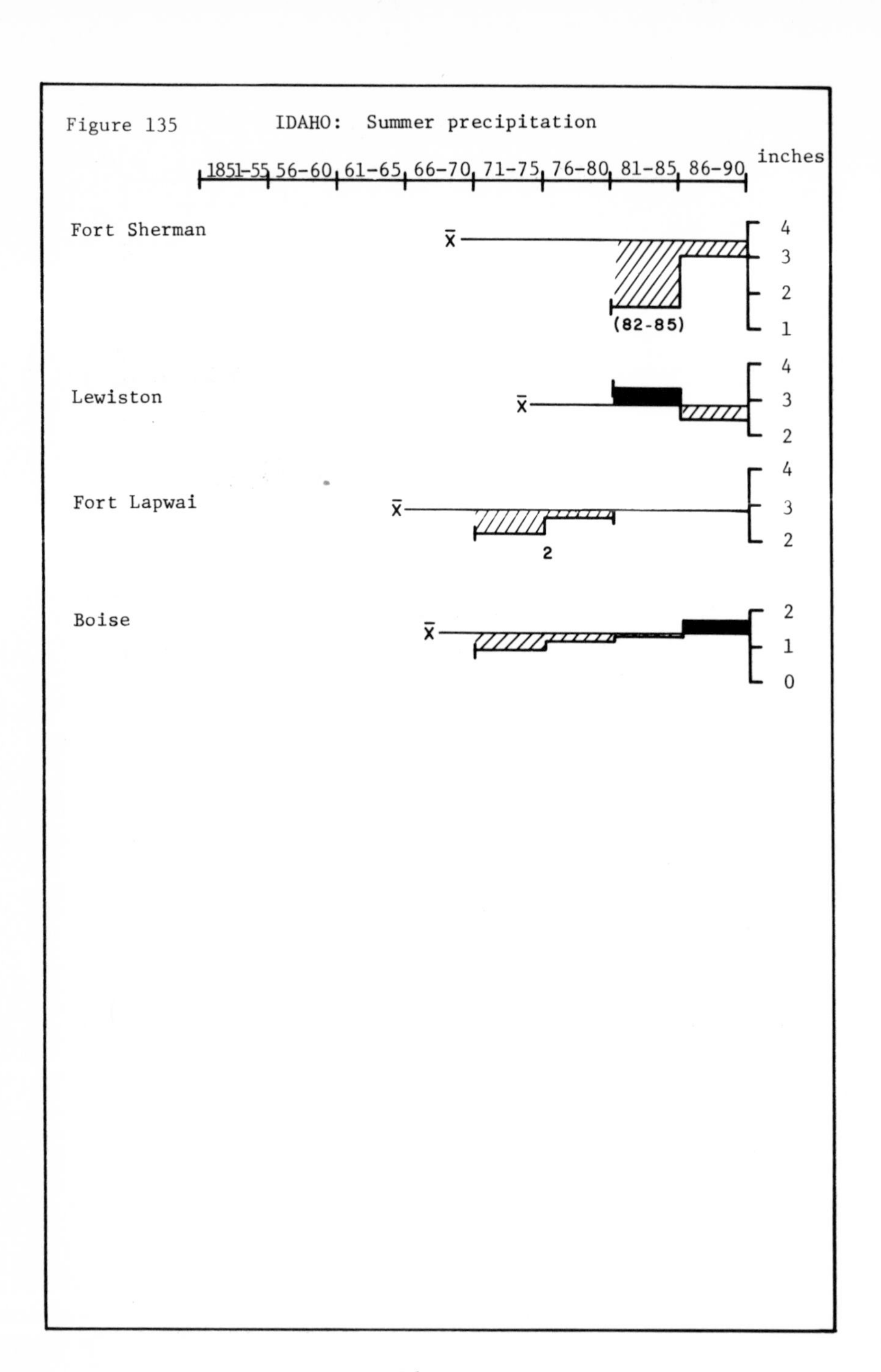
Figure 135
IDAHO: Summer precipitation
1851-55
56-60
61-65
66-70
71-75
76-80
81-85
86-90
inches
Fort Sherman
x̄
(82-85)
4
3
2
1
Lewiston
x̄
4
3
2
Fort Lapwai
x̄
2
4
3
2
Boise
x̄
2
1
0

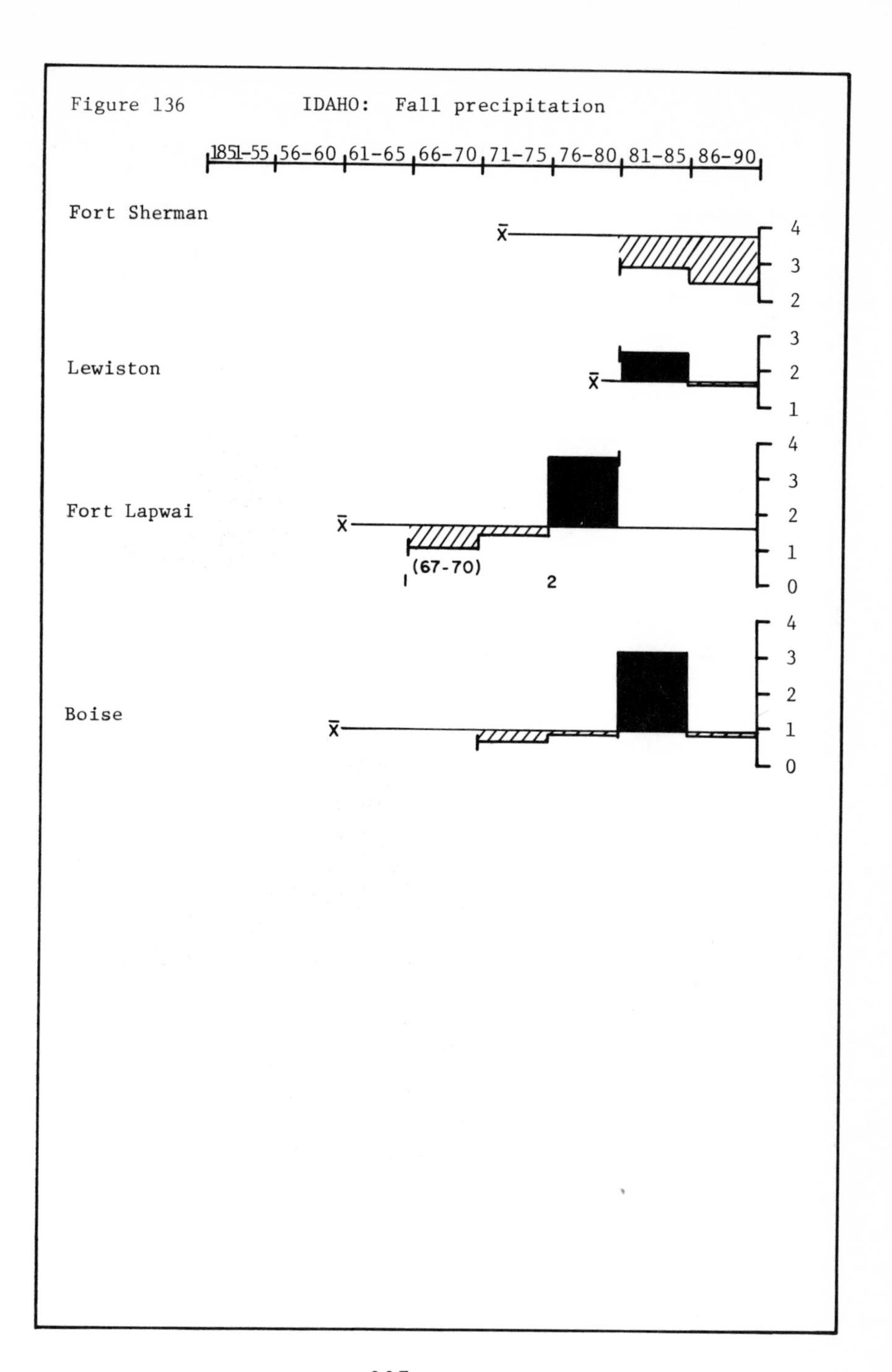
Figure 136
IDAHO: Fall precipitation
1851-55
56-60
61-65
66-70
71-75
76-80
81-85
86-90
Fort Sherman
Lewiston
Fort Lapwai
(67-70)
1
2
Boise
4
3
2
1
0

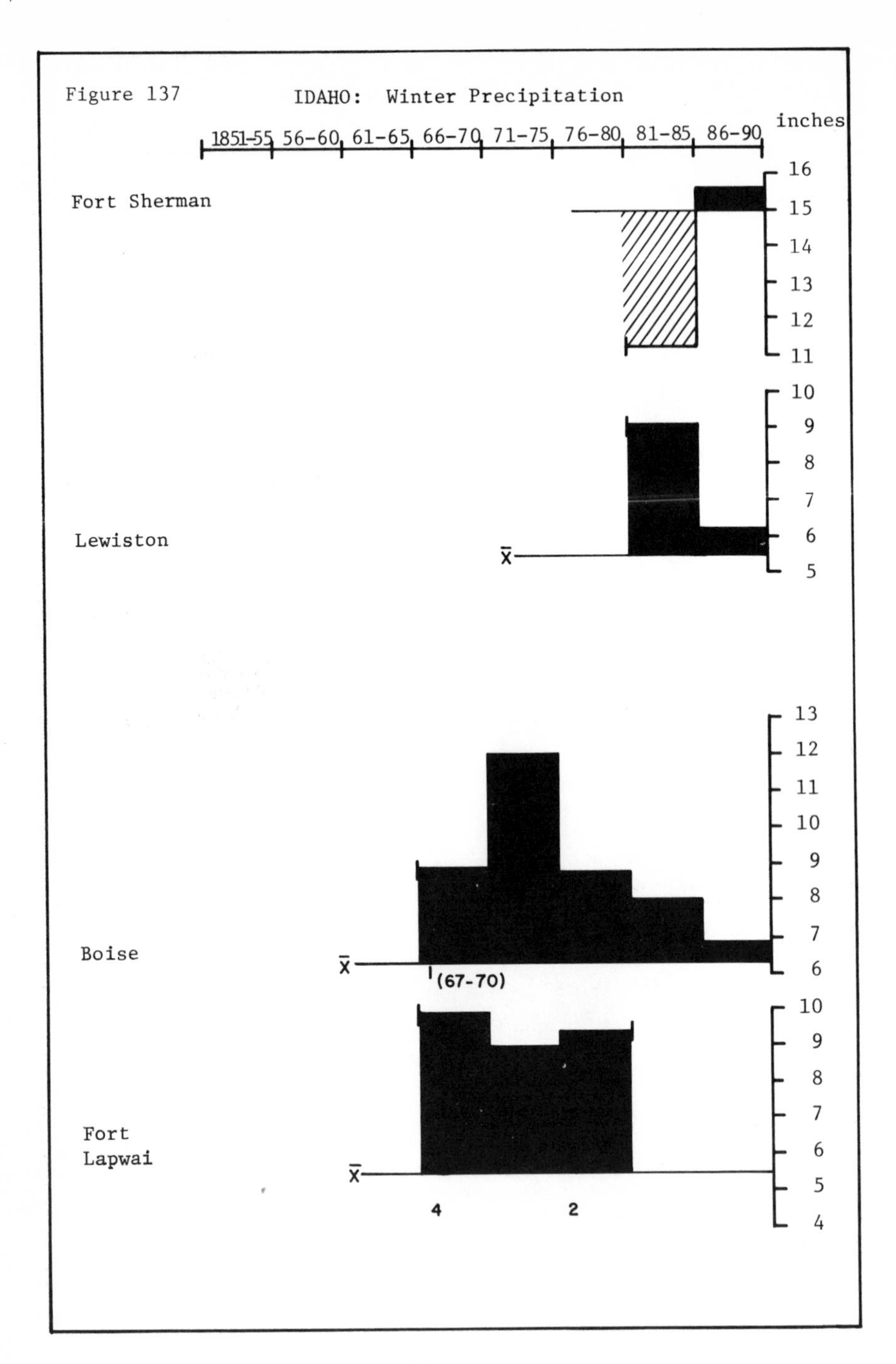
Figure 137
IDAHO: Winter Precipitation
1851-55
56-60
61-65
66-70
71-75
76-80
81-85
86-90
inches
Fort Sherman
16
15
14
13
12
11
Lewiston
10
9
8
7
6
5
Boise
13
12
11
10
9
8
7
6
(67-70)
Fort Lapwai
10
9
8
7
6
5
4
4
2

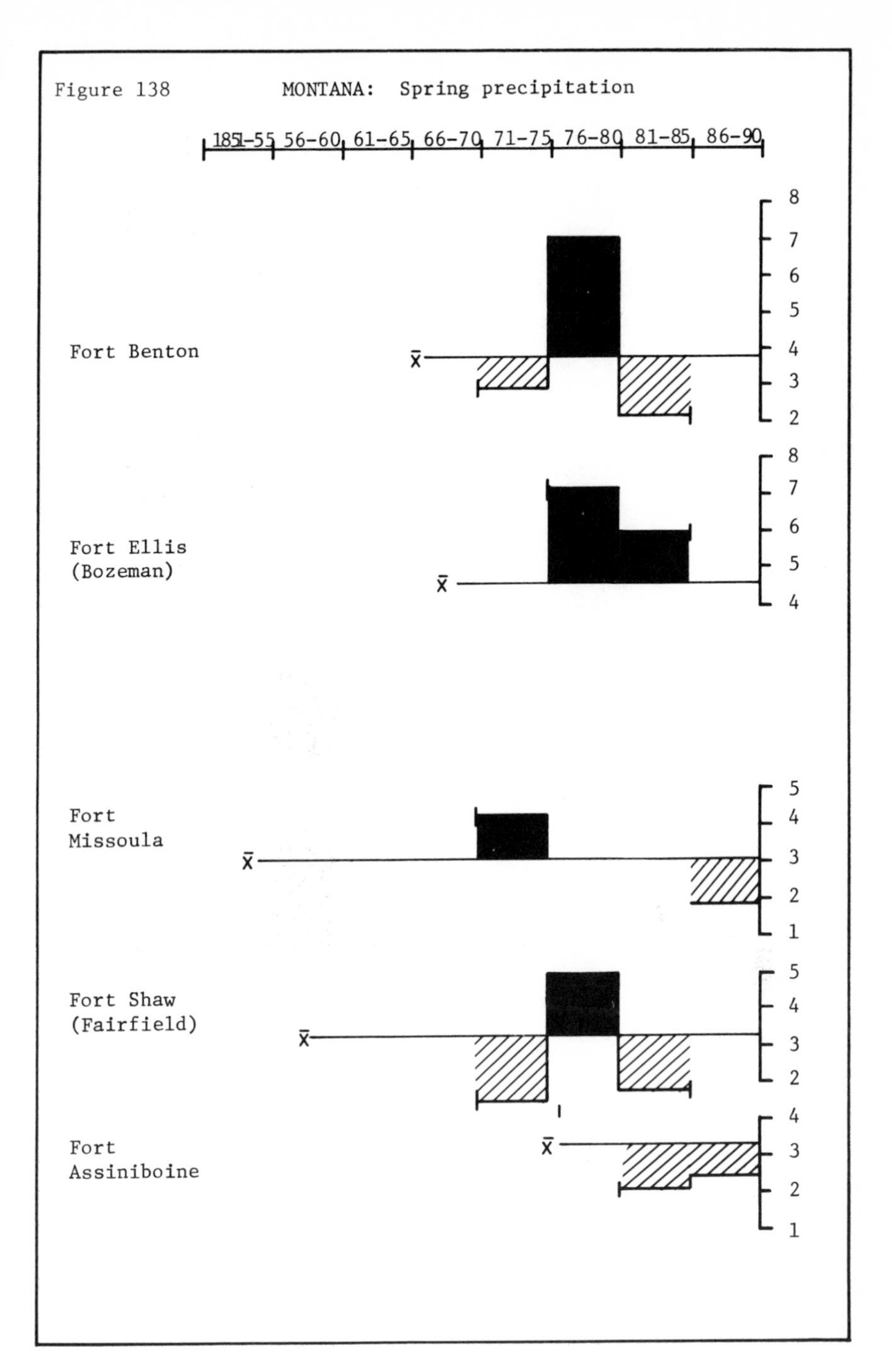
Figure 138
MONTANA: Spring precipitation
1851-55
56-60
61-65
66-70
71-75
76-80
81-85
86-90
Fort Benton
Fort Ellis
(Bozeman)
Fort
Missoula
Fort Shaw
(Fairfield)
Fort
Assiniboine
x̄
8
7
6
5
4
3
2
1

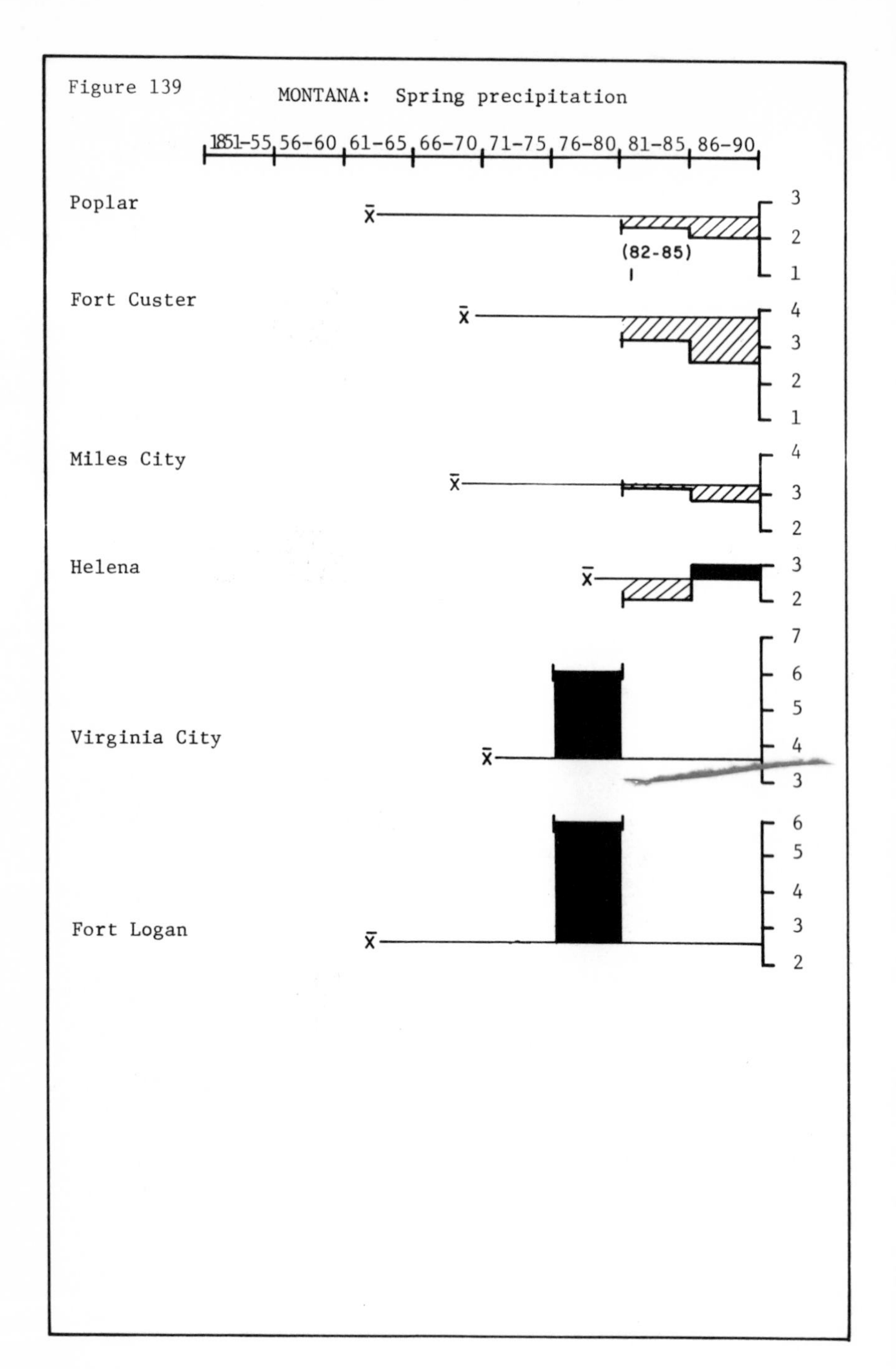

Figure 139
MONTANA: Spring precipitation
1851-55
56-60
61-65
66-70
71-75
76-80
81-85
86-90
Poplar
(82-85)
Fort Custer
Miles City
Helena
Virginia City
Fort Logan

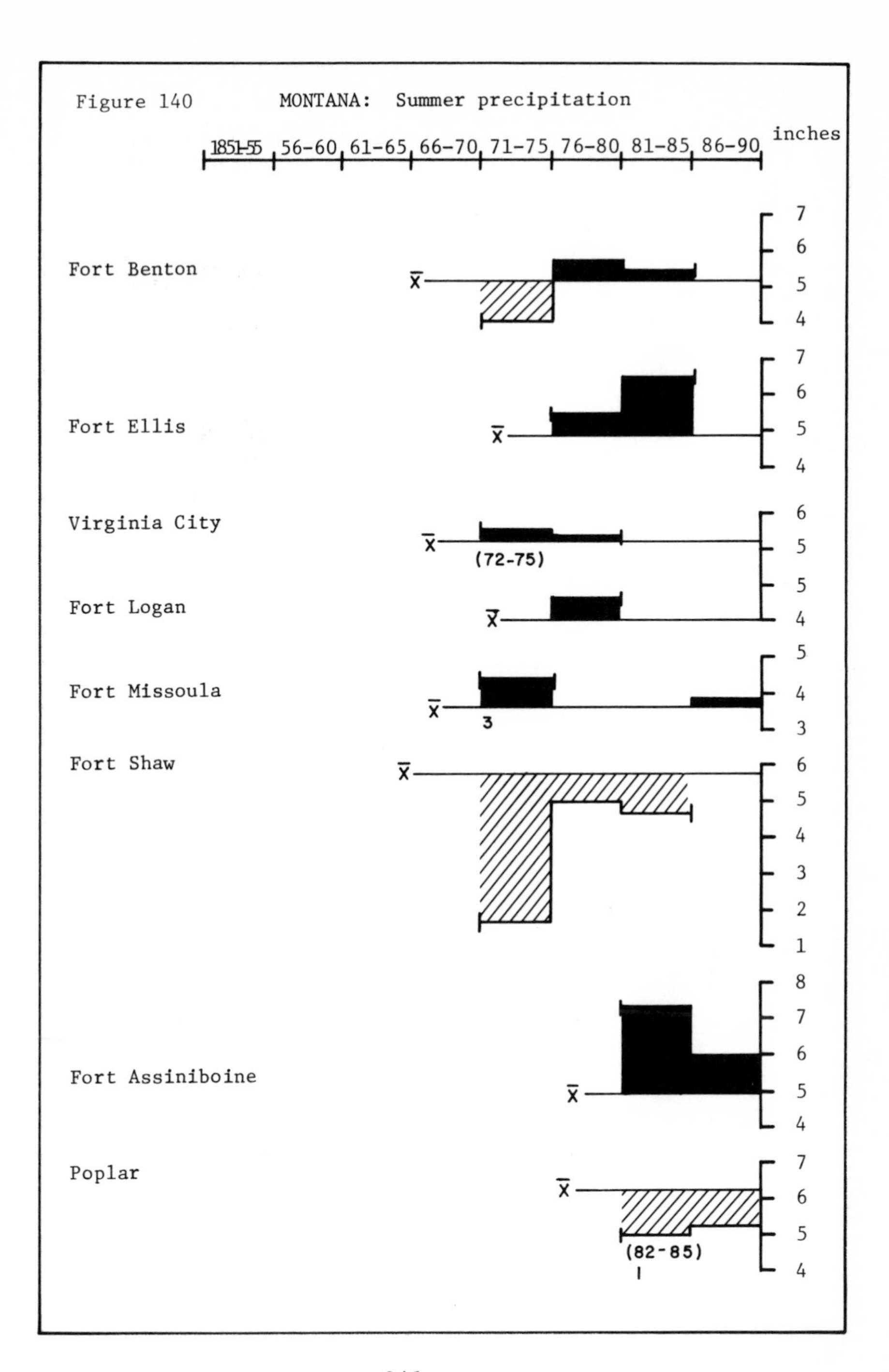
Figure 140 MONTANA: Summer precipitation
1851-55
56-60
61-65
66-70
71-75
76-80
81-85
86-90
inches
Fort Benton
Fort Ellis
Virginia City
(72-75)
Fort Logan
Fort Missoula
3
Fort Shaw
Fort Assiniboine
Poplar
(82-85)
1
X̄

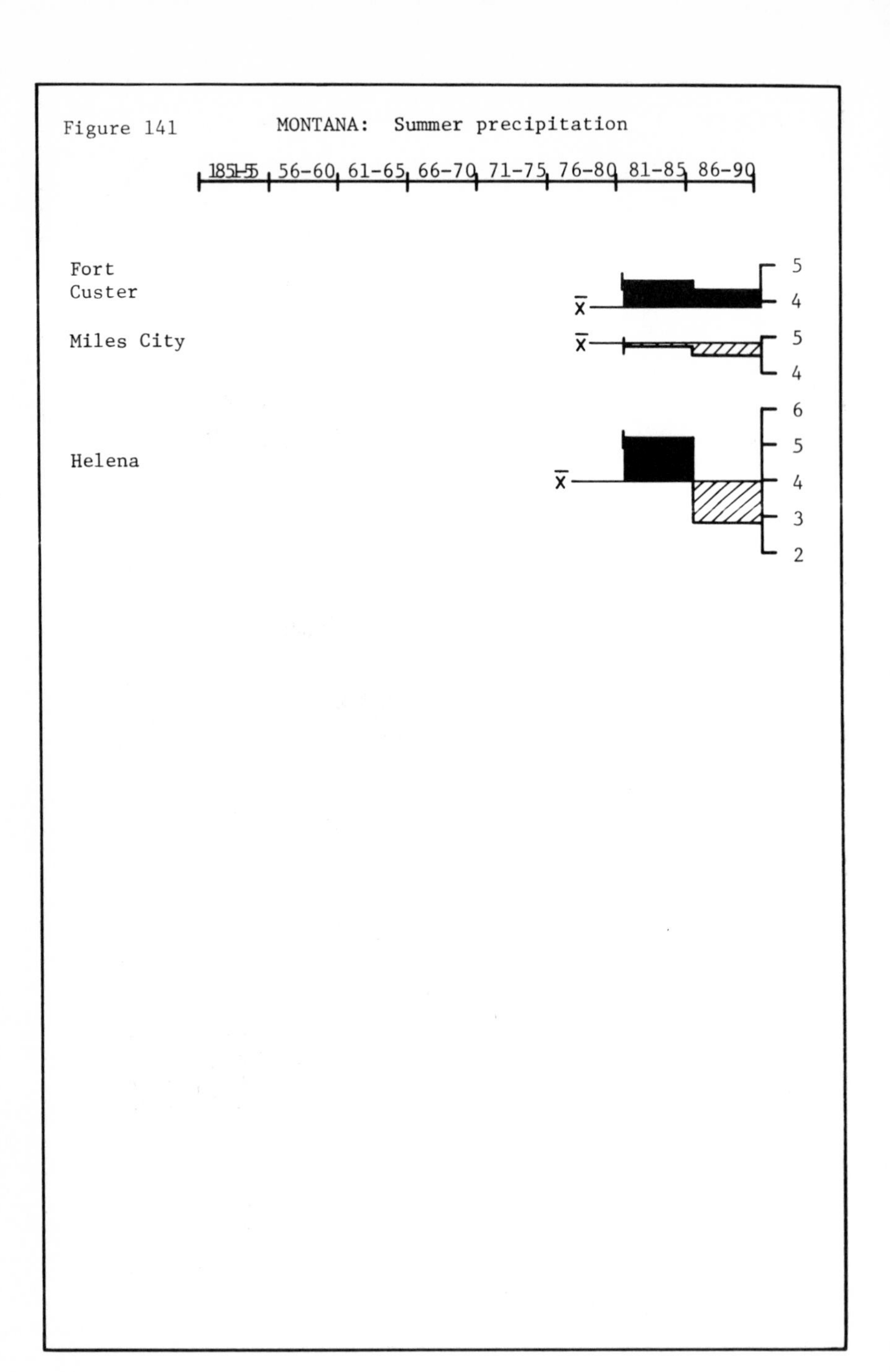
Figure 141
MONTANA: Summer precipitation
1851-55
56-60
61-65
66-70
71-75
76-80
81-85
86-90
Fort Custer
Miles City
Helena
5
4
5
4
6
5
4
3
2

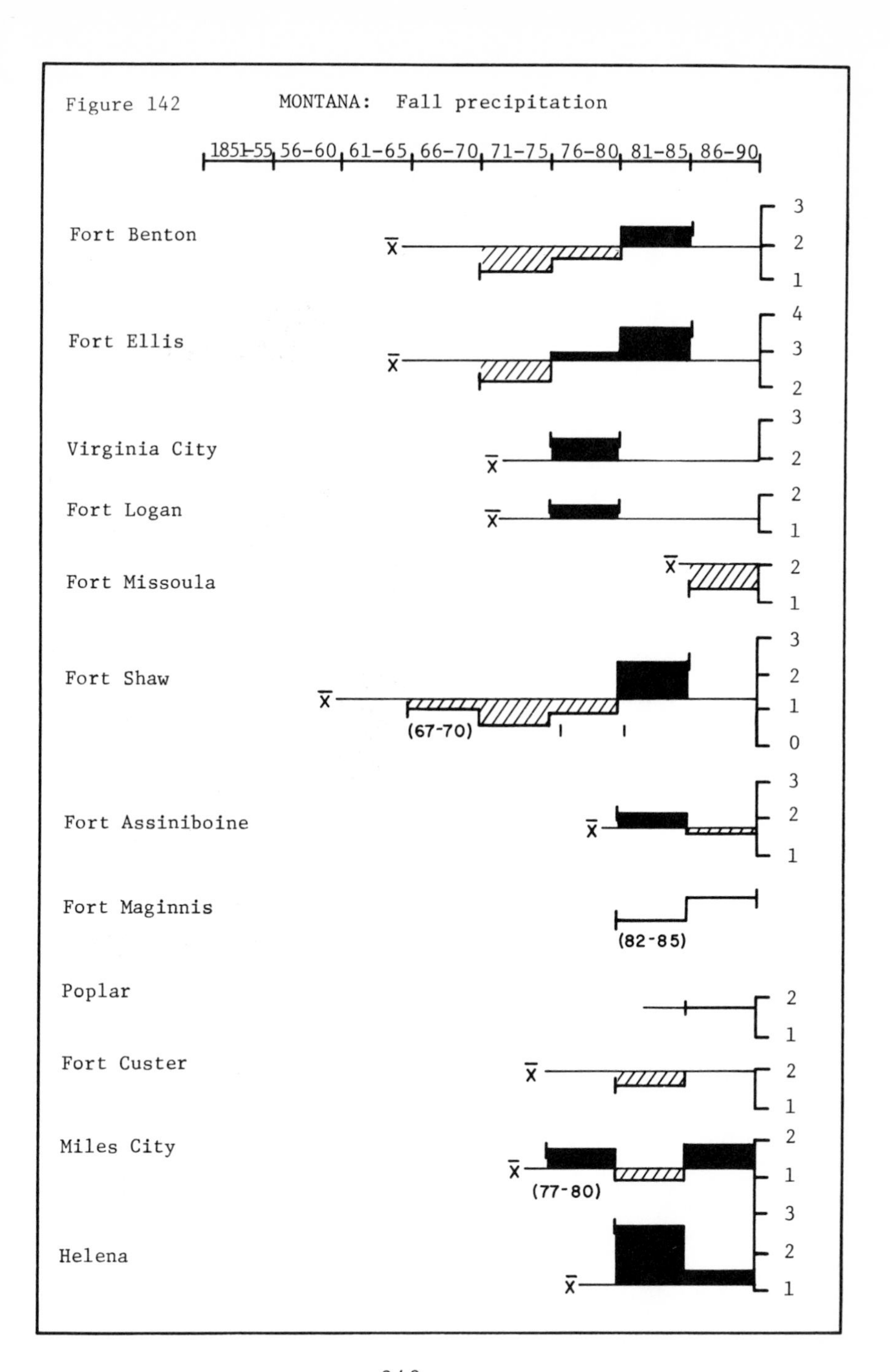
Figure 142
MONTANA: Fall precipitation
1851-55
56-60
61-65
66-70
71-75
76-80
81-85
86-90
Fort Benton
Fort Ellis
Virginia City
Fort Logan
Fort Missoula
Fort Shaw
(67-70)
Fort Assiniboine
Fort Maginnis
(82-85)
Poplar
Fort Custer
Miles City
(77-80)
Helena

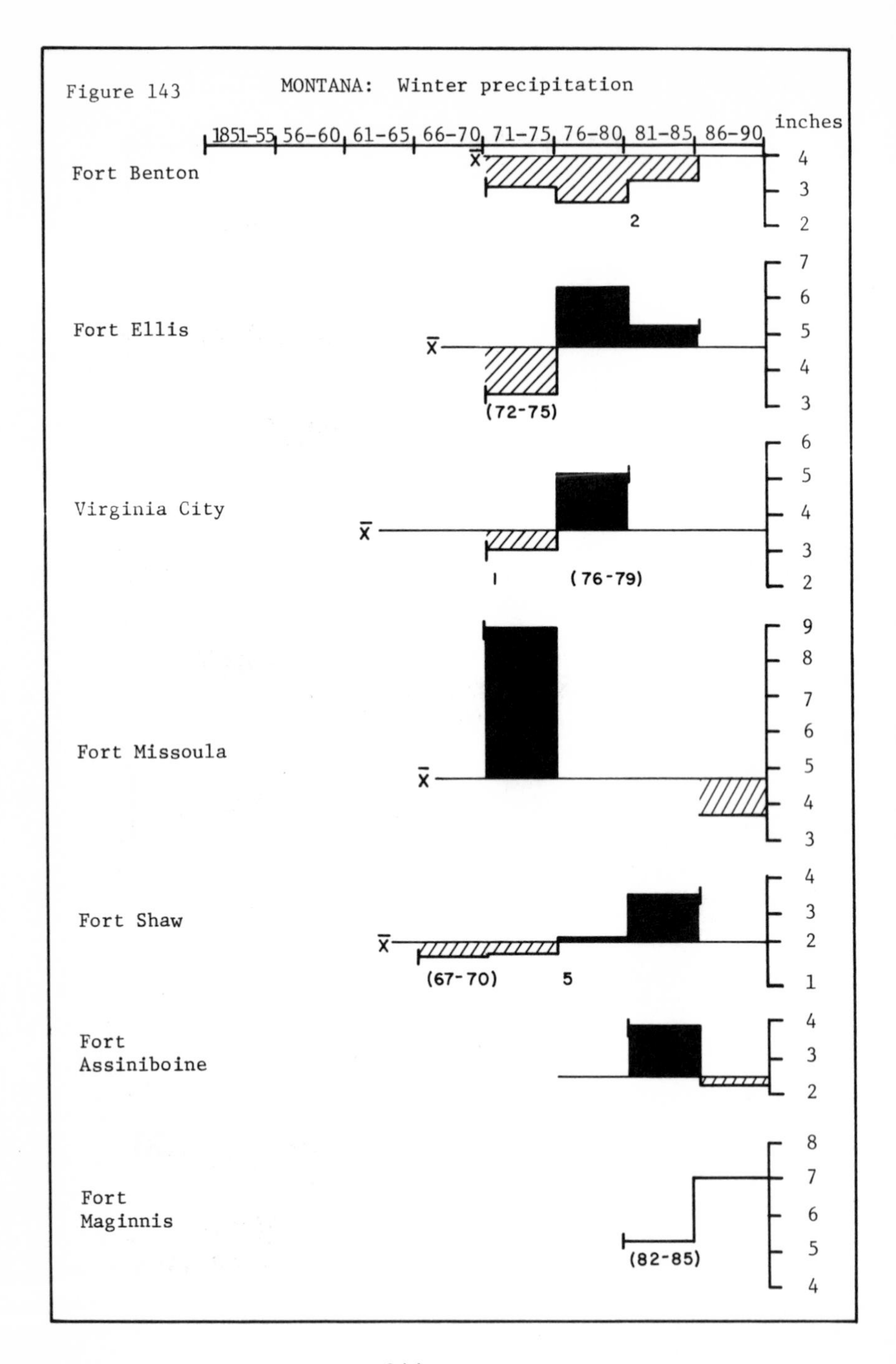
Figure 143
MONTANA: Winter precipitation
1851-55
56-60
61-65
66-70
71-75
76-80
81-85
86-90
inches
Fort Benton
Fort Ellis
(72-75)
Virginia City
(76-79)
Fort Missoula
Fort Shaw
(67-70)
Fort
Assiniboine
Fort
Maginnis
(82-85)

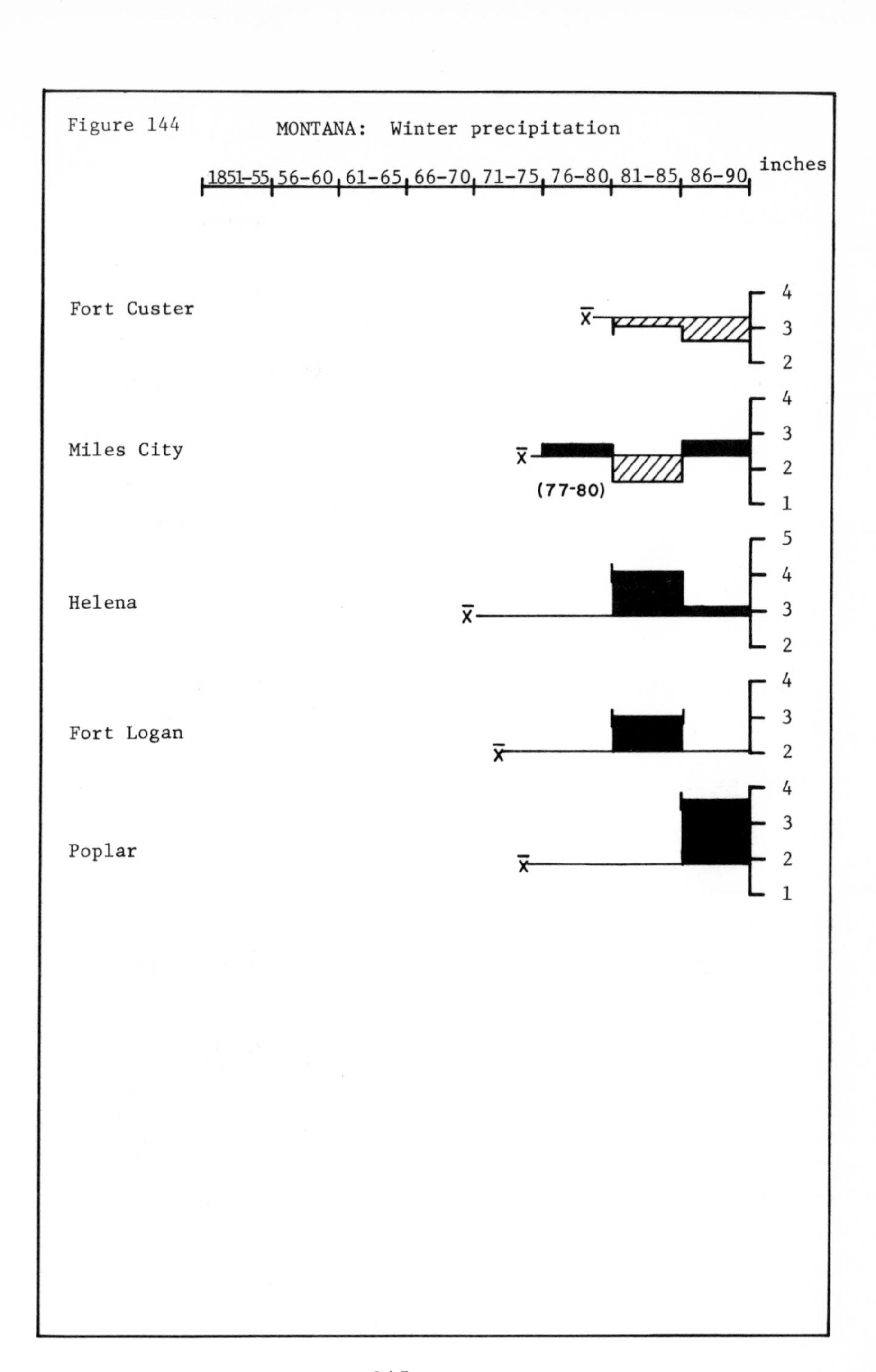

Figure 144
MONTANA: Winter precipitation
1851-55
56-60
61-65
66-70
71-75
76-80
81-85
86-90
inches
Fort Custer
Miles City
(77-80)
Helena
Fort Logan
Poplar
4
3
2
4
3
2
1
5
4
3
2
4
3
2
4
3
2
1

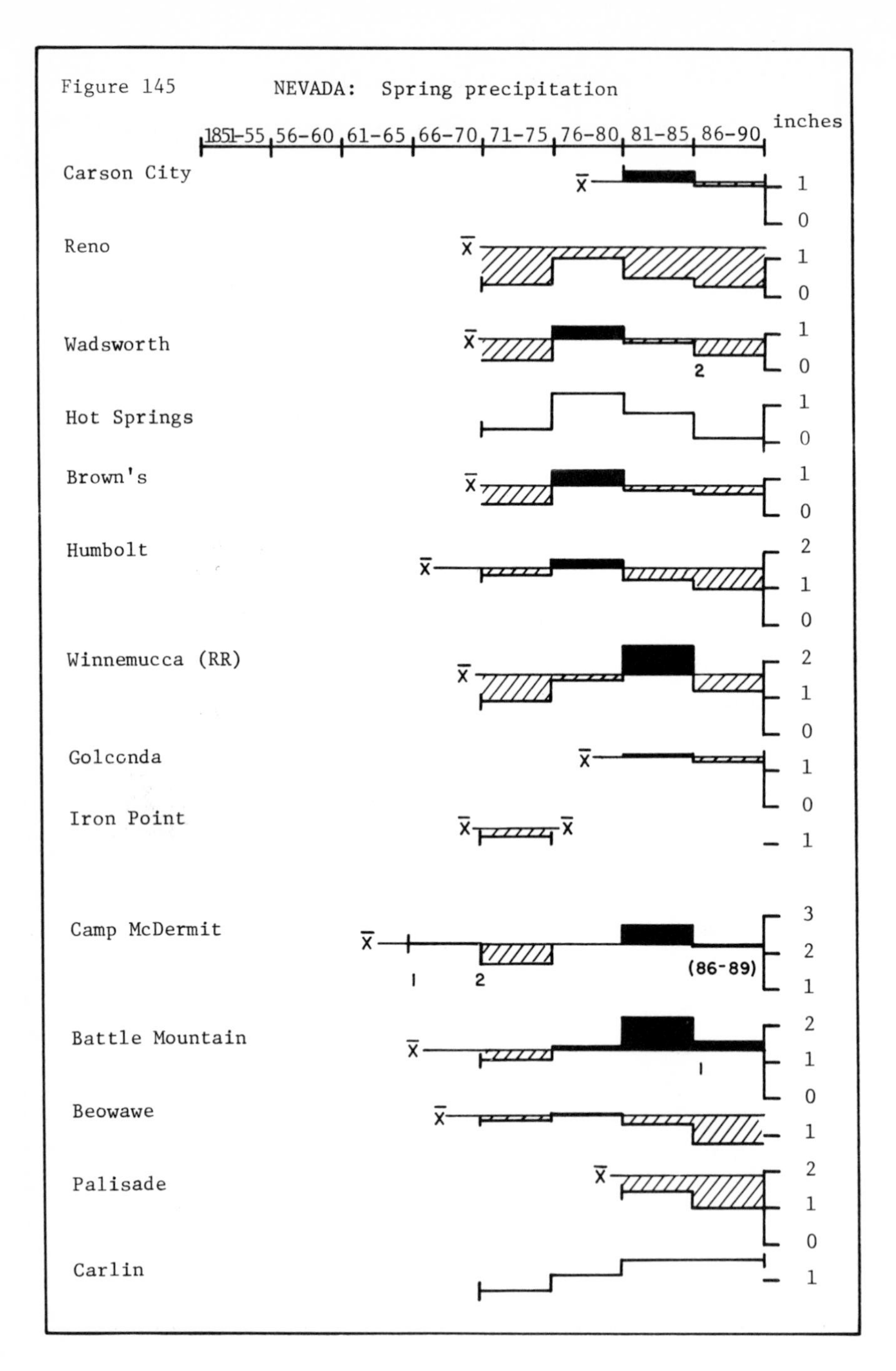
Figure 145
NEVADA: Spring precipitation
1851-55
56-60
61-65
66-70
71-75
76-80
81-85
86-90
inches
Carson City
Reno
Wadsworth
Hot Springs
Brown's
Humbolt
Winnemucca (RR)
Golconda
Iron Point
Camp McDermit
(86-89)
Battle Mountain
Beowawe
Palisade
Carlin

Figure 146 NEVADA: Spring Precipitation

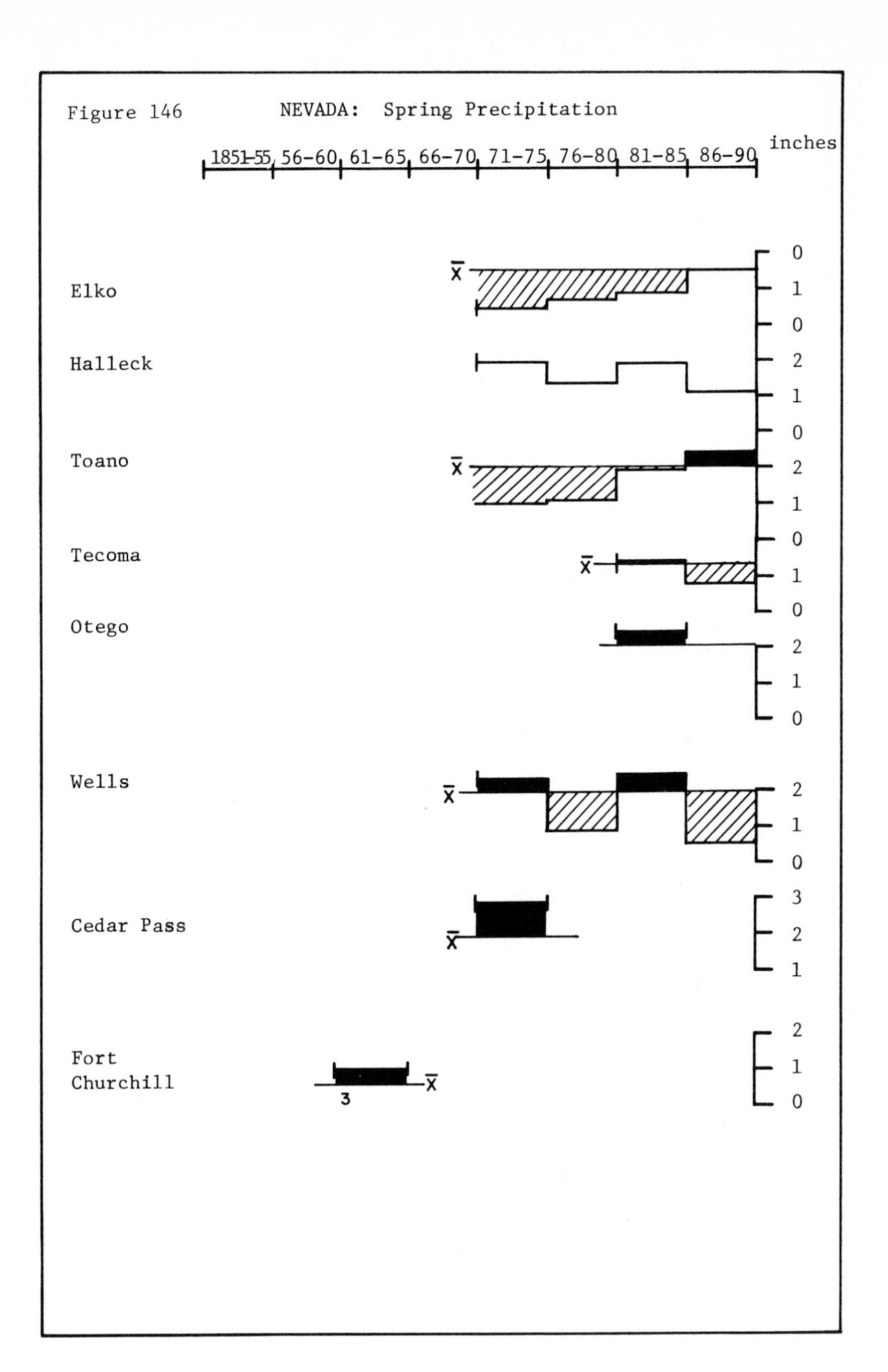

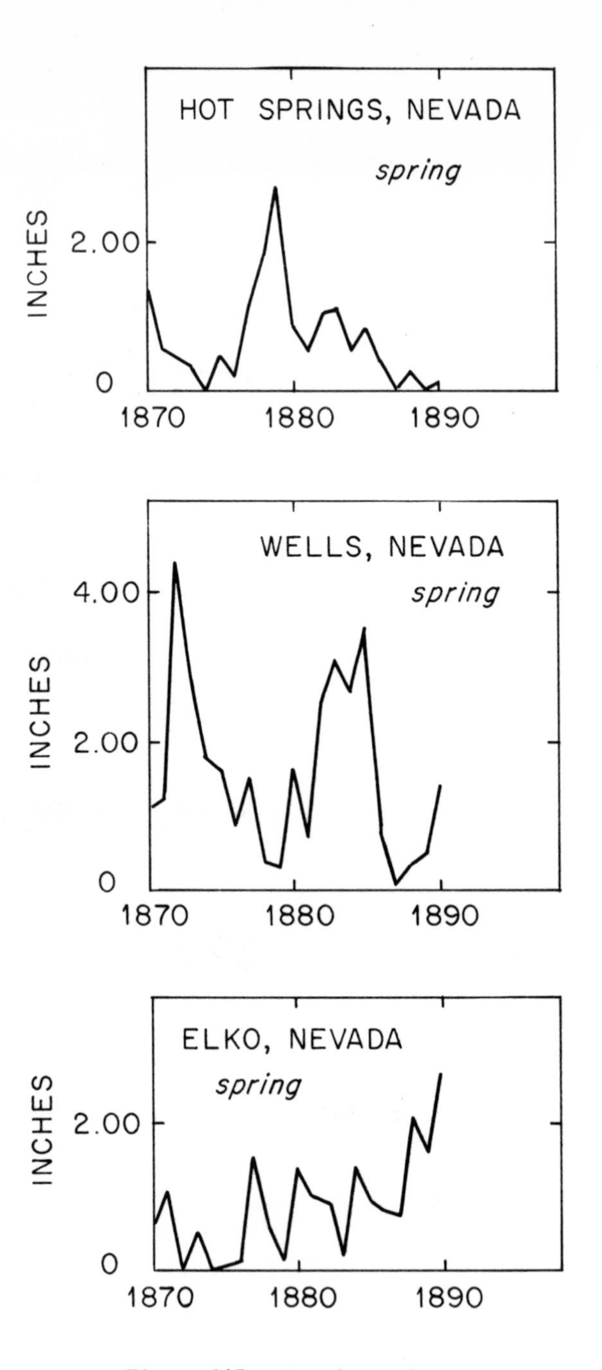

Figure 147: Nevada spring precipitation

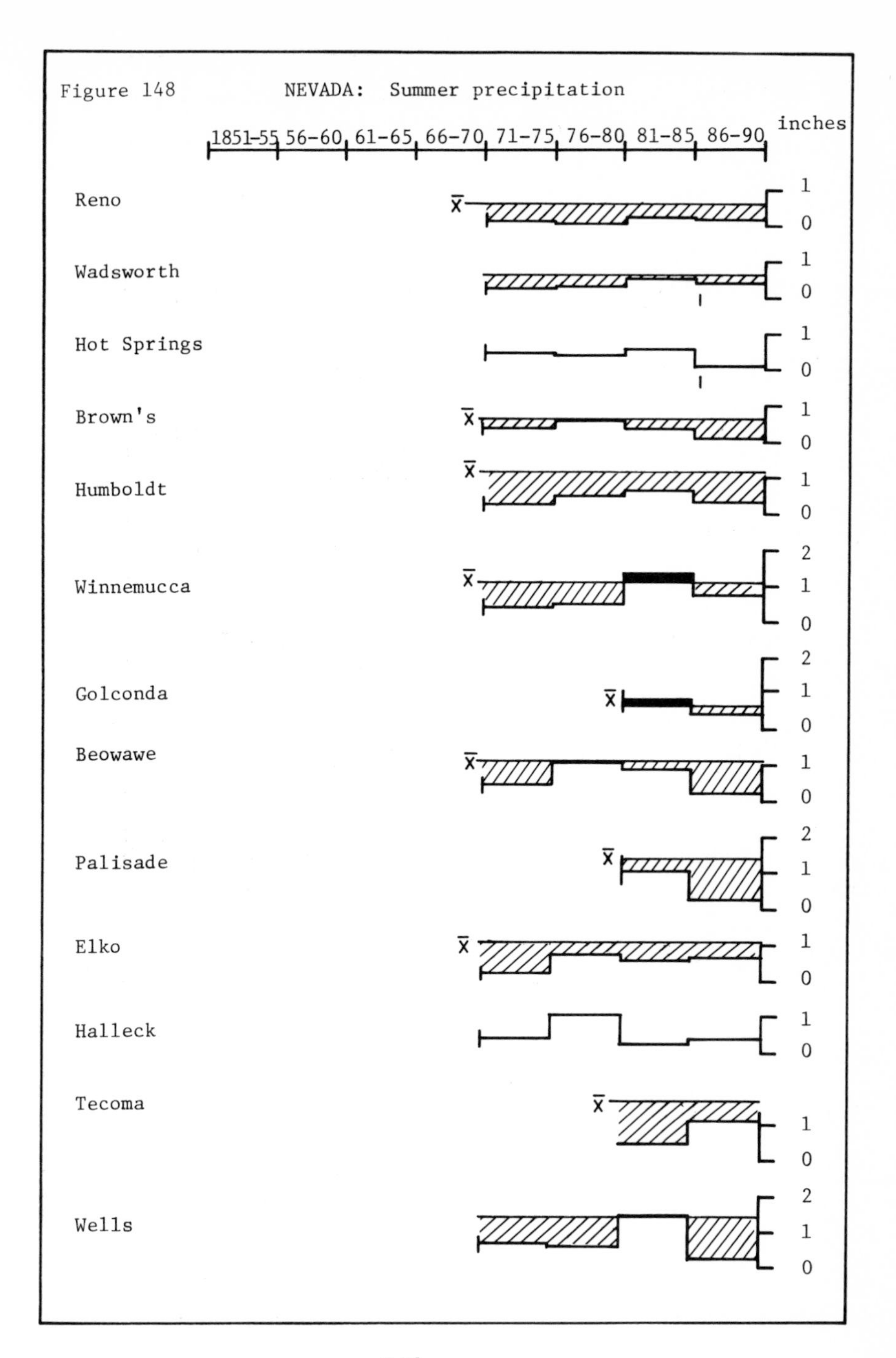
Figure 148
NEVADA: Summer precipitation
1851-55
56-60
61-65
66-70
71-75
76-80
81-85
86-90
inches
Reno
Wadsworth
Hot Springs
Brown's
Humboldt
Winnemucca
Golconda
Beowawe
Palisade
Elko
Halleck
Tecoma
Wells
x̄
2
1
0

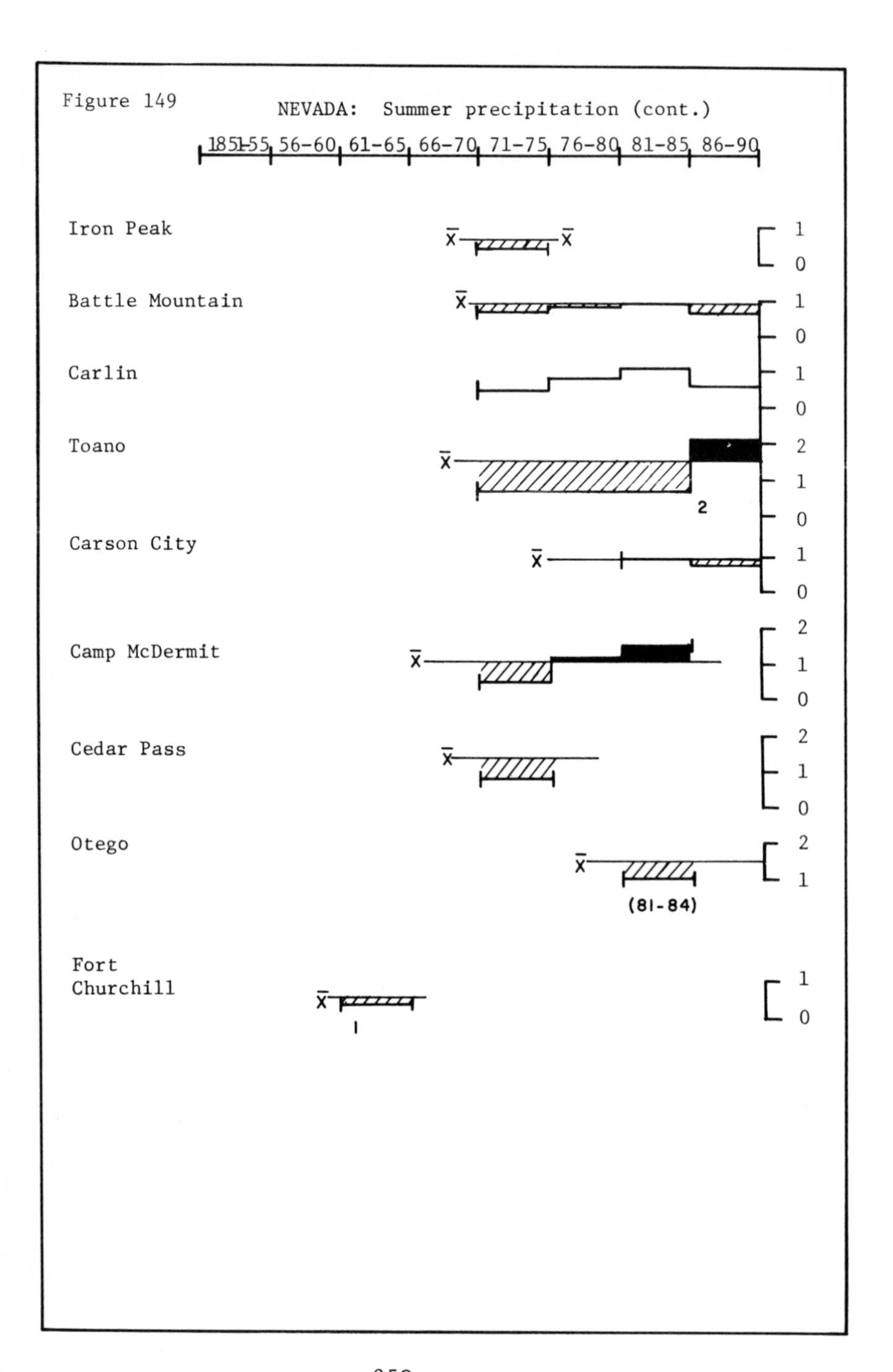

Figure 149
NEVADA: Summer precipitation (cont.)
1851-55
56-60
61-65
66-70
71-75
76-80
81-85
86-90
Iron Peak
Battle Mountain
Carlin
Toano
Carson City
Camp McDermit
Cedar Pass
Otego
(81-84)
Fort Churchill

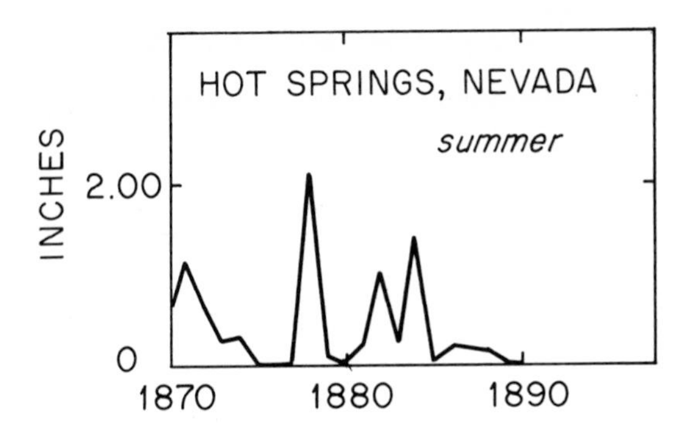

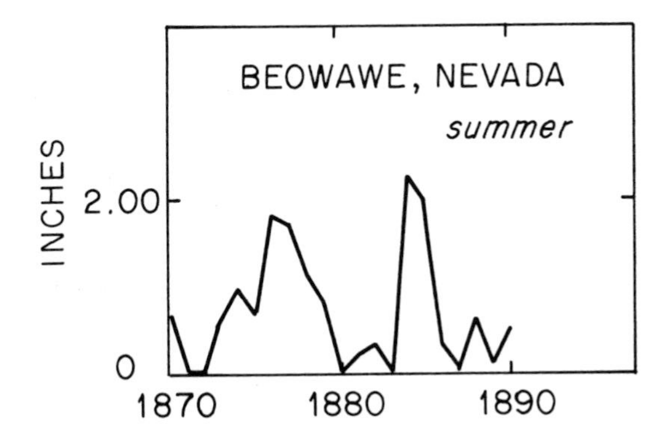

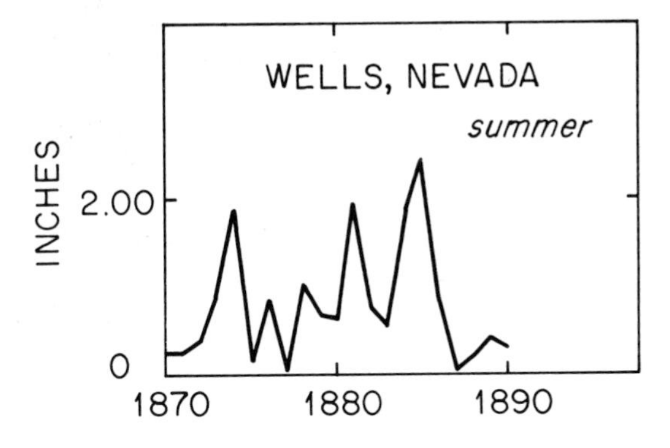

Figure 150: Nevada summer precipitation

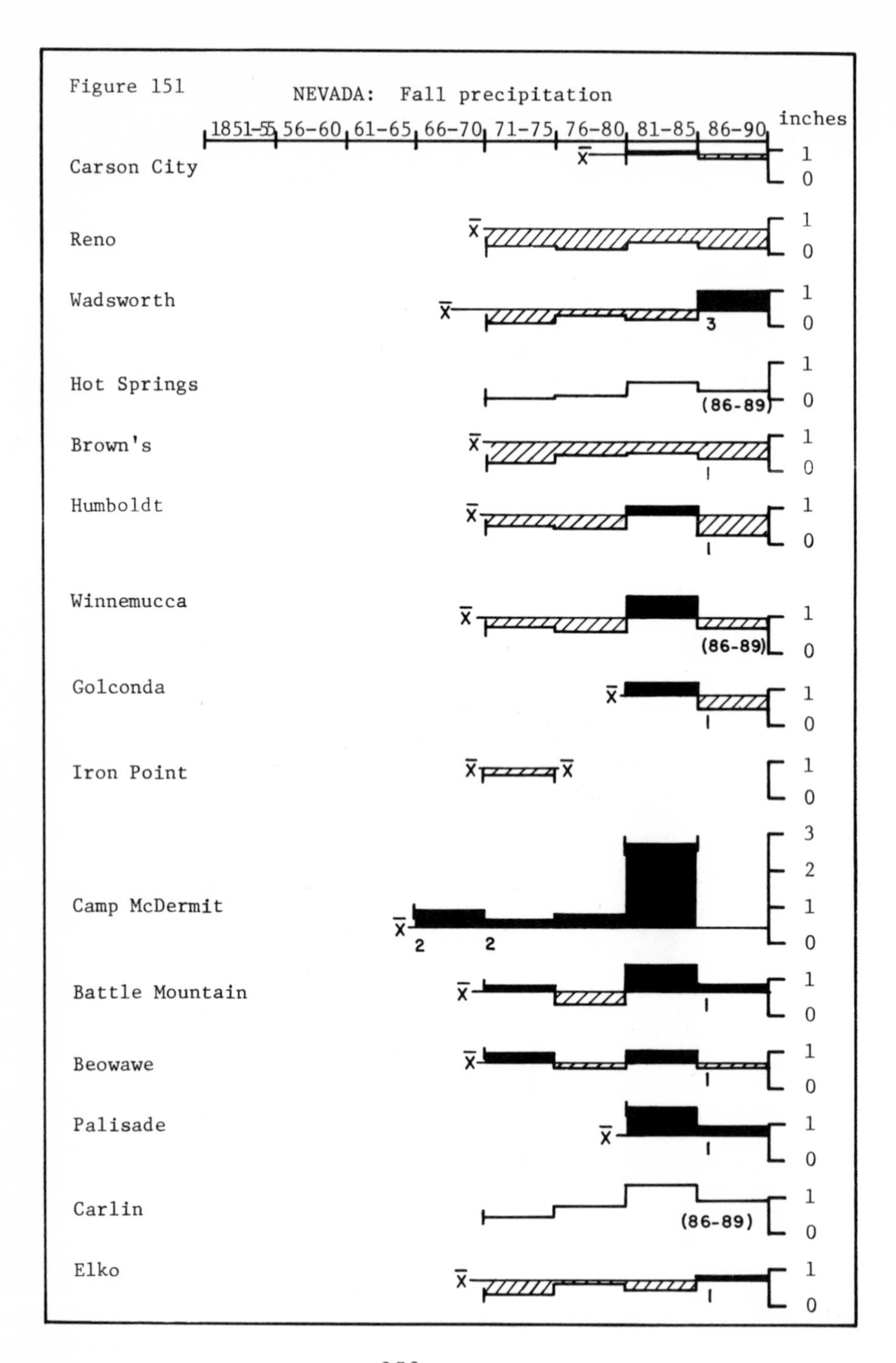

Figure 151
NEVADA: Fall precipitation
1851-55
56-60
61-65
66-70
71-75
76-80
81-85
86-90
inches
Carson City
Reno
Wadsworth
Hot Springs
(86-89)
Brown's
Humboldt
Winnemucca
(86-89)
Golconda
Iron Point
Camp McDermit
Battle Mountain
Beowawe
Palisade
Carlin
(86-89)
Elko

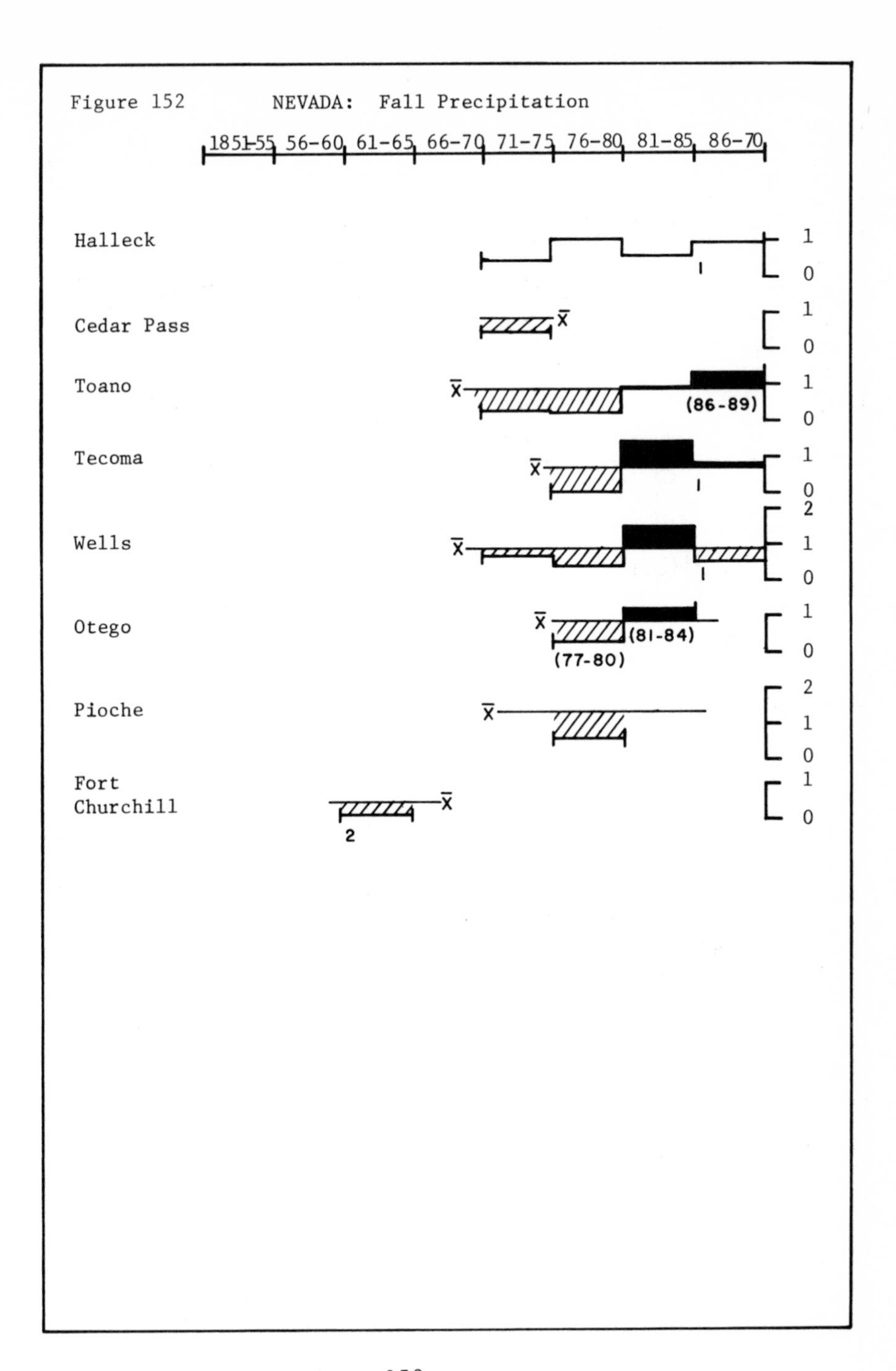
Figure 152
NEVADA: Fall Precipitation
1851-55
56-60
61-65
66-70
71-75
76-80
81-85
86-70
Halleck
Cedar Pass
Toano
(86-89)
Tecoma
Wells
Otego
(81-84)
(77-80)
Pioche
Fort Churchill
2
1
0

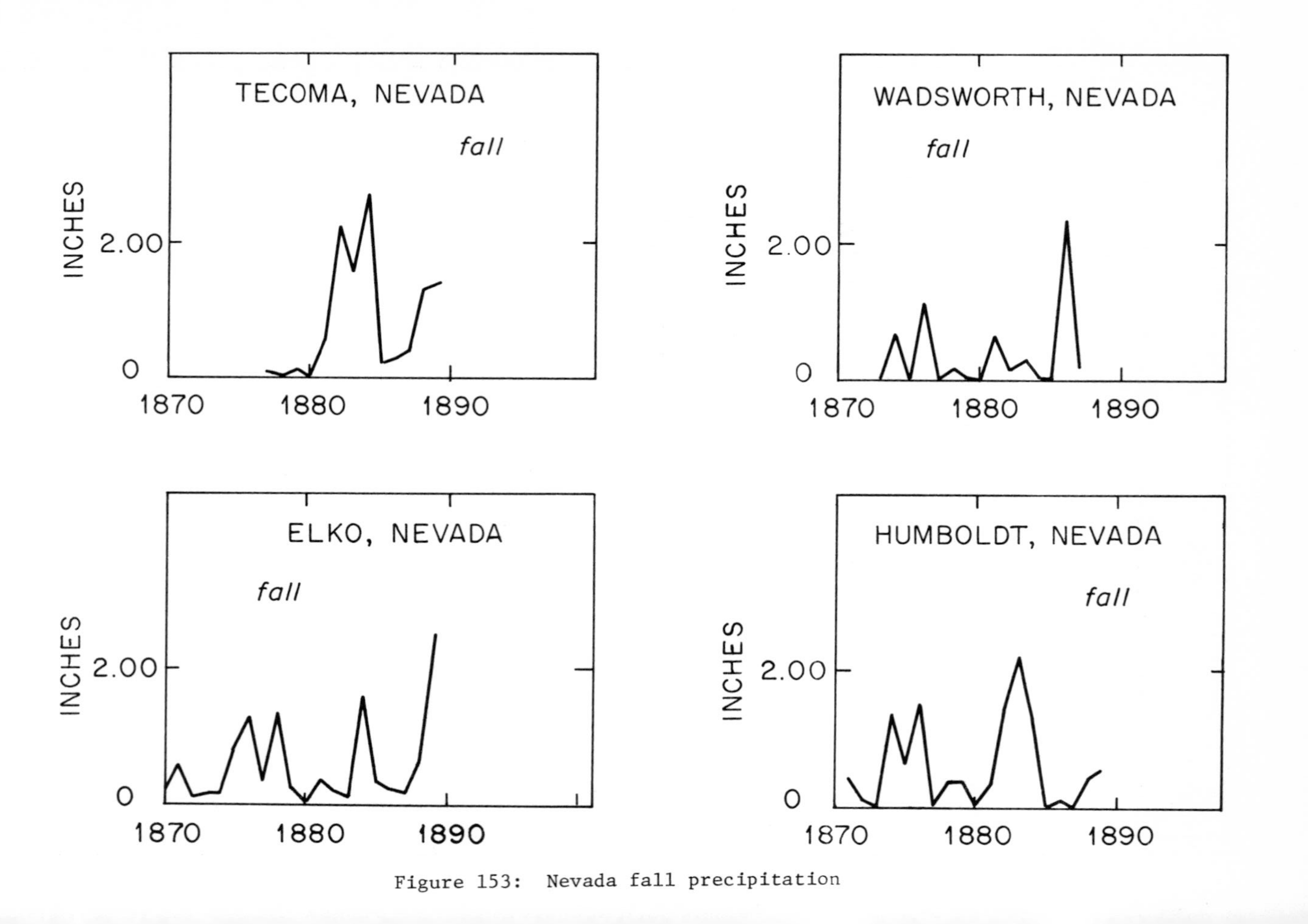

Figure 153: Nevada fall precipitation

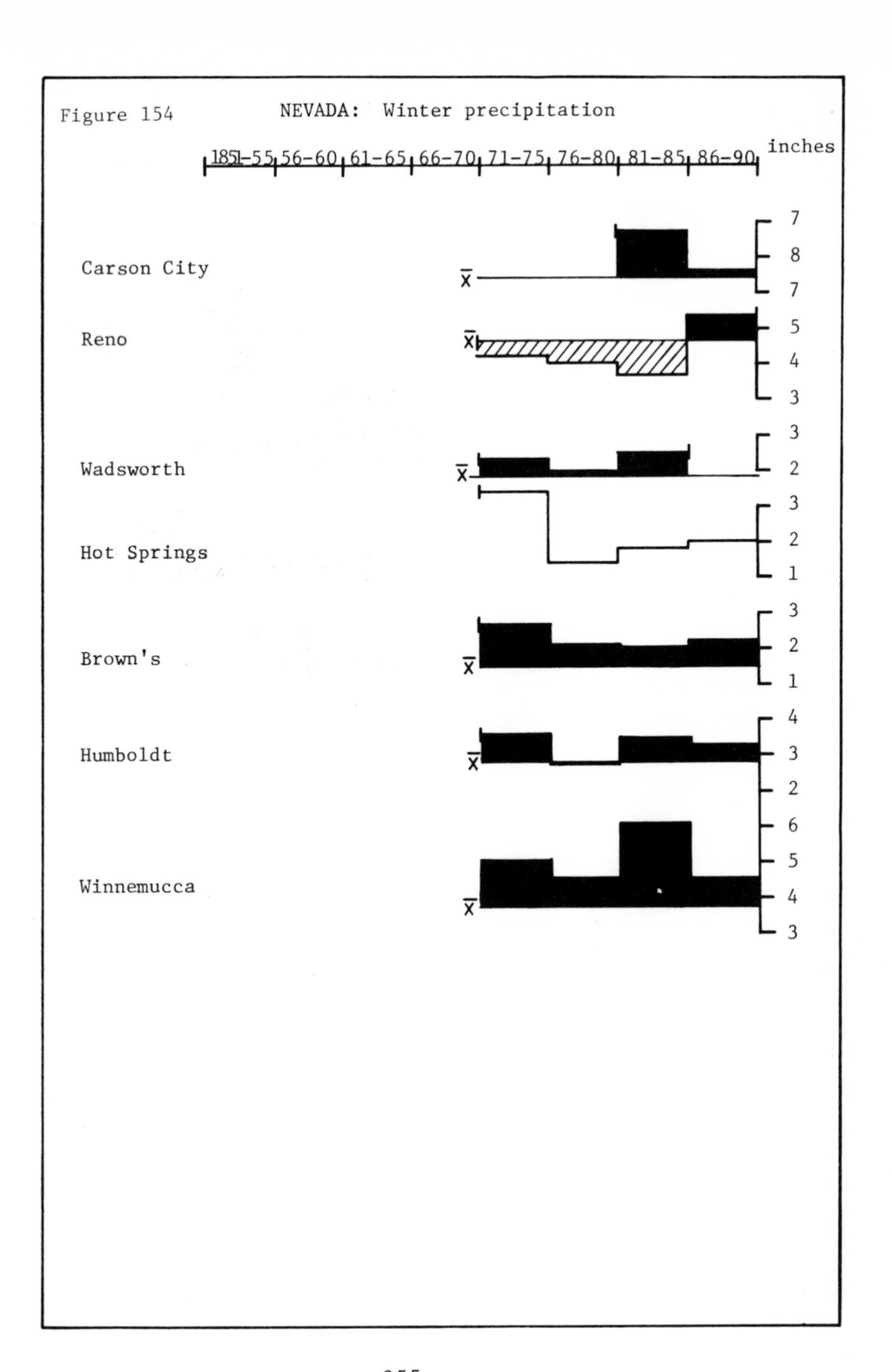
Figure 154
NEVADA: Winter precipitation
1851-55 56-60 61-65 66-70 71-75 76-80 81-85 86-90
inches
Carson City
Reno
Wadsworth
Hot Springs
Brown's
Humboldt
Winnemucca
7
8
7
5
4
3
3
2
3
2
1
3
2
1
4
3
2
6
5
4
3

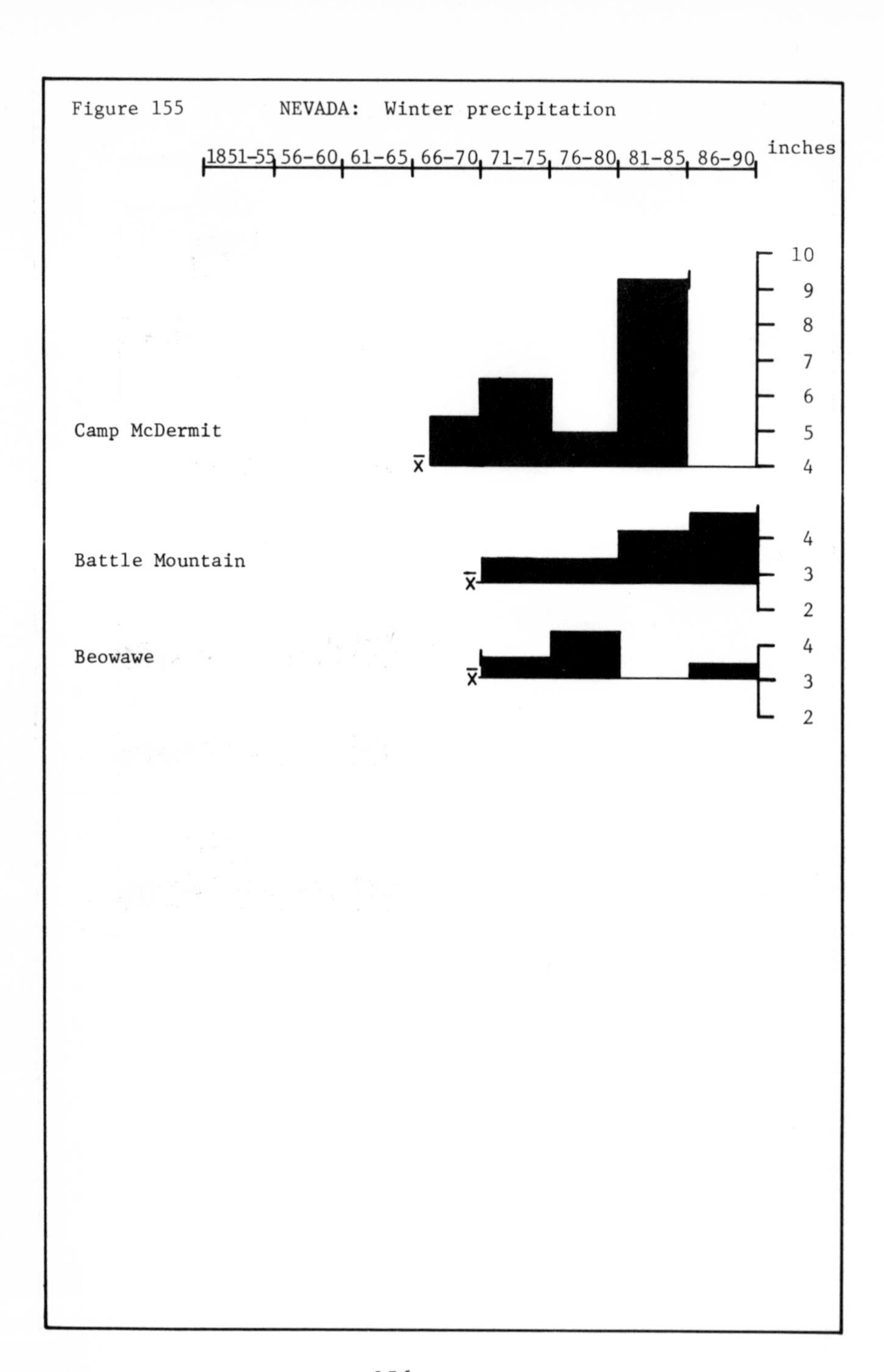
Figure 155
NEVADA: Winter precipitation
1851-55
56-60
61-65
66-70
71-75
76-80
81-85
86-90
inches
10
9
8
7
6
5
4
Camp McDermit
x̄
4
3
2
Battle Mountain
x̄
4
3
2
Beowawe
x̄

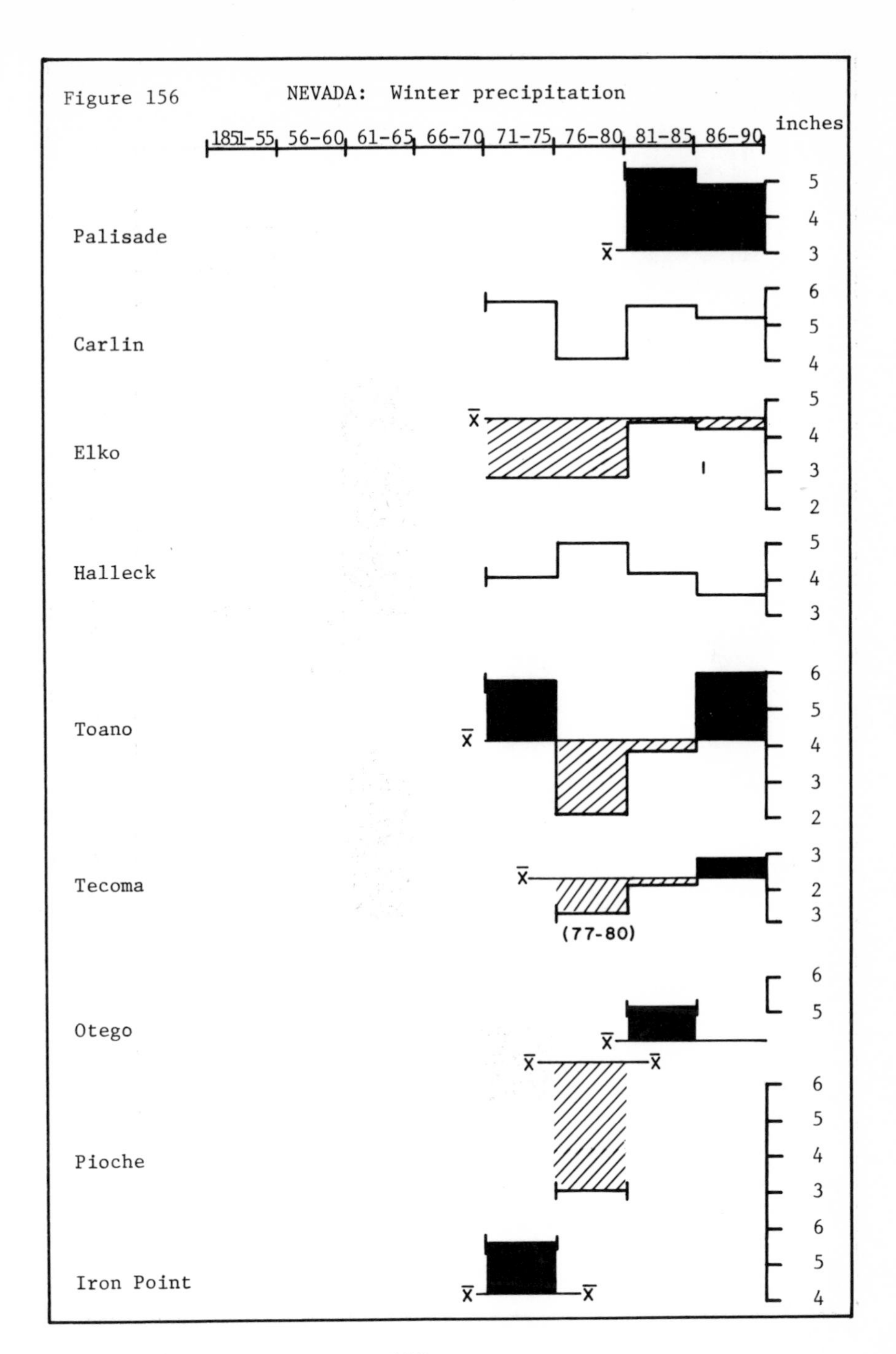

Figure 156
NEVADA: Winter precipitation
1851-55
56-60
61-65
66-70
71-75
76-80
81-85
86-90
inches
Palisade
Carlin
Elko
Halleck
Toano
Tecoma
(77-80)
Otego
Pioche
Iron Point

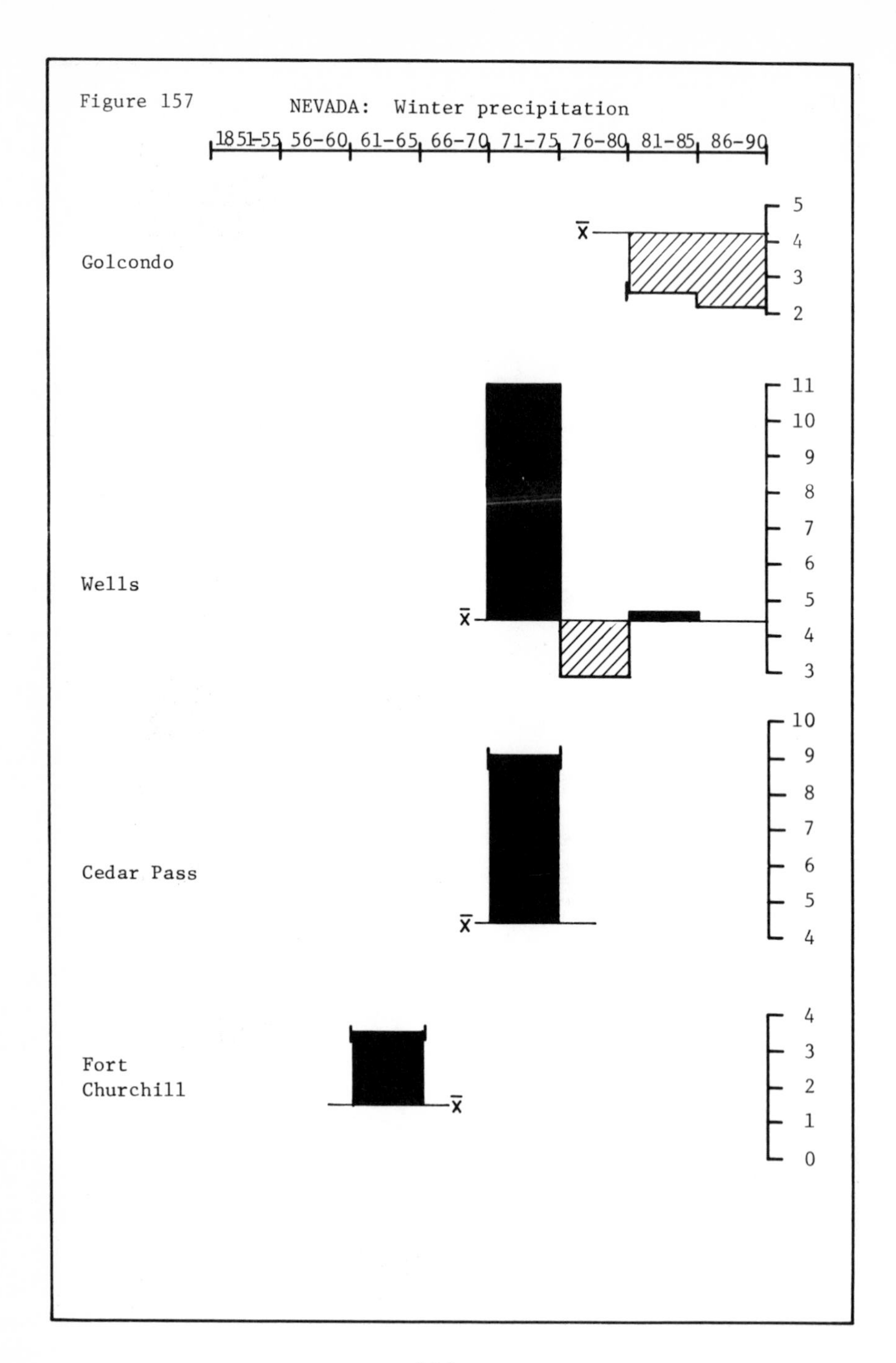
Figure 157
NEVADA: Winter precipitation
1851-55
56-60
61-65
66-70
71-75
76-80
81-85
86-90
Golcondo
Wells
Cedar Pass
Fort Churchill
x̄
5
4
3
2
11
10
9
8
7
6
5
4
3
10
9
8
7
6
5
4
4
3
2
1
0

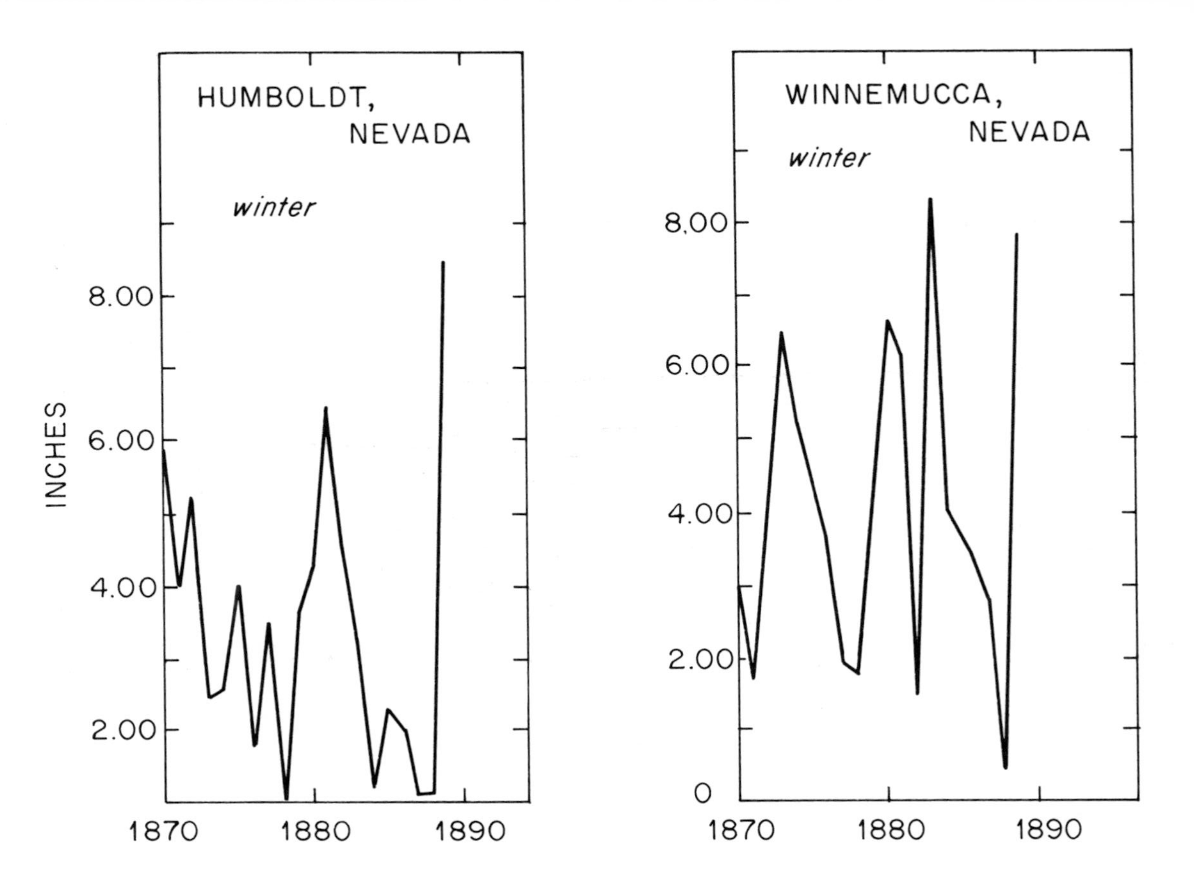

Figure 158: Nevada winter precipitation

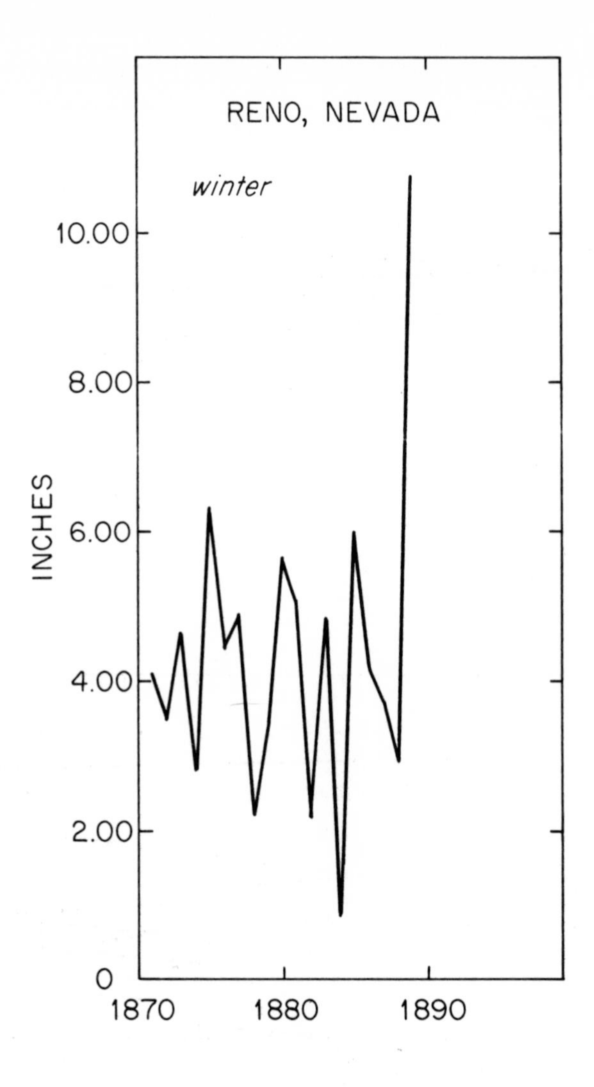

Figure 159: Nevada winter precipitation

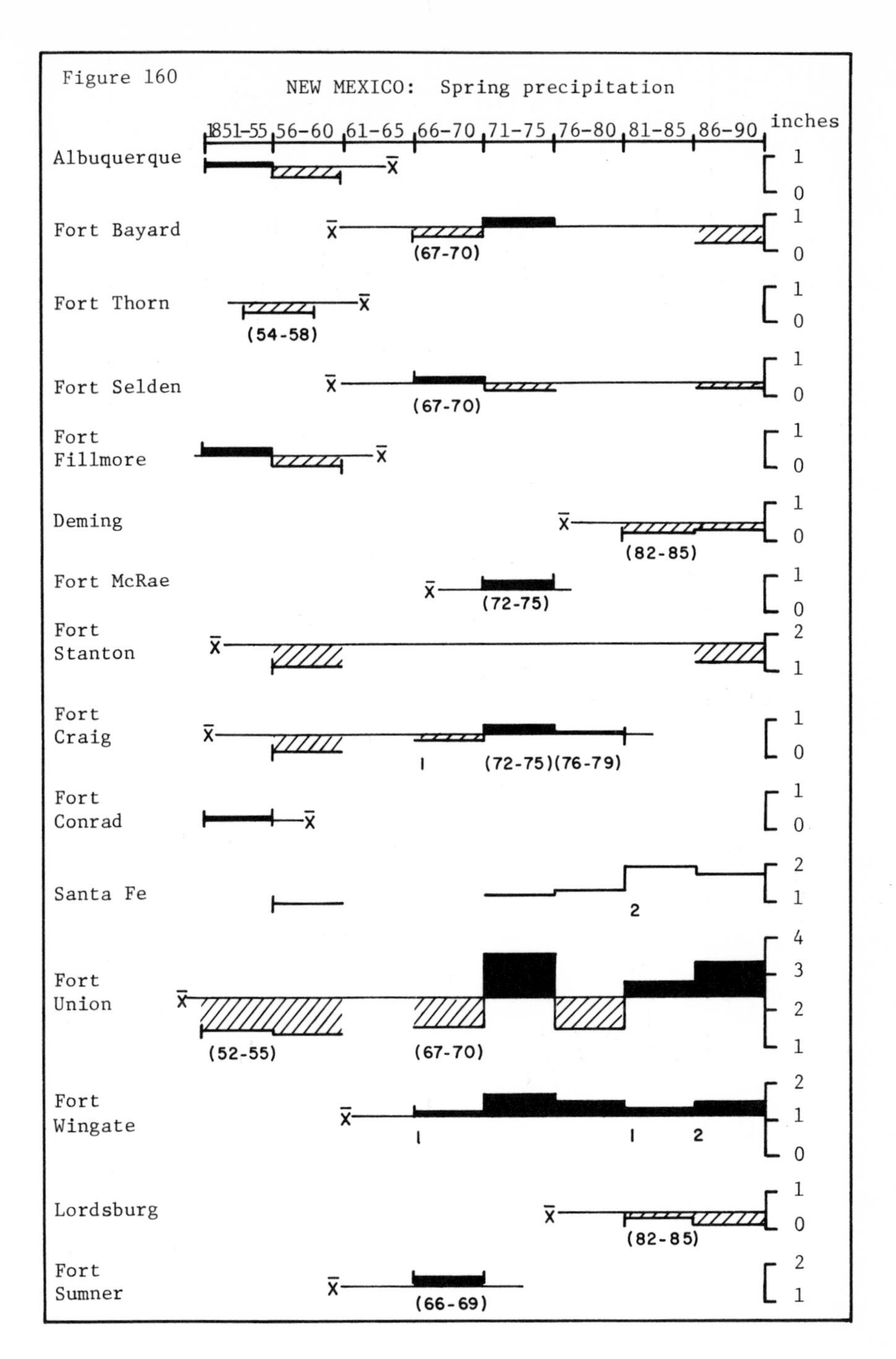

Figure 160
NEW MEXICO: Spring precipitation
1851-55
56-60
61-65
66-70
71-75
76-80
81-85
86-90
inches
Albuquerque
Fort Bayard
(67-70)
Fort Thorn
(54-58)
Fort Selden
(67-70)
Fort Fillmore
Deming
(82-85)
Fort McRae
(72-75)
Fort Stanton
Fort Craig
(72-75)(76-79)
Fort Conrad
Santa Fe
Fort Union
(52-55)
(67-70)
Fort Wingate
Lordsburg
(82-85)
Fort Sumner
(66-69)

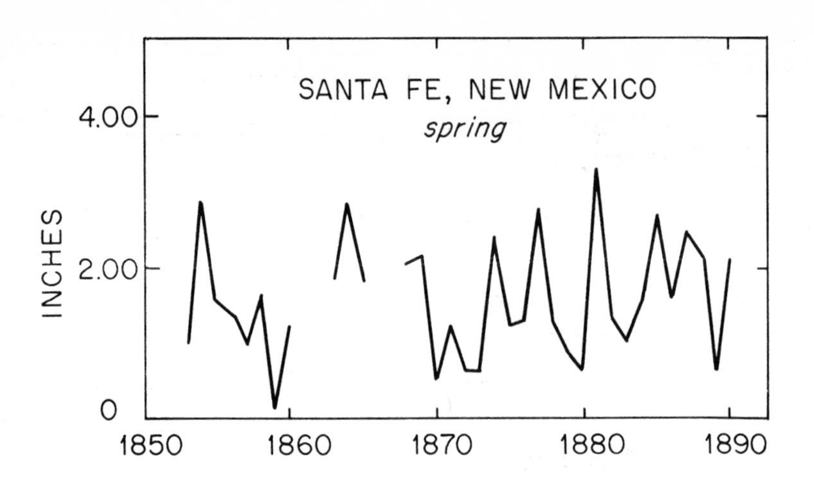

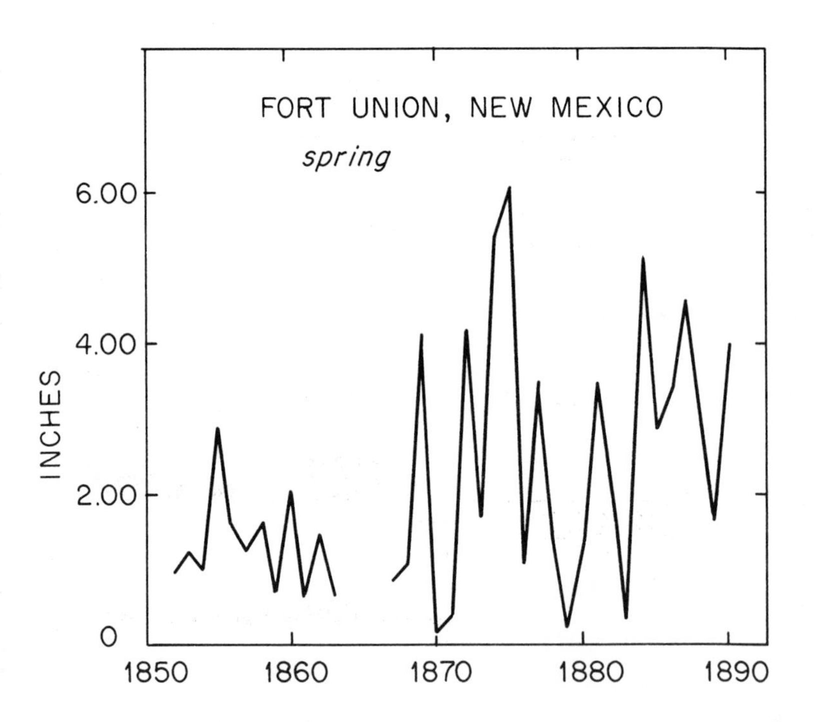

Figure 161: New Mexico spring precipitation

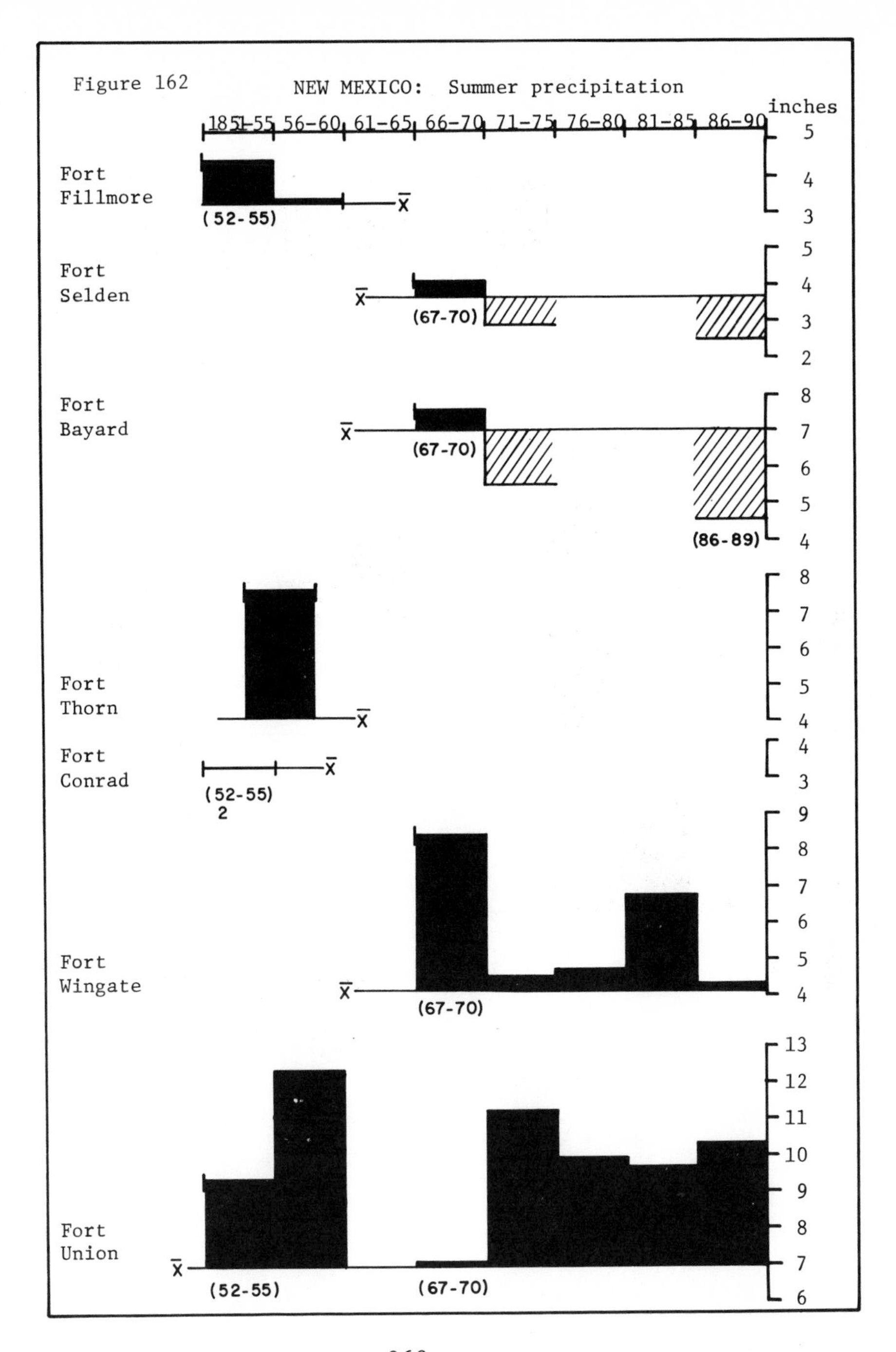

Figure 162
NEW MEXICO: Summer precipitation
inches
1851-55 56-60 61-65 66-70 71-75 76-80 81-85 86-90
Fort Fillmore
(52-55)
Fort Selden
(67-70)
Fort Bayard
(67-70)
(86-89)
Fort Thorn
Fort Conrad
(52-55)
2
Fort Wingate
(67-70)
Fort Union
(52-55)
(67-70)

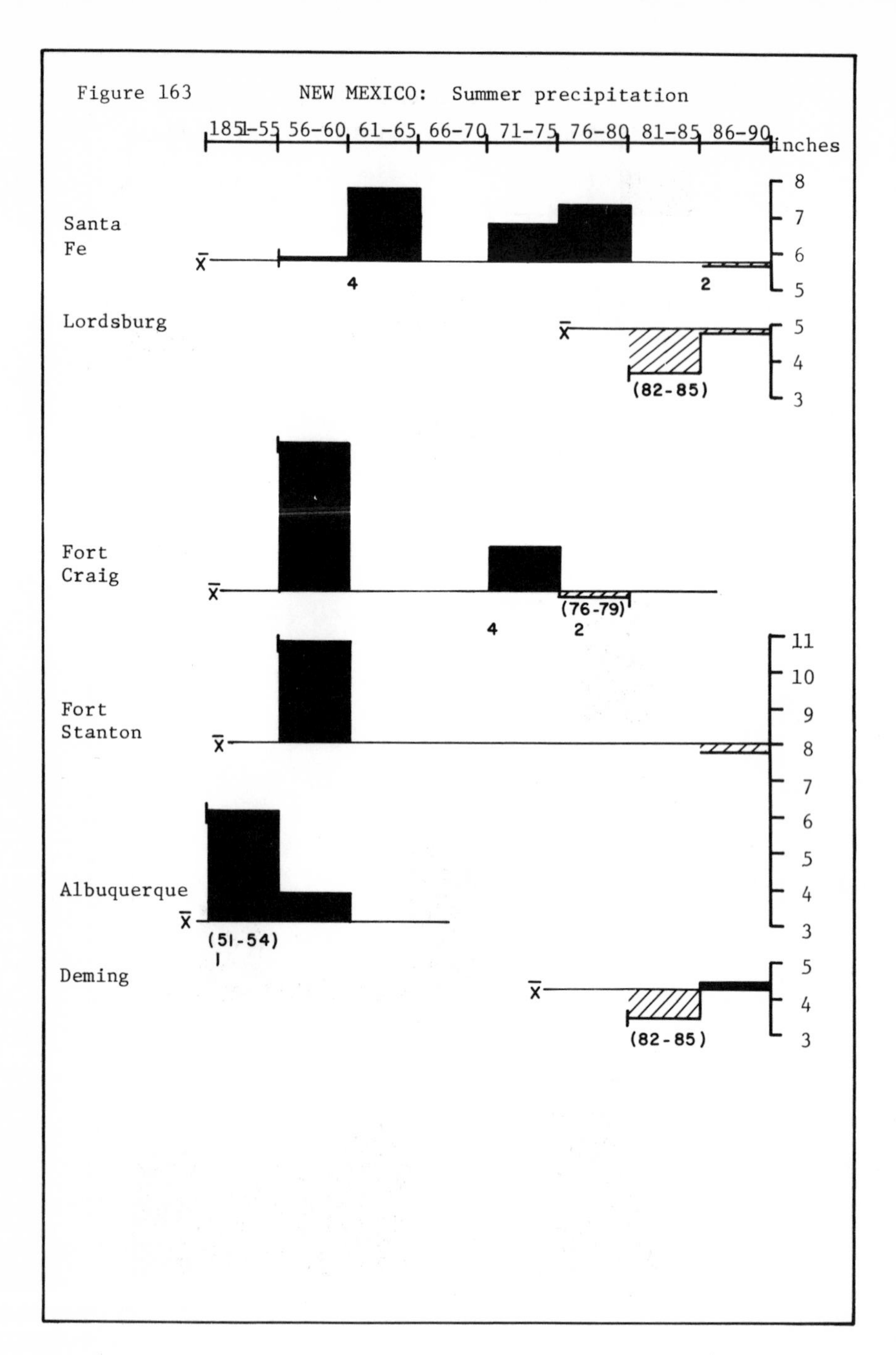
Figure 163
NEW MEXICO: Summer precipitation
1851-55
56-60
61-65
66-70
71-75
76-80
81-85
86-90
inches
Santa Fe
4
2
8
7
6
5
Lordsburg
(82-85)
5
4
3
Fort Craig
(76-79)
4
2
Fort Stanton
11
10
9
8
7
6
5
4
3
Albuquerque
(51-54)
1
Deming
(82-85)
5
4
3

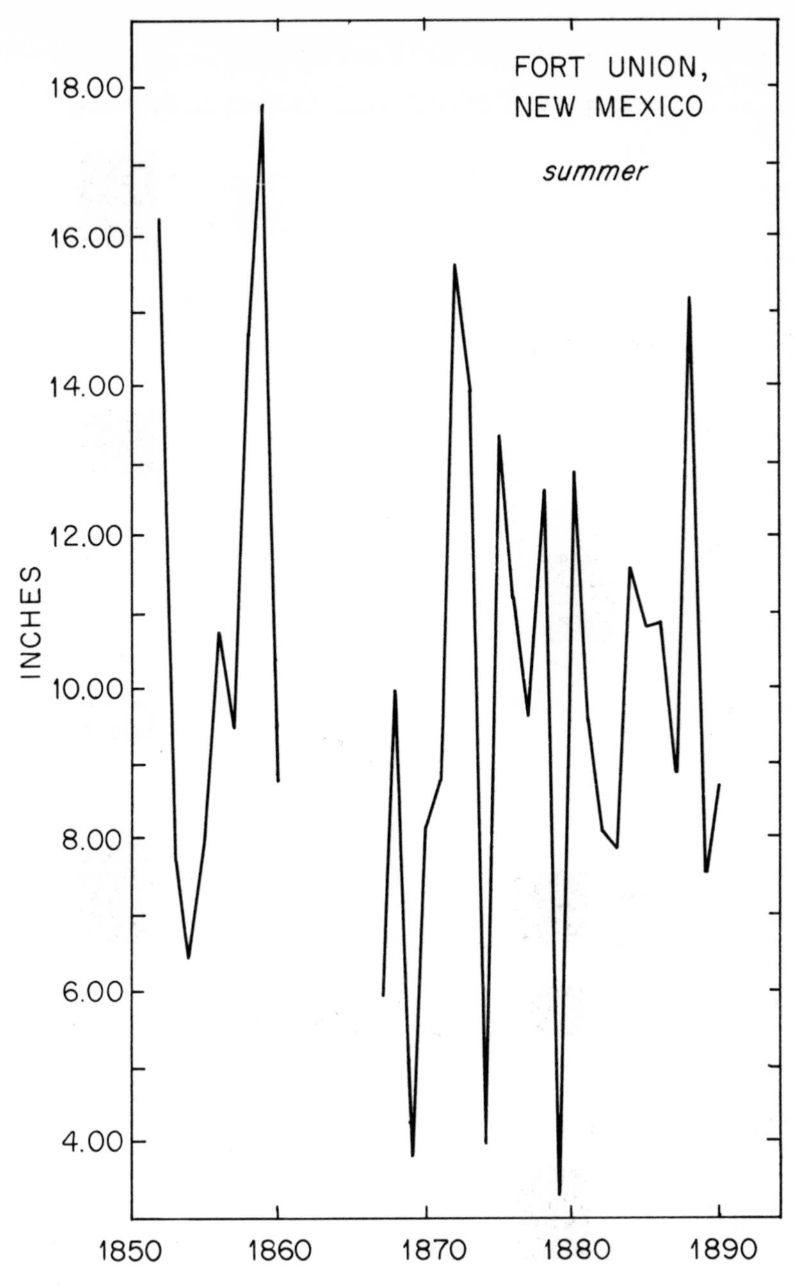

Figure 164: New Mexico summer precipitation

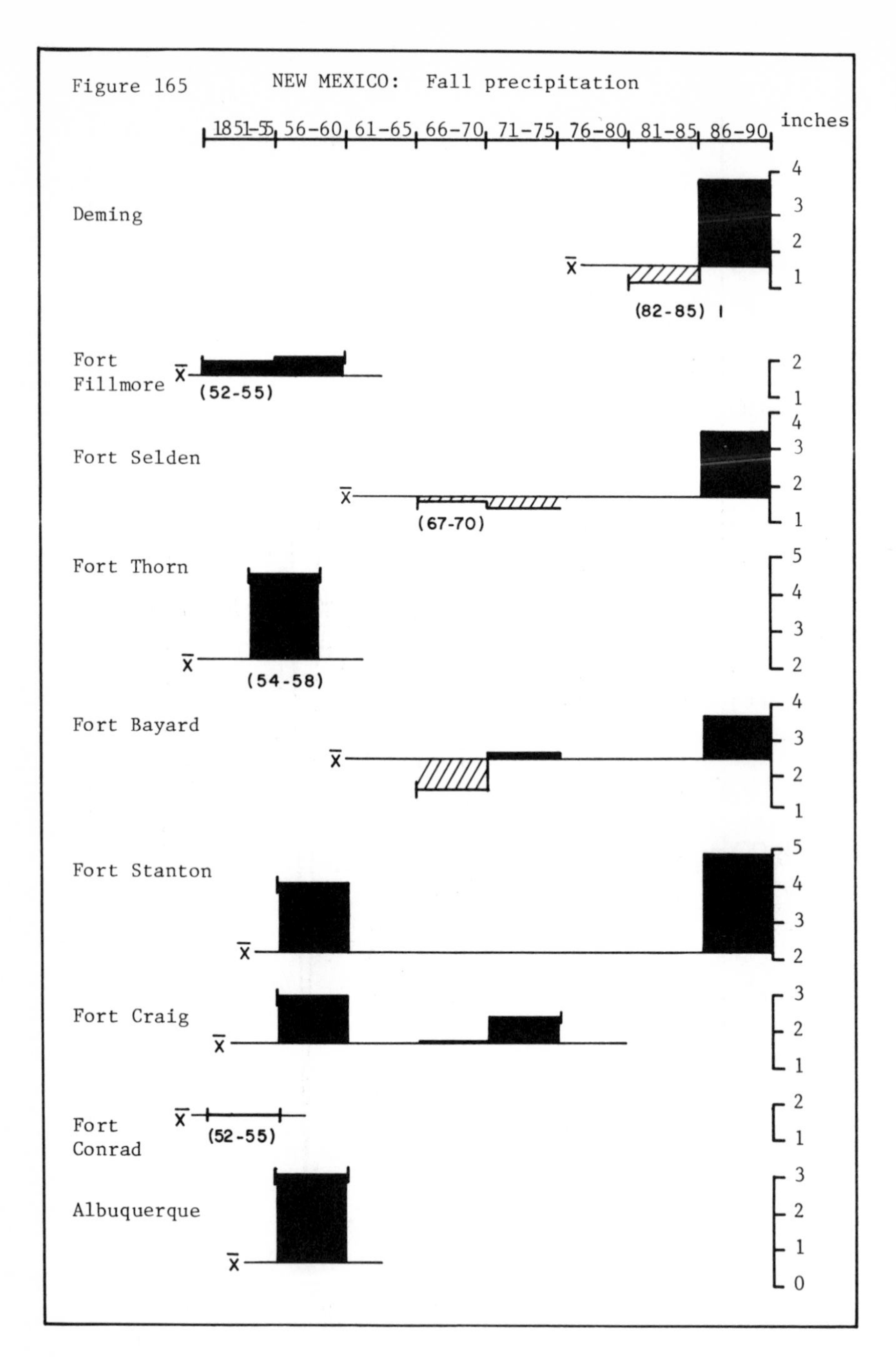
Figure 165
NEW MEXICO: Fall precipitation
1851-55
56-60
61-65
66-70
71-75
76-80
81-85
86-90
inches
Deming
(82-85)
Fort Fillmore
(52-55)
Fort Selden
(67-70)
Fort Thorn
(54-58)
Fort Bayard
Fort Stanton
Fort Craig
Fort Conrad
(52-55)
Albuquerque

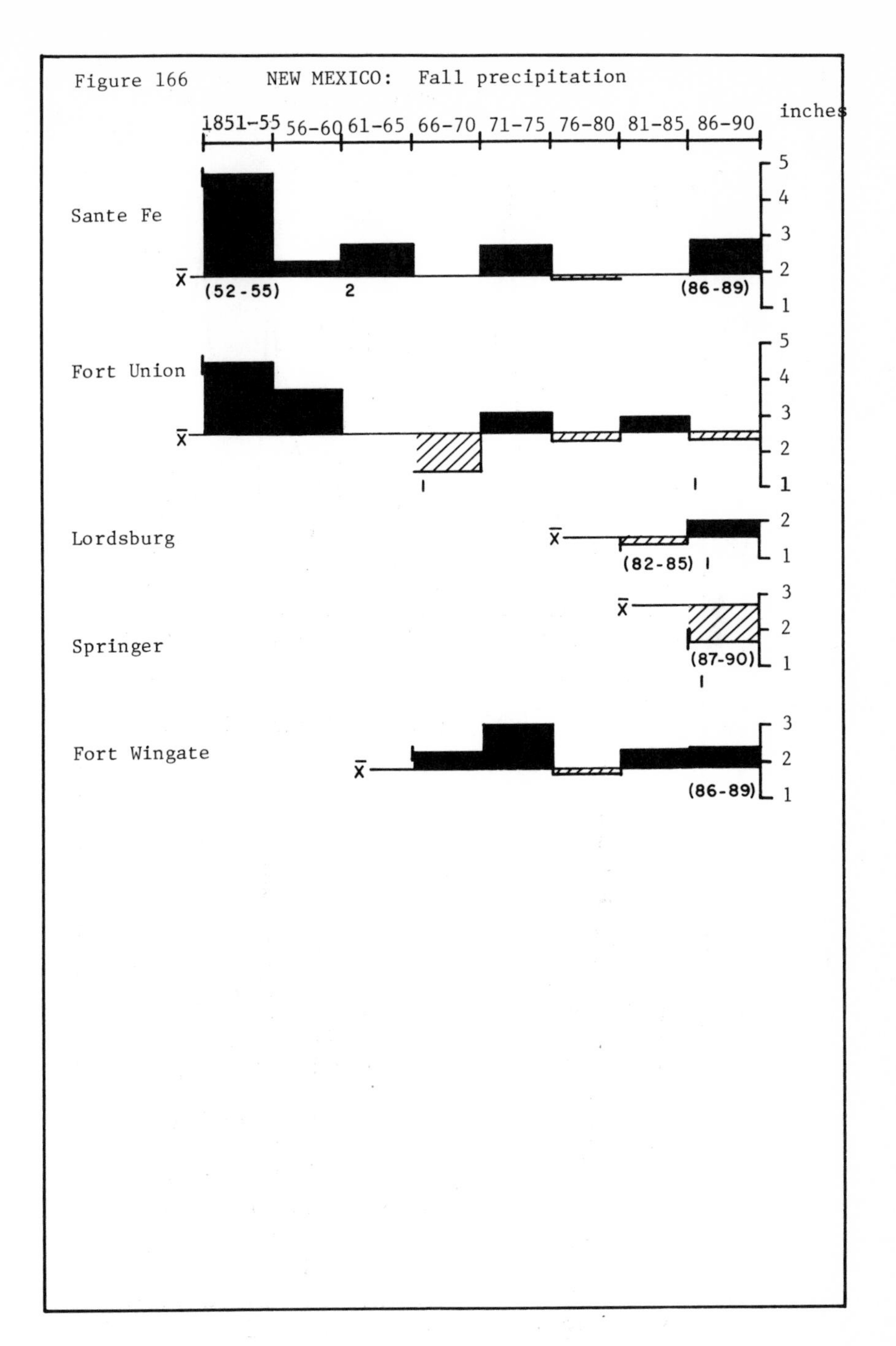
Figure 166
NEW MEXICO: Fall precipitation
1851-55
56-60
61-65
66-70
71-75
76-80
81-85
86-90
inches
Sante Fe
(52-55)
2
(86-89)
Fort Union
Lordsburg
(82-85)
Springer
(87-90)
Fort Wingate
(86-89)
5
4
3
2
1

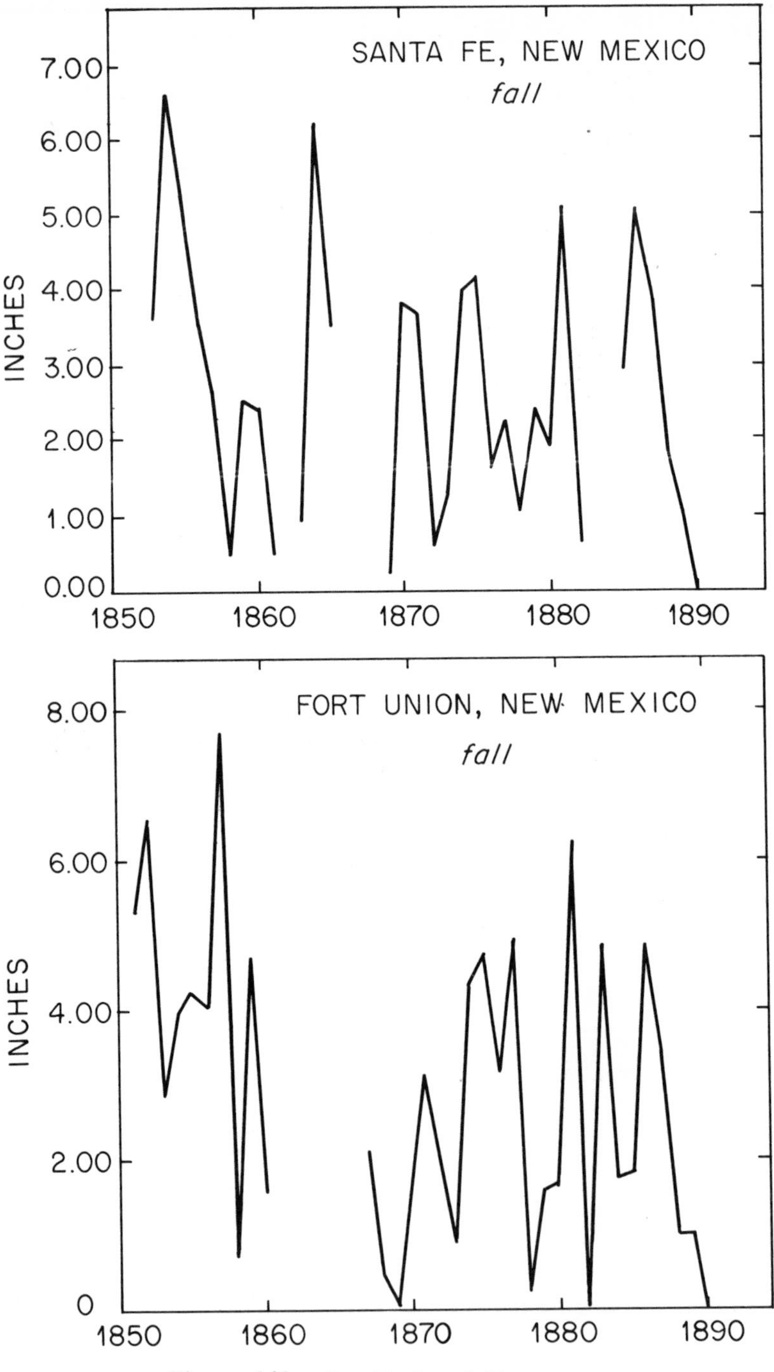

Figure 167: New Mexico fall precipitation

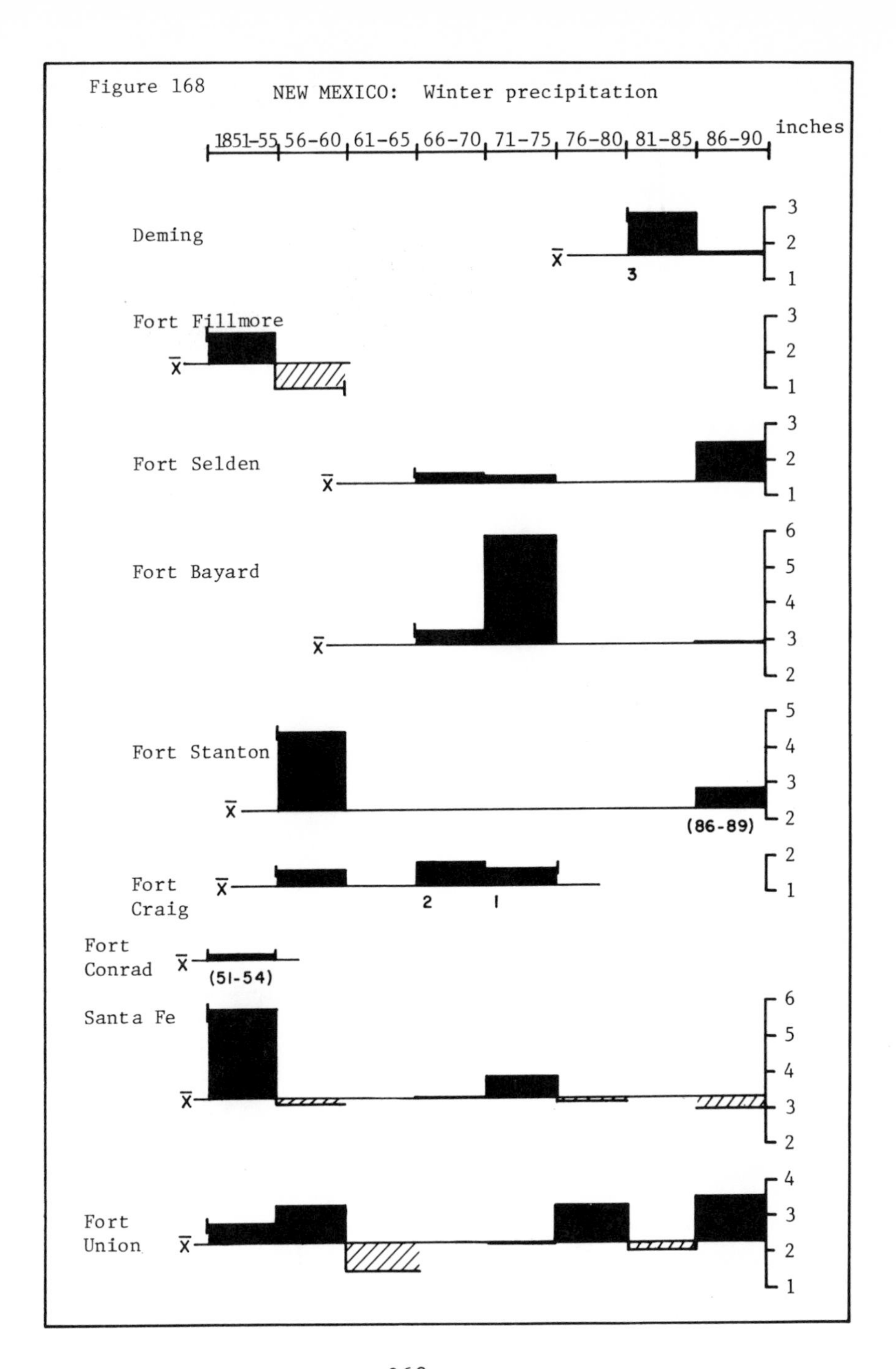

Figure 168
NEW MEXICO: Winter precipitation
1851-55
56-60
61-65
66-70
71-75
76-80
81-85
86-90
inches
Deming
3
Fort Fillmore
Fort Selden
Fort Bayard
Fort Stanton
(86-89)
Fort Craig
2
1
Fort Conrad
(51-54)
Santa Fe
Fort Union

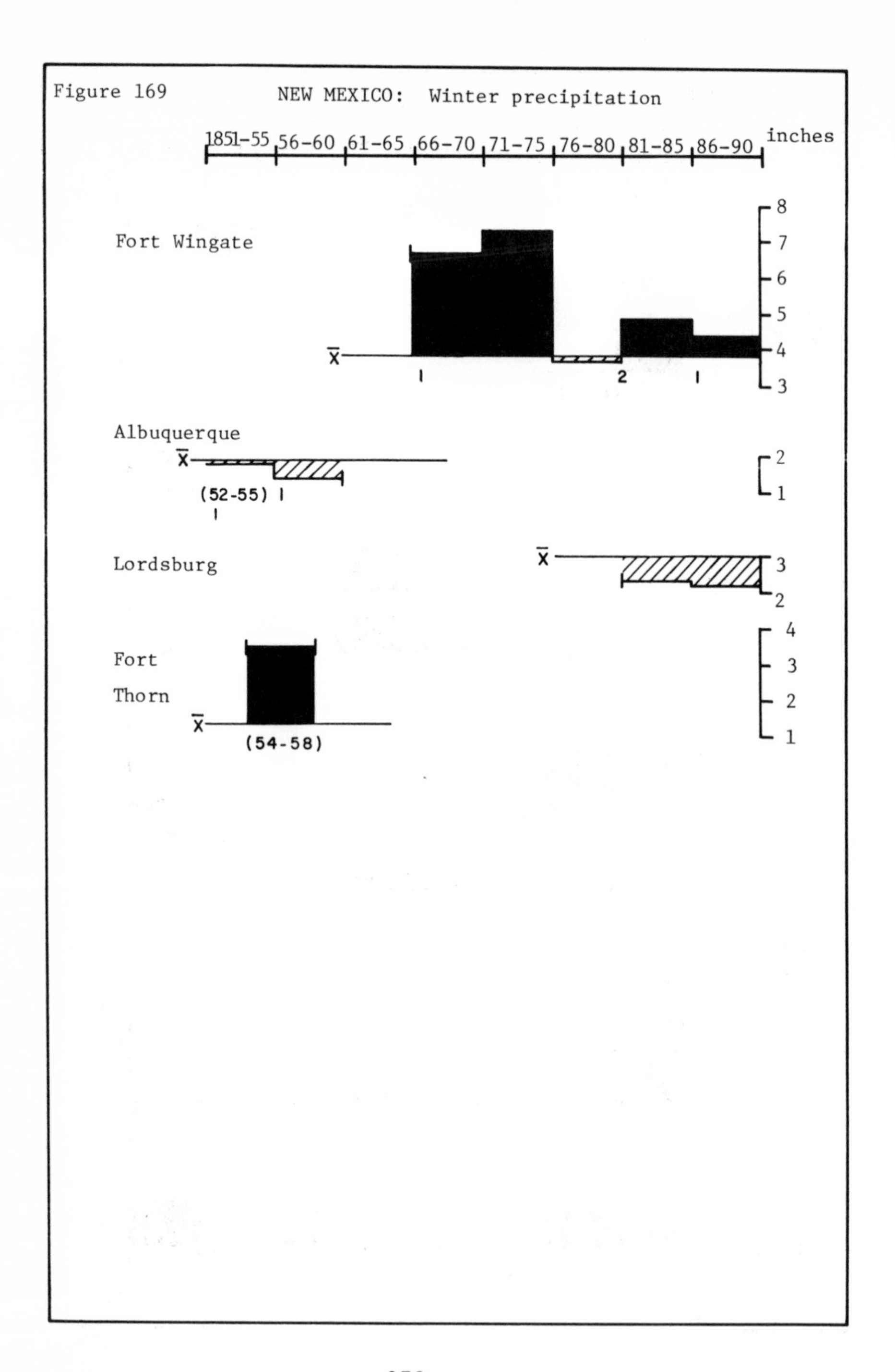
Figure 169
NEW MEXICO: Winter precipitation
1851-55
56-60
61-65
66-70
71-75
76-80
81-85
86-90
inches
Fort Wingate
Albuquerque
(52-55)
Lordsburg
Fort
Thorn
(54-58)

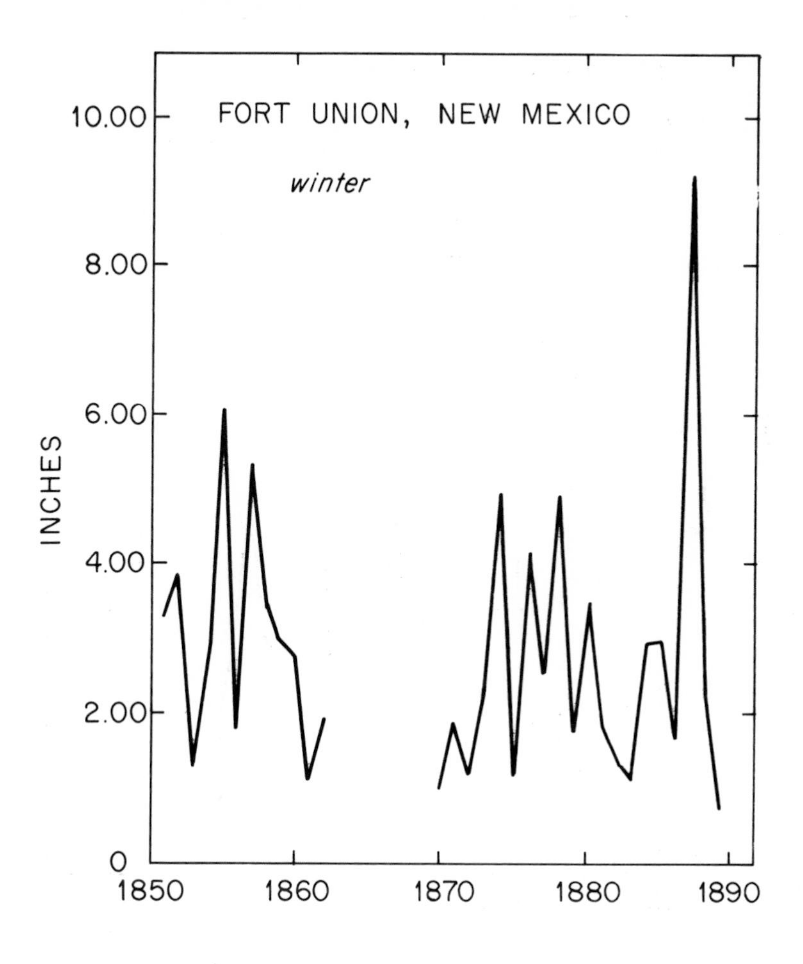

Figure 170: New Mexico winter precipitation

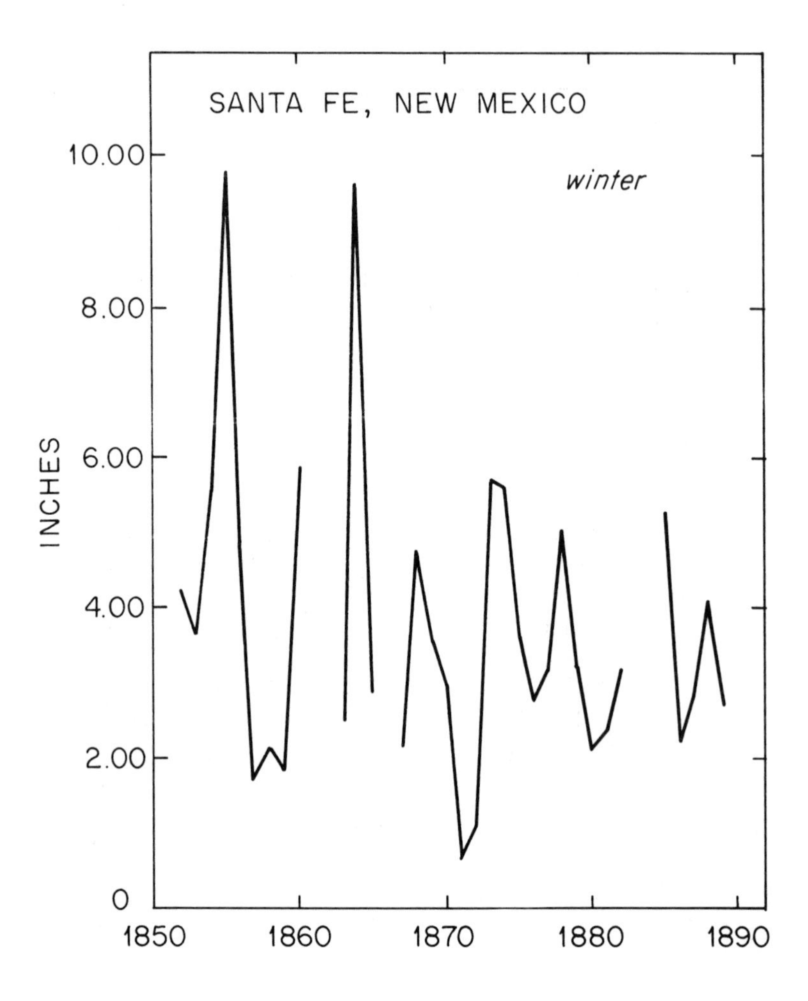

Figure 170: New Mexico winter precipitation

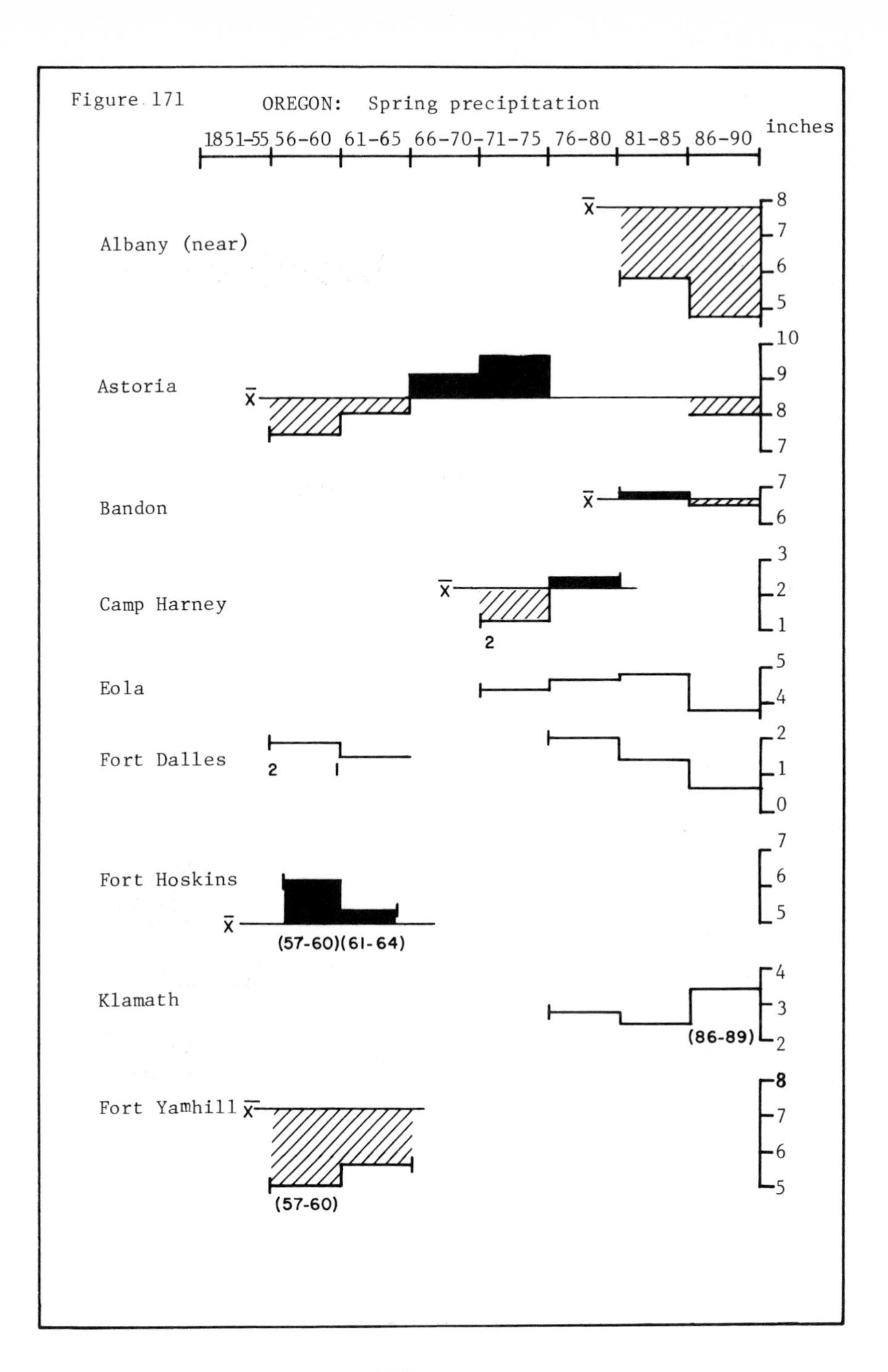
Figure 171
OREGON: Spring precipitation
1851-55
56-60
61-65
66-70
71-75
76-80
81-85
86-90
inches
Albany (near)
Astoria
Bandon
Camp Harney
Eola
Fort Dalles
Fort Hoskins
(57-60)(61-64)
Klamath
(86-89)
Fort Yamhill
(57-60)

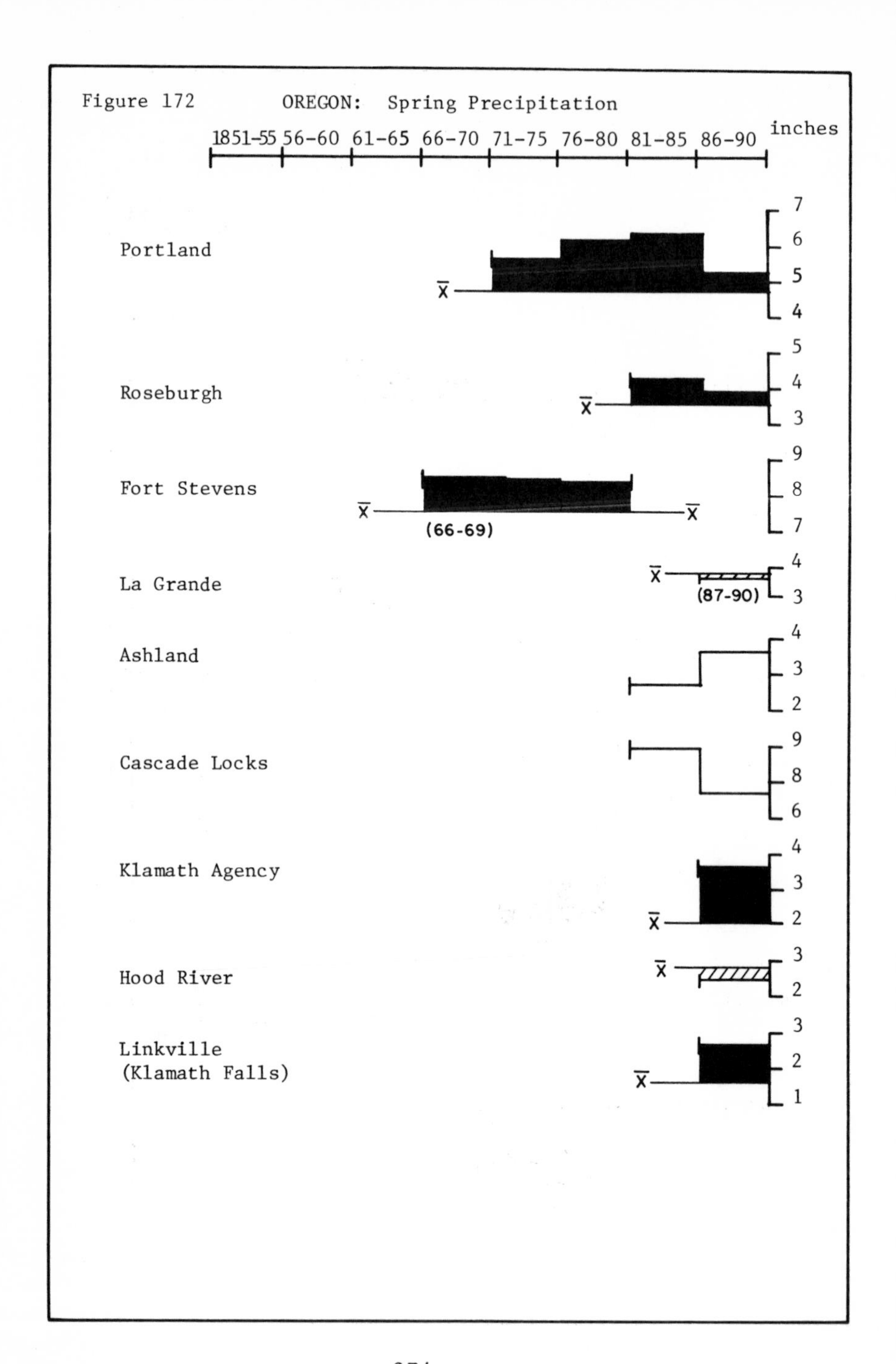

Figure 172 OREGON: Spring Precipitation
1851-55
56-60
61-65
66-70
71-75
76-80
81-85
86-90
inches
Portland
Roseburgh
Fort Stevens
(66-69)
La Grande
(87-90)
Ashland
Cascade Locks
Klamath Agency
Hood River
Linkville
(Klamath Falls)

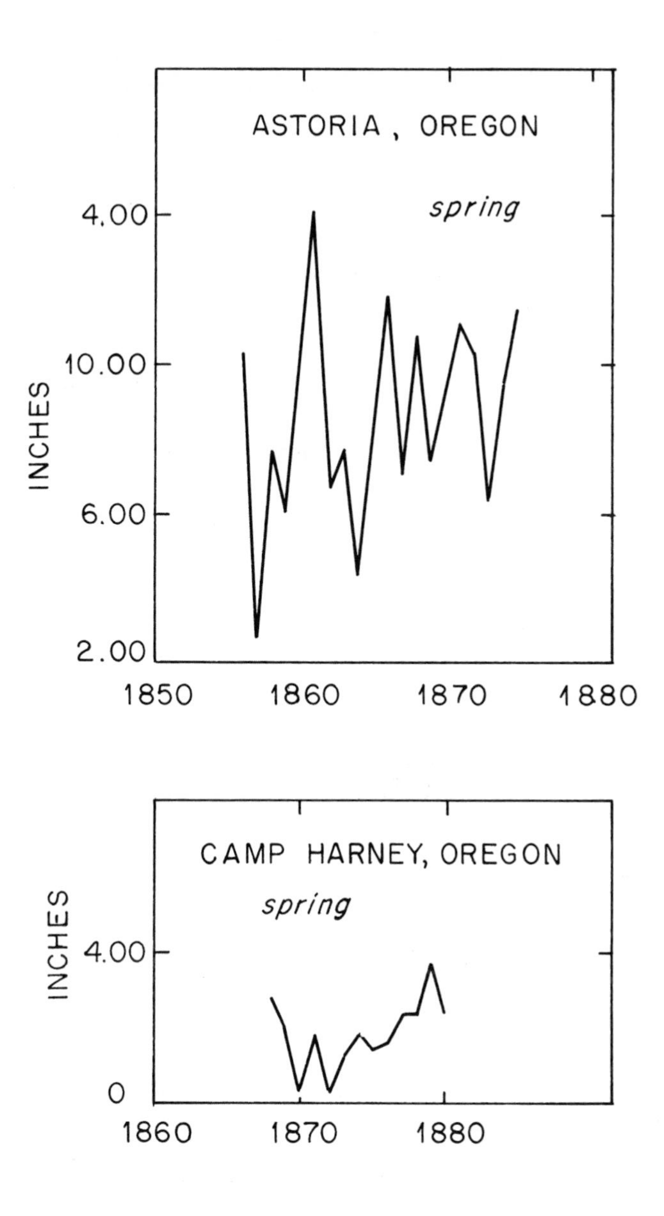

Figure 173: Oregon spring precipitation

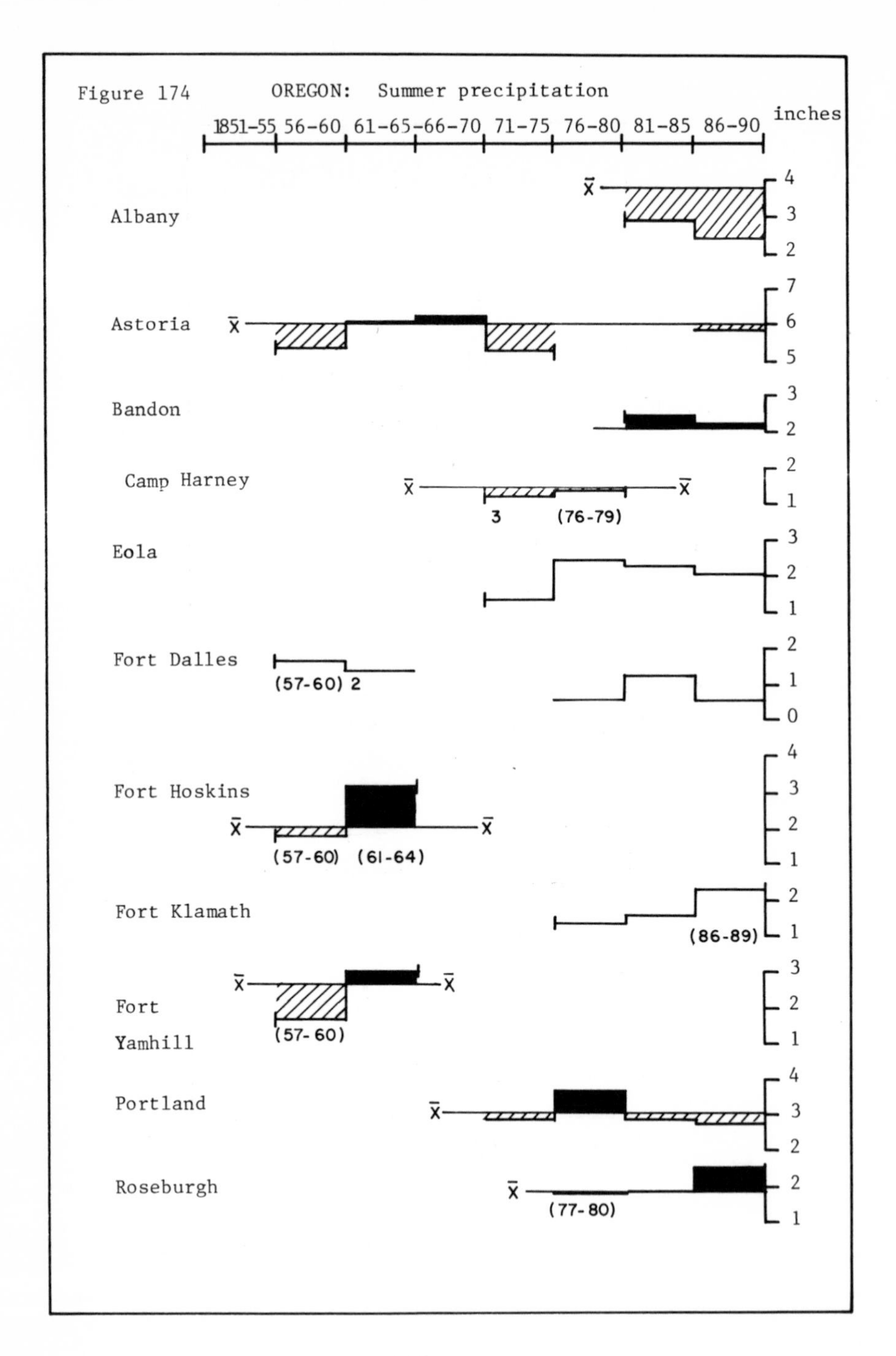
Figure 174 OREGON: Summer precipitation
1851-55
56-60
61-65
66-70
71-75
76-80
81-85
86-90
inches
Albany
Astoria
Bandon
Camp Harney
3
(76-79)
Eola
Fort Dalles
(57-60) 2
Fort Hoskins
(57-60)
(61-64)
Fort Klamath
(86-89)
Fort
Yamhill
(57-60)
Portland
Roseburgh
(77-80)

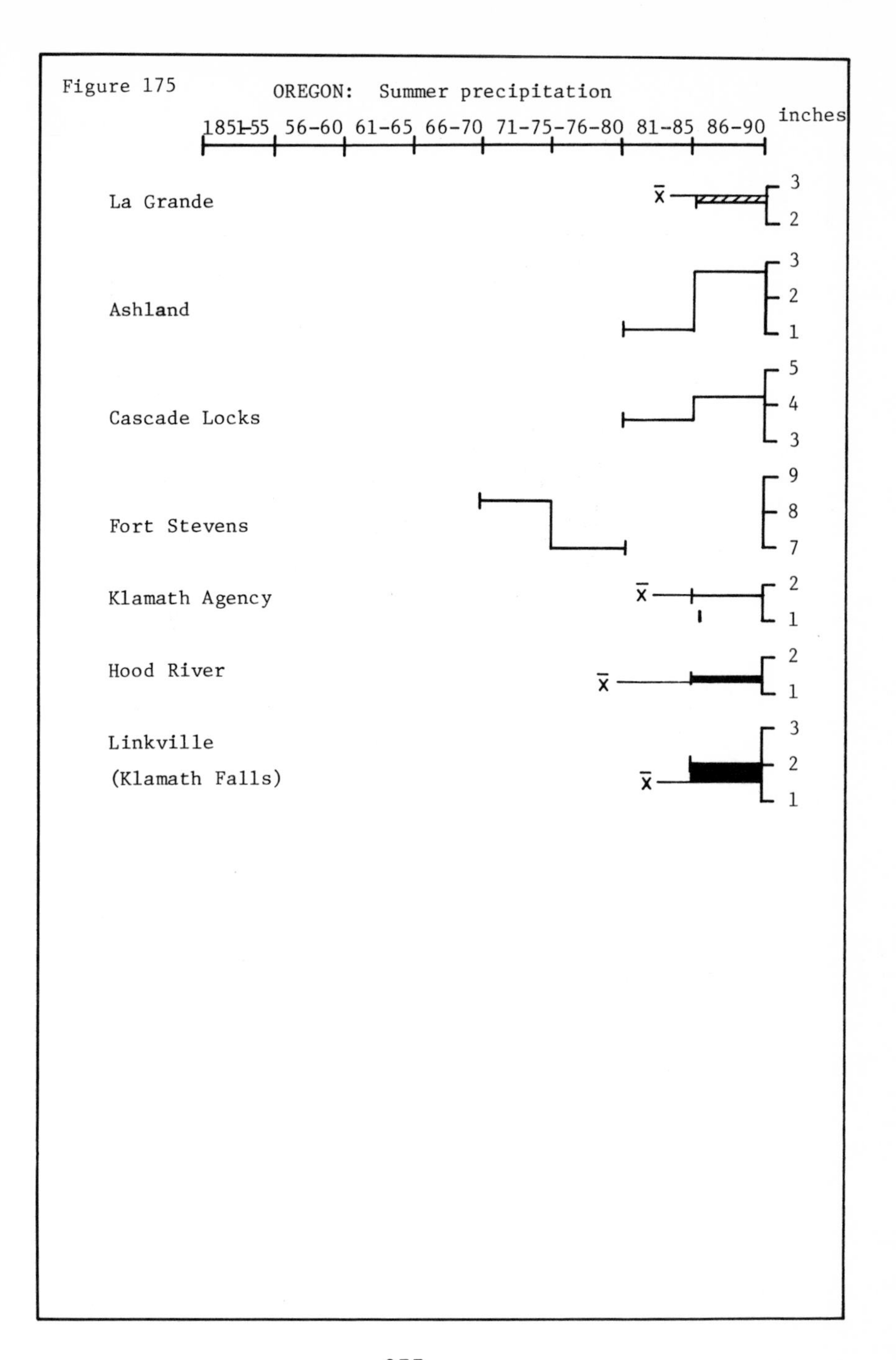

Figure 175
OREGON: Summer precipitation
1851-55
56-60
61-65
66-70
71-75
76-80
81-85
86-90
inches
La Grande
Ashland
Cascade Locks
Fort Stevens
Klamath Agency
Hood River
Linkville
(Klamath Falls)
3
2
3
2
1
5
4
3
9
8
7
2
1
2
1
3
2
1

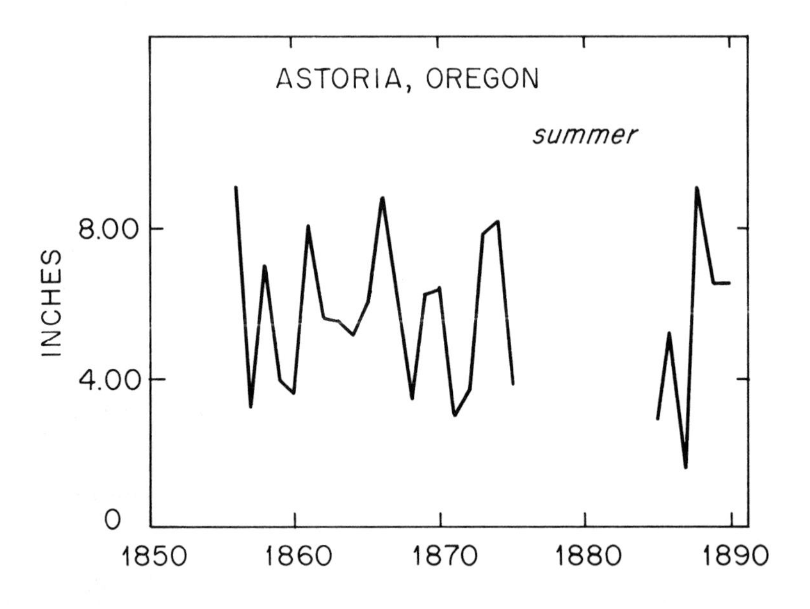

Figure 176: Oregon summer precipitation

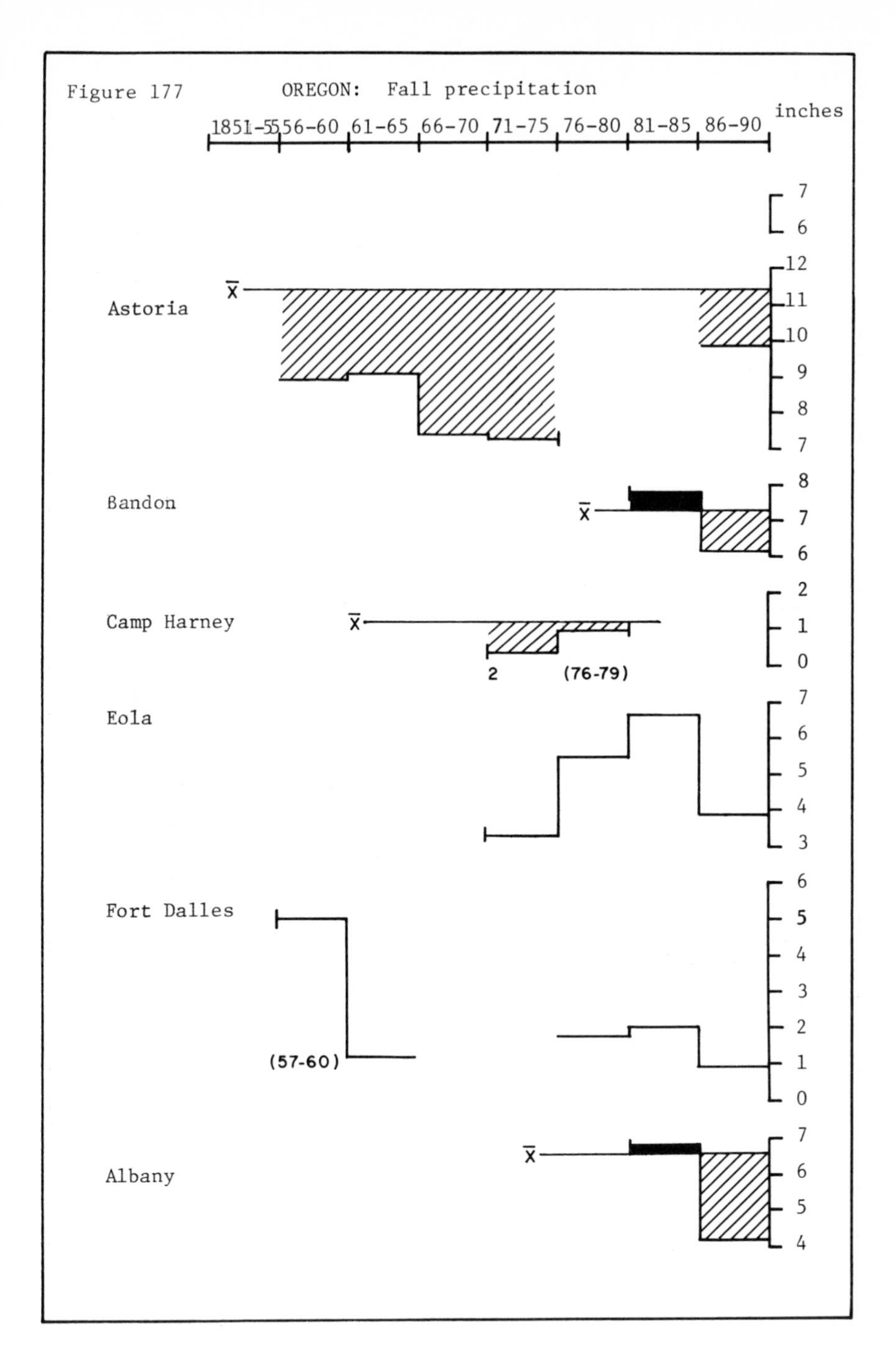
Figure 177
OREGON: Fall precipitation
1851-55
56-60
61-65
66-70
71-75
76-80
81-85
86-90
inches
Astoria
Bandon
Camp Harney
Eola
Fort Dalles
Albany
2
(76-79)
(57-60)
7
6
12
11
10
9
8
7
8
7
6
2
1
0
7
6
5
4
3
6
5
4
3
2
1
0
7
6
5
4

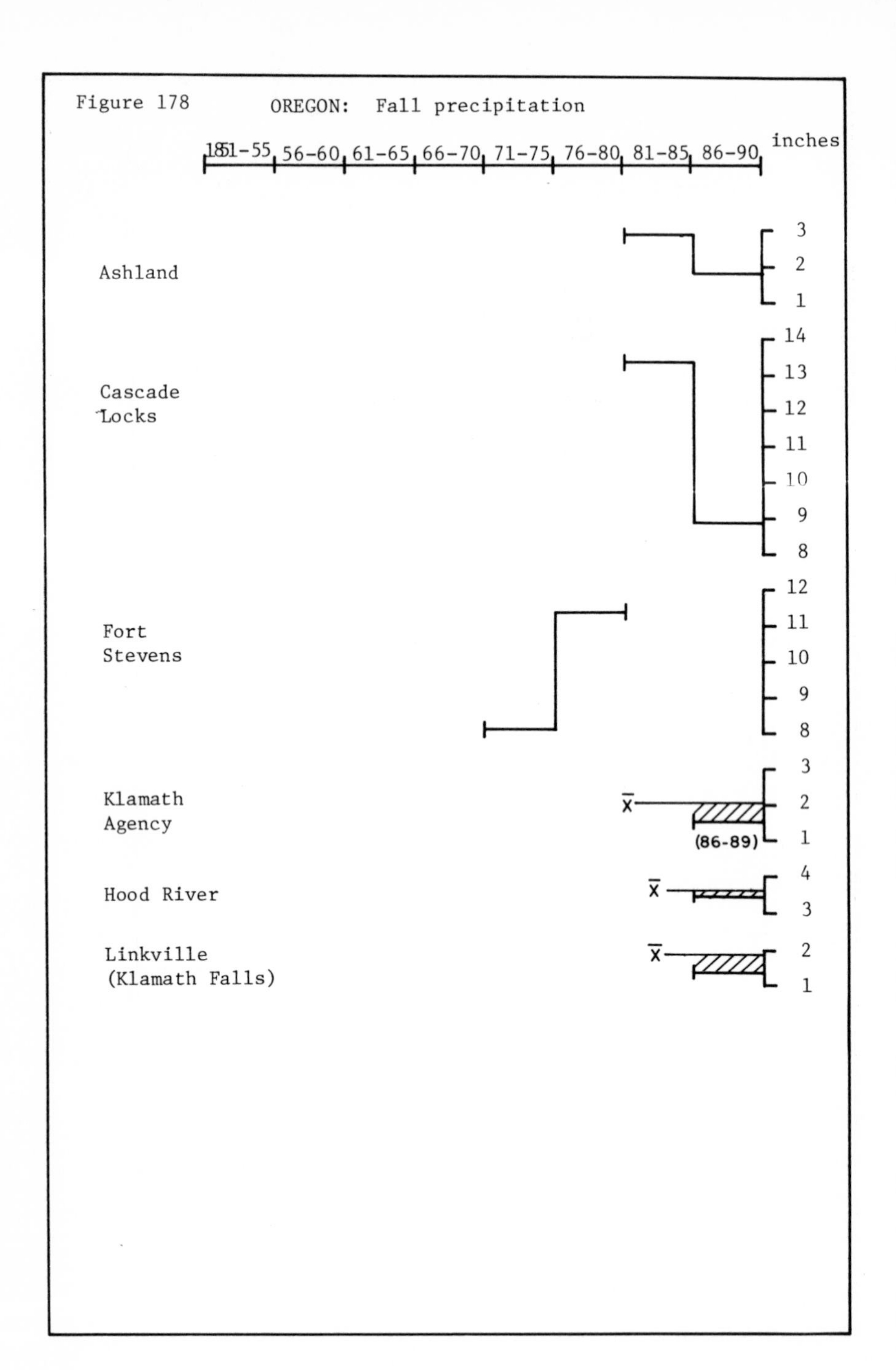
Figure 178
OREGON: Fall precipitation
1851-55
56-60
61-65
66-70
71-75
76-80
81-85
86-90
inches
Ashland
3
2
1
Cascade Locks
14
13
12
11
10
9
8
Fort Stevens
12
11
10
9
8
Klamath Agency
X̄
(86-89)
3
2
1
Hood River
X̄
4
3
Linkville (Klamath Falls)
X̄
2
1

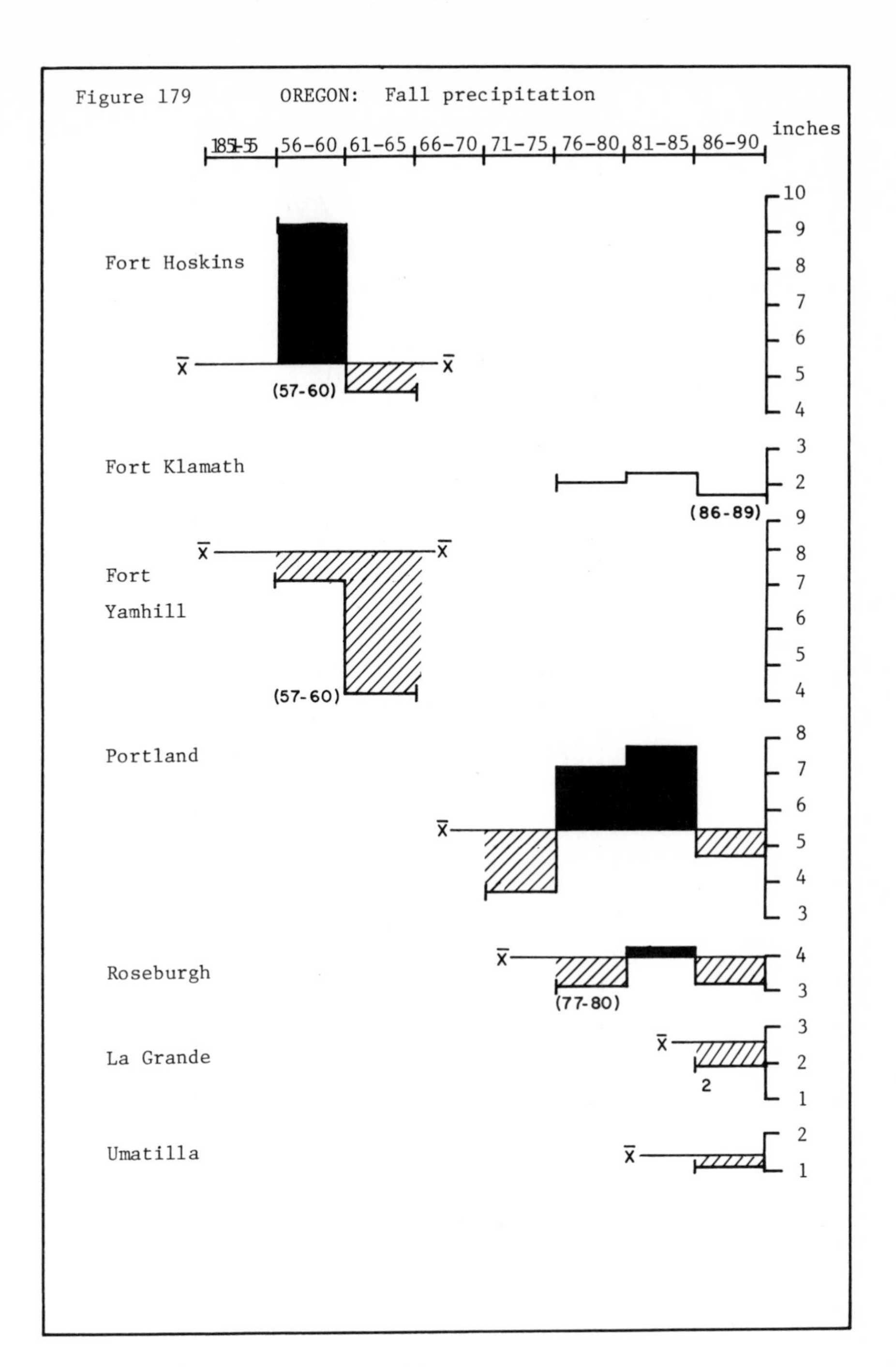

Figure 179 OREGON: Fall precipitation

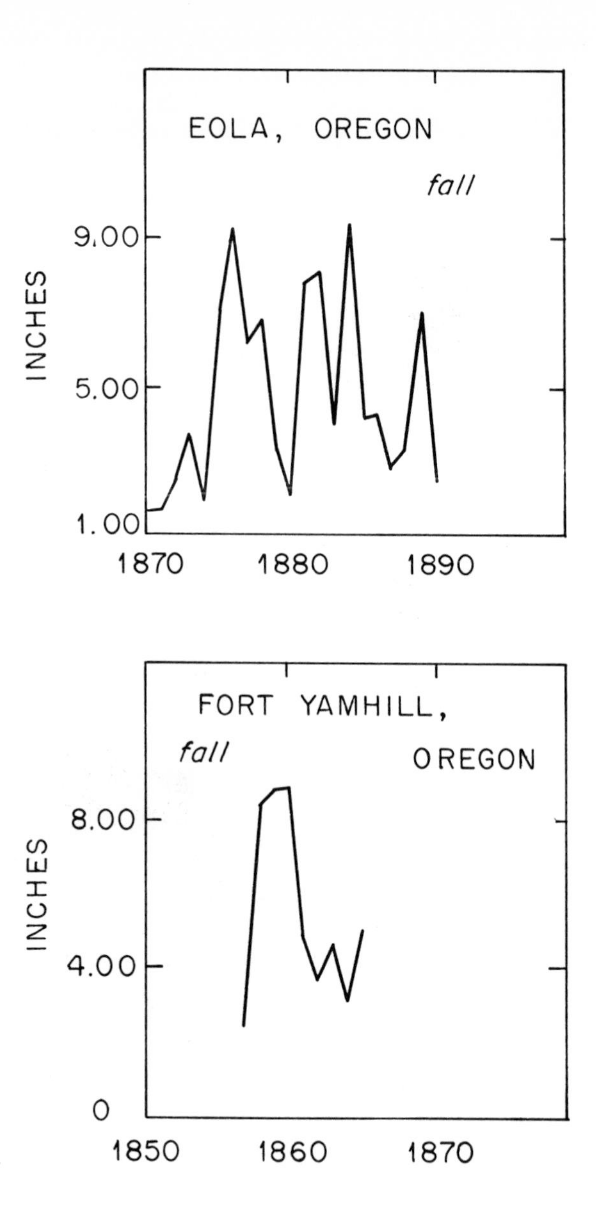

Figure 180: Oregon fall precipitation

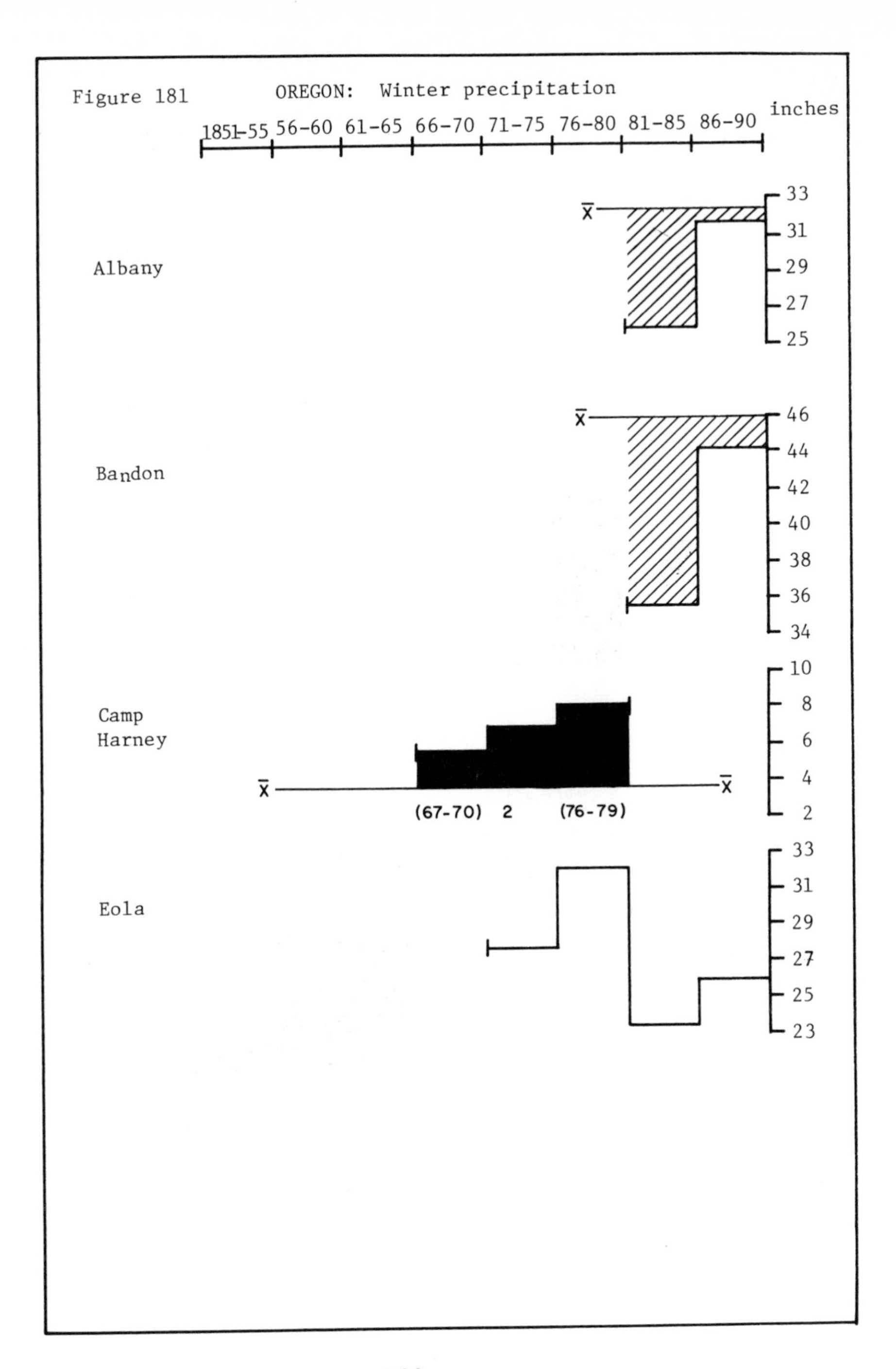
Figure 181
OREGON: Winter precipitation
1851-55
56-60
61-65
66-70
71-75
76-80
81-85
86-90
inches
Albany
33
31
29
27
25
Bandon
46
44
42
40
38
36
34
Camp Harney
10
8
6
4
2
(67-70)
2
(76-79)
Eola
33
31
29
27
25
23

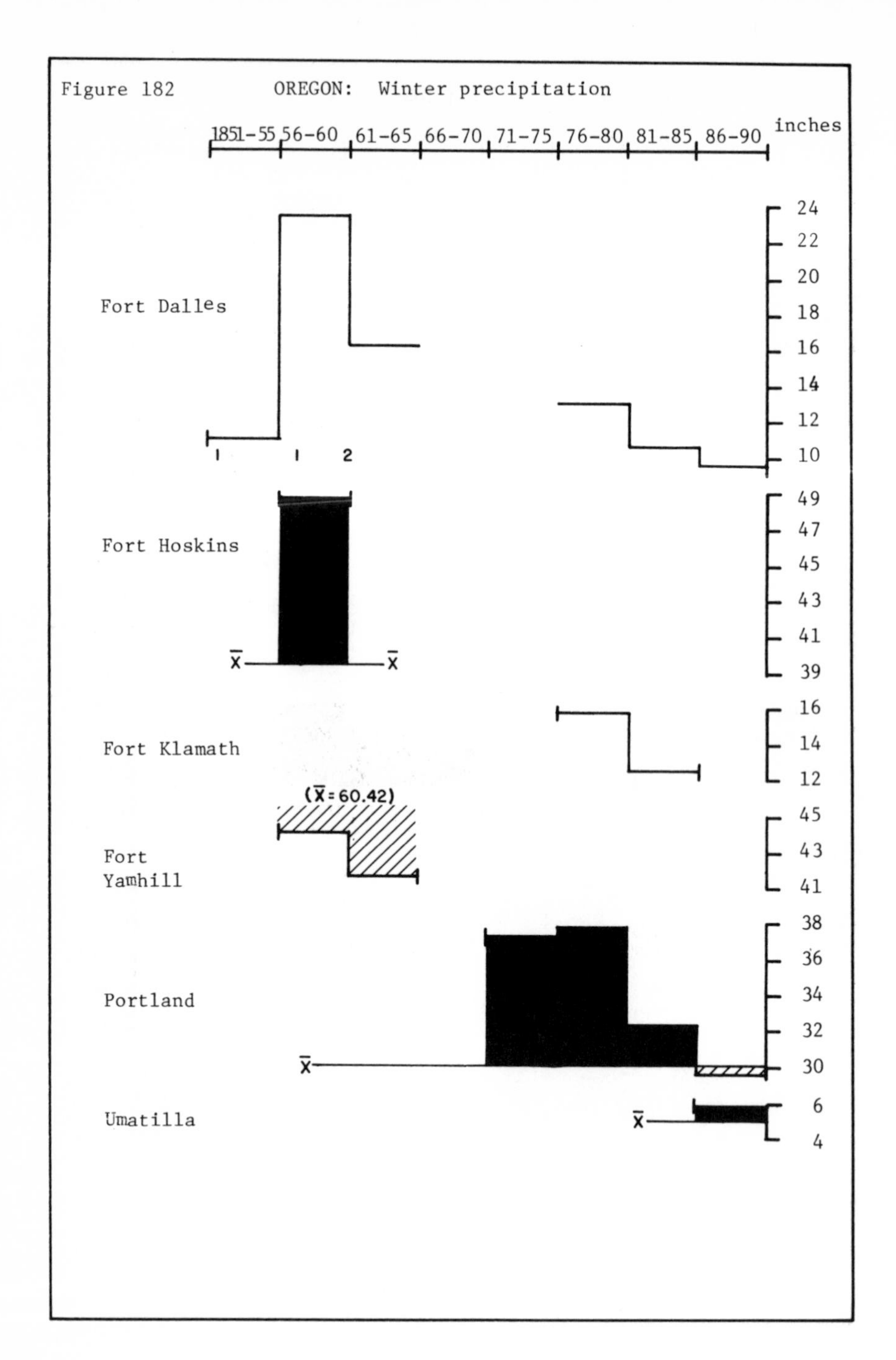
Figure 182
OREGON: Winter precipitation
1851-55
56-60
61-65
66-70
71-75
76-80
81-85
86-90
inches
Fort Dalles
24
22
20
18
16
14
12
10
1
1
2
Fort Hoskins
49
47
45
43
41
39
X̄
X̄
Fort Klamath
16
14
12
(X̄=60.42)
Fort
Yamhill
45
43
41
Portland
38
36
34
32
30
X̄
Umatilla
6
4
X̄

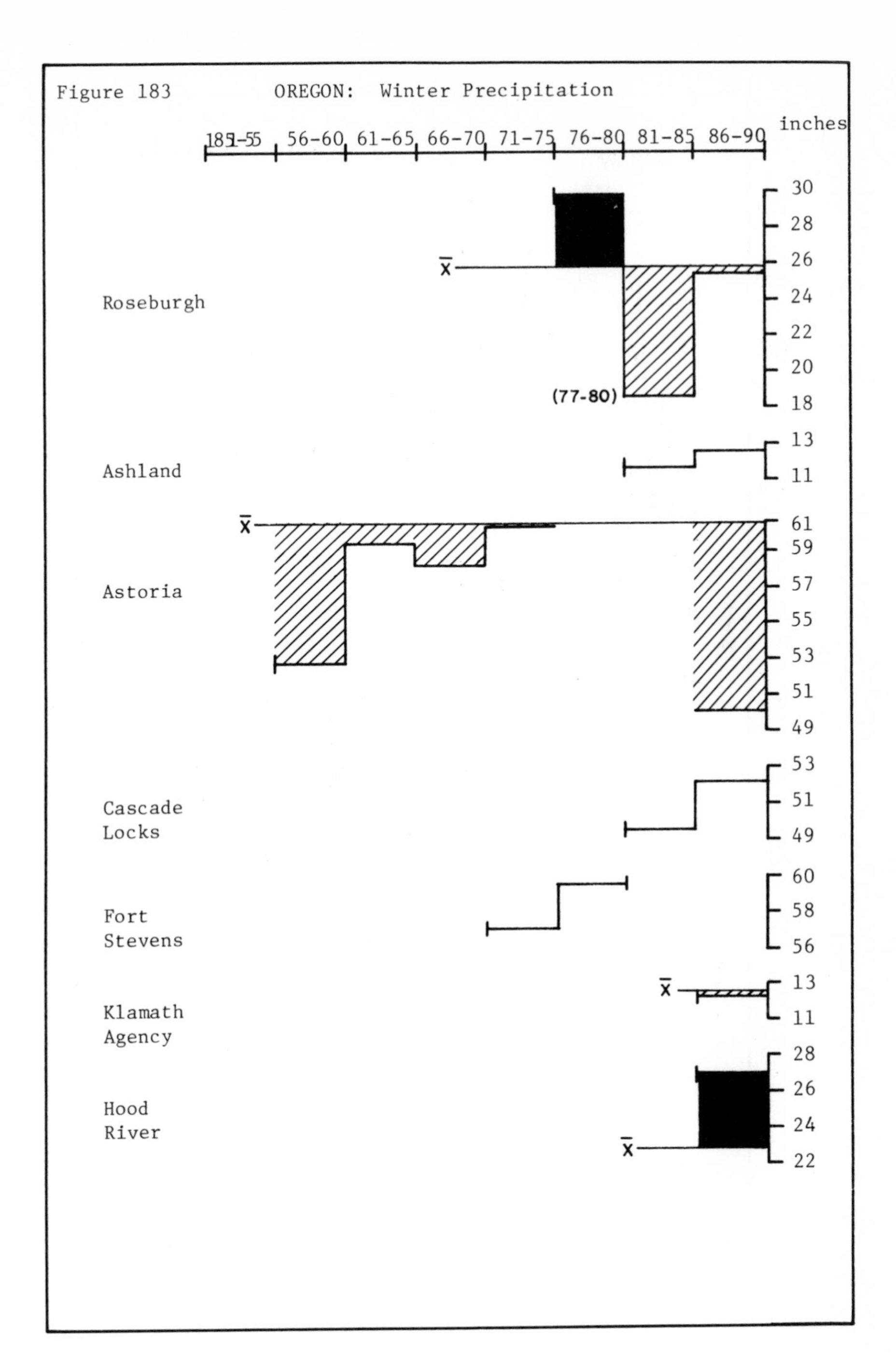
Figure 183 OREGON: Winter Precipitation
1851-55
56-60
61-65
66-70
71-75
76-80
81-85
86-90
inches
Roseburgh
(77-80)
30
28
26
24
22
20
18
Ashland
13
11
Astoria
61
59
57
55
53
51
49
Cascade Locks
53
51
49
Fort Stevens
60
58
56
Klamath Agency
13
11
Hood River
28
26
24
22

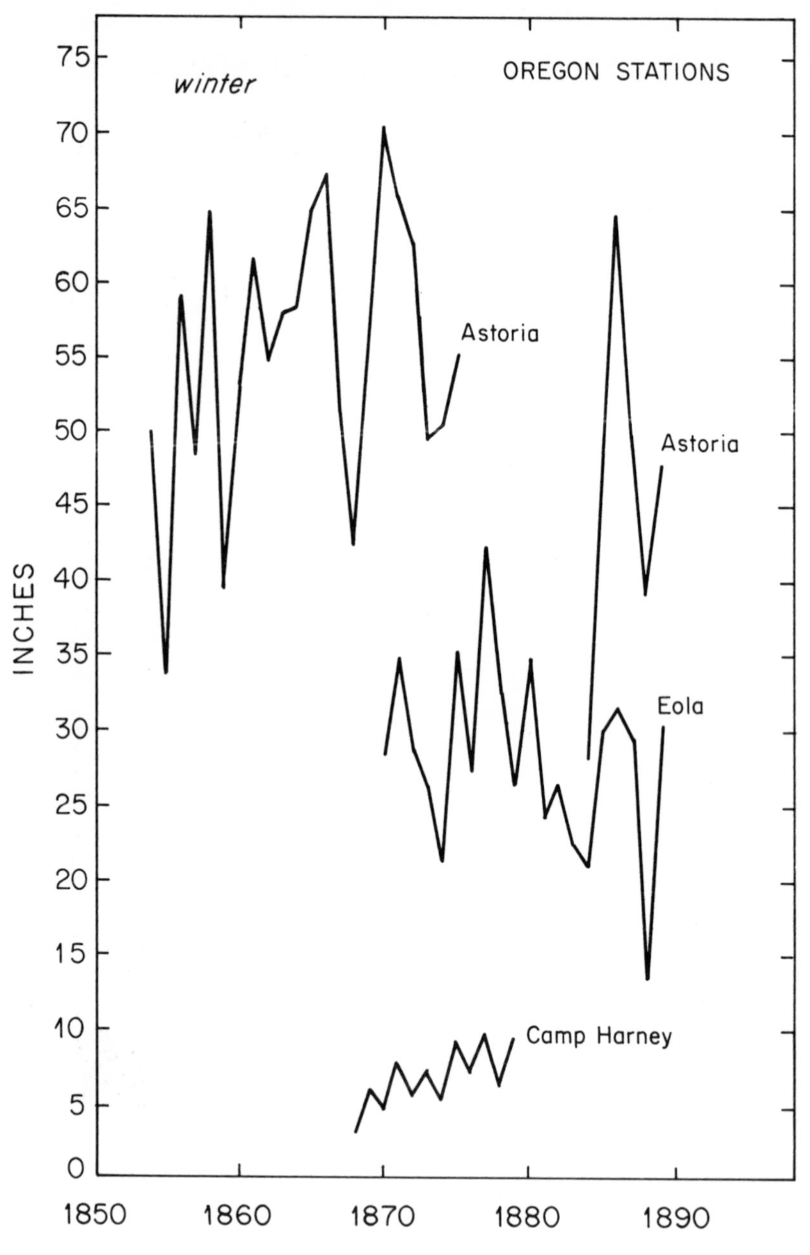

Figure 184: Oregon fall precipitation

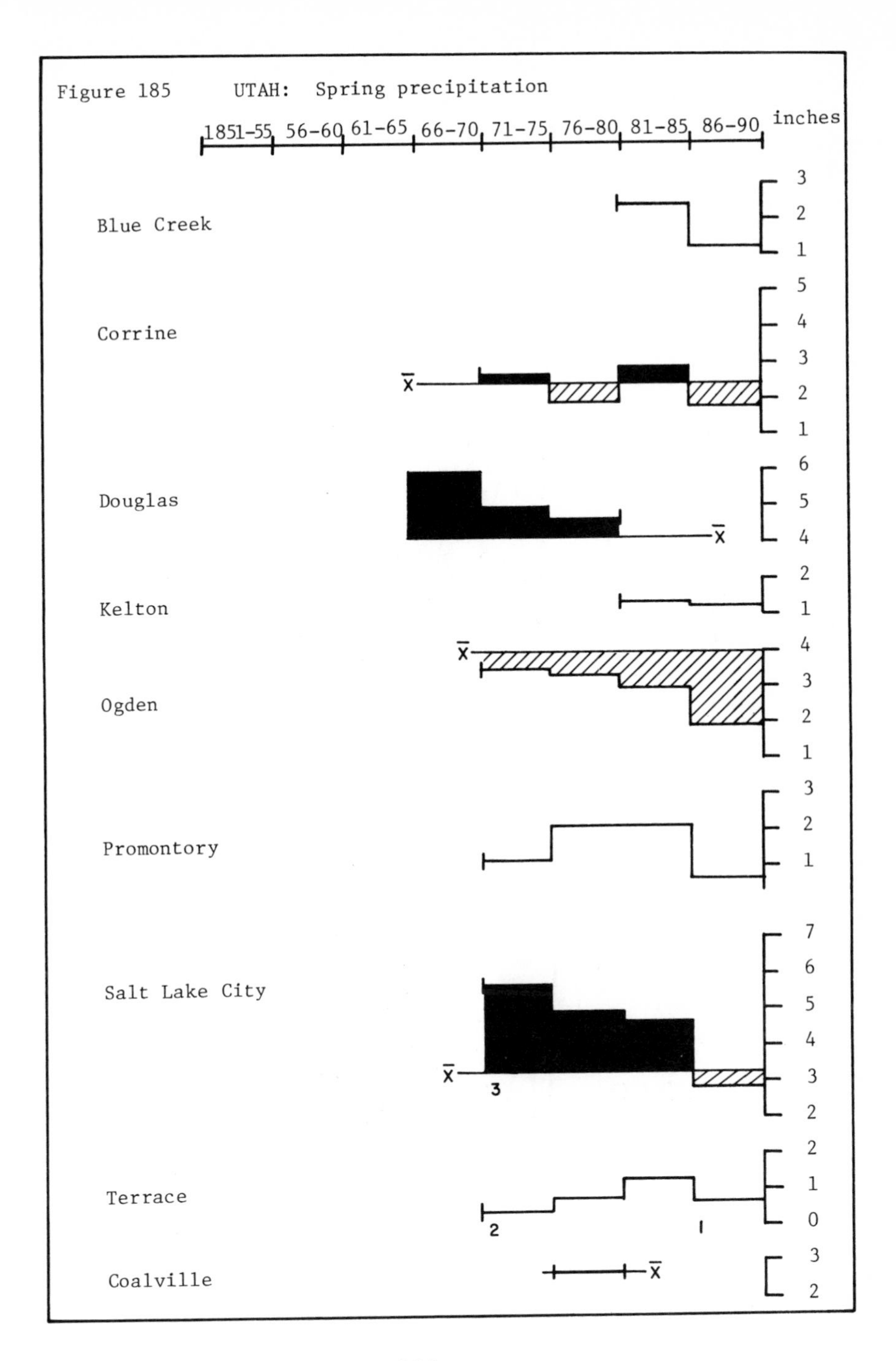

Figure 185 UTAH: Spring precipitation
1851-55
56-60
61-65
66-70
71-75
76-80
81-85
86-90
inches
Blue Creek
Corrine
Douglas
Kelton
Ogden
Promontory
Salt Lake City
Terrace
Coalville

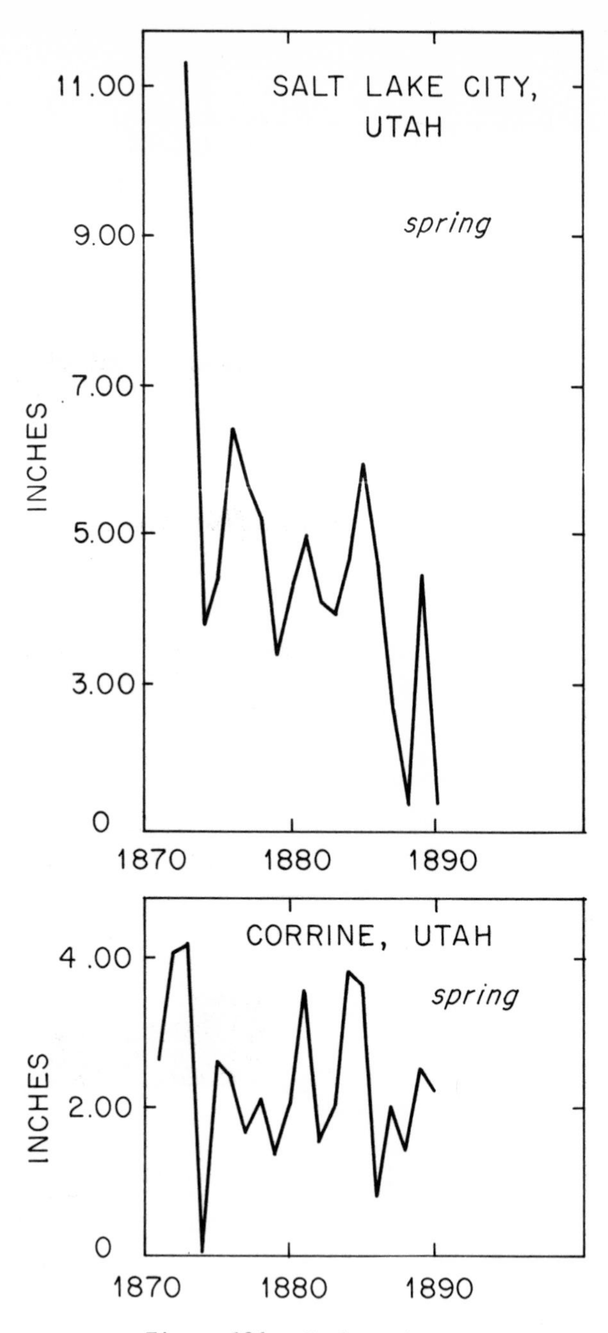

Figure 186: Utah spring precipitation

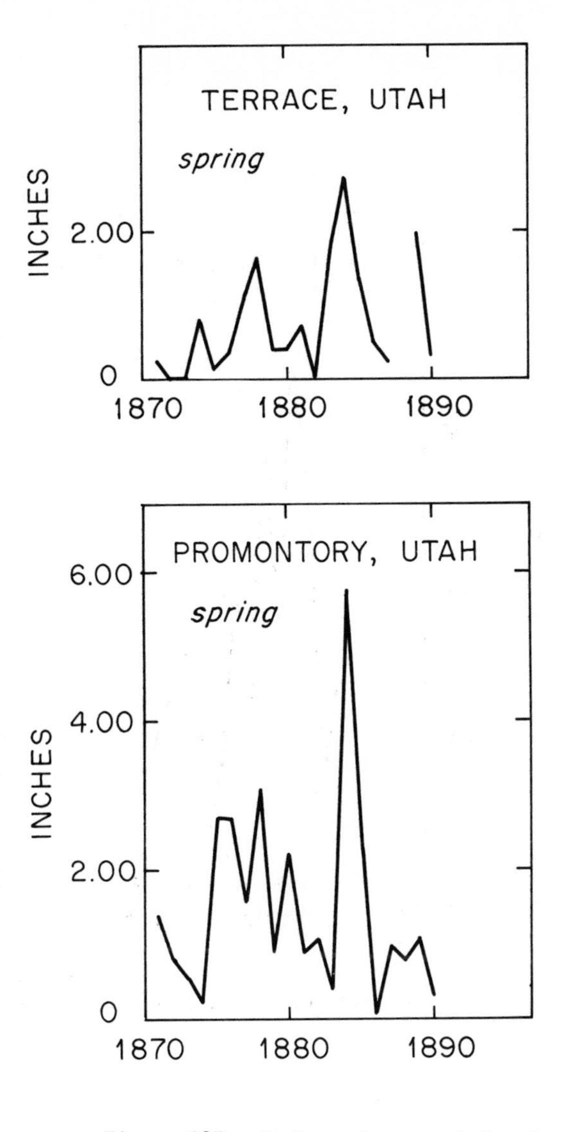

Figure 187: Utah spring precipitation

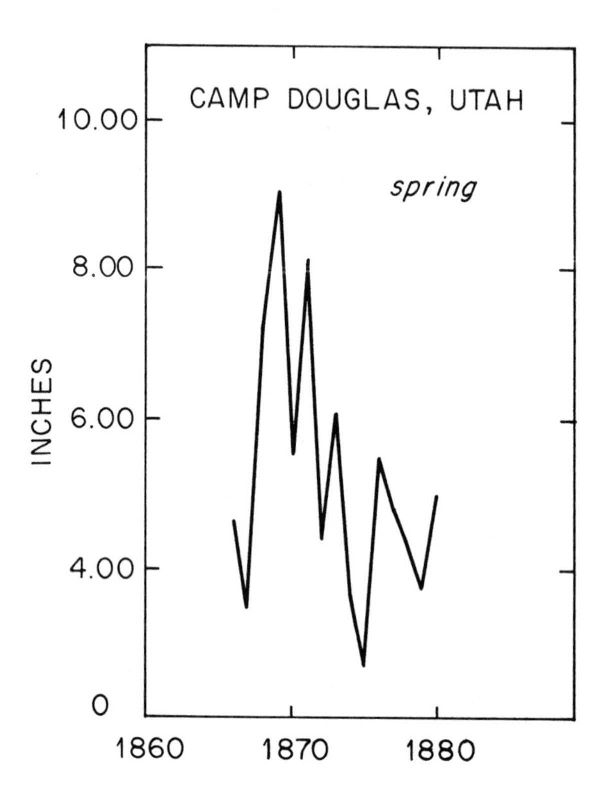

Figure 187: Utah spring precipitation

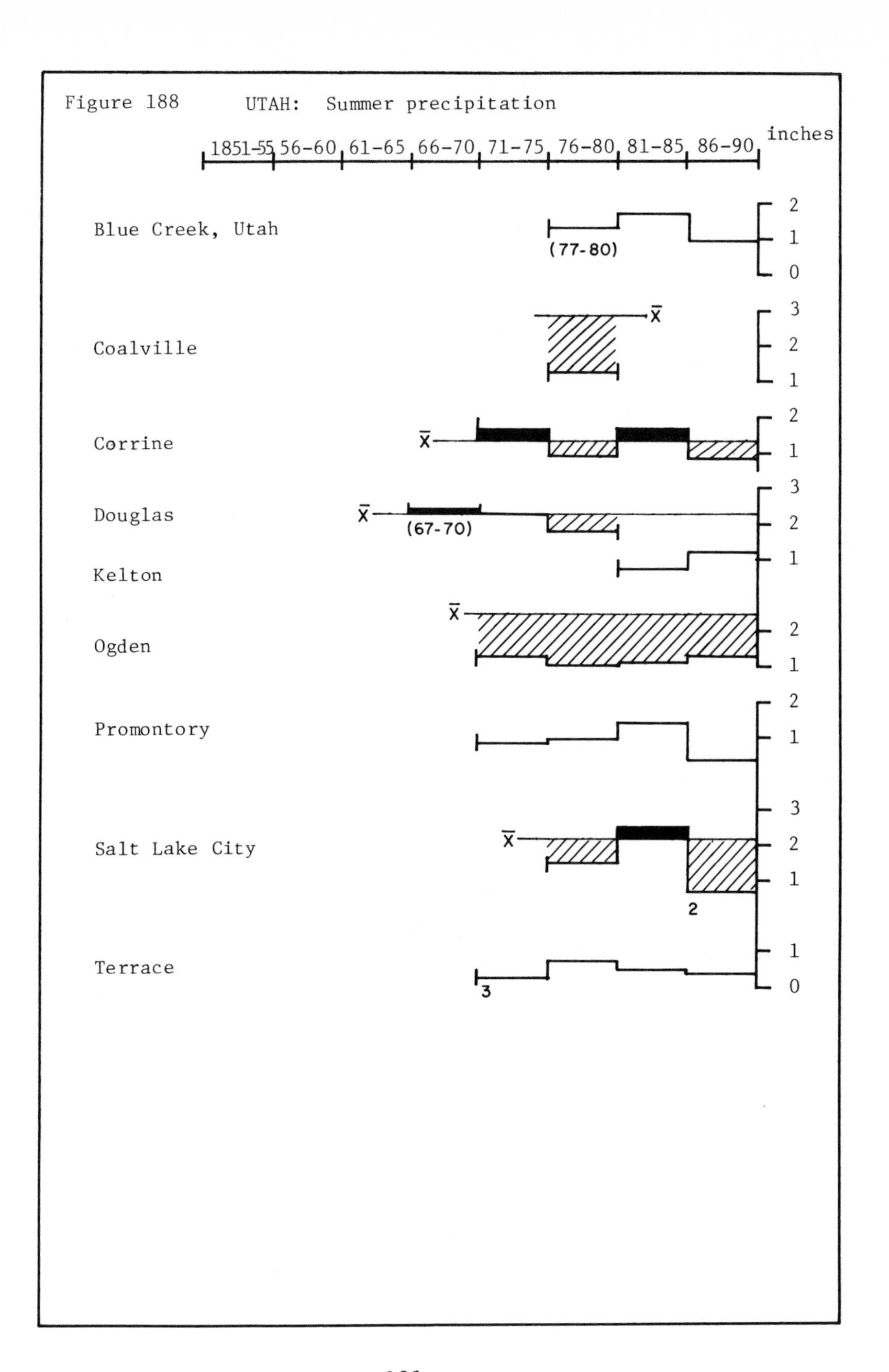

Figure 188 UTAH: Summer precipitation
1851-55
56-60
61-65
66-70
71-75
76-80
81-85
86-90
inches
Blue Creek, Utah
(77-80)
Coalville
Corrine
Douglas
(67-70)
Kelton
Ogden
Promontory
Salt Lake City
Terrace
X̄
2
3
0
1

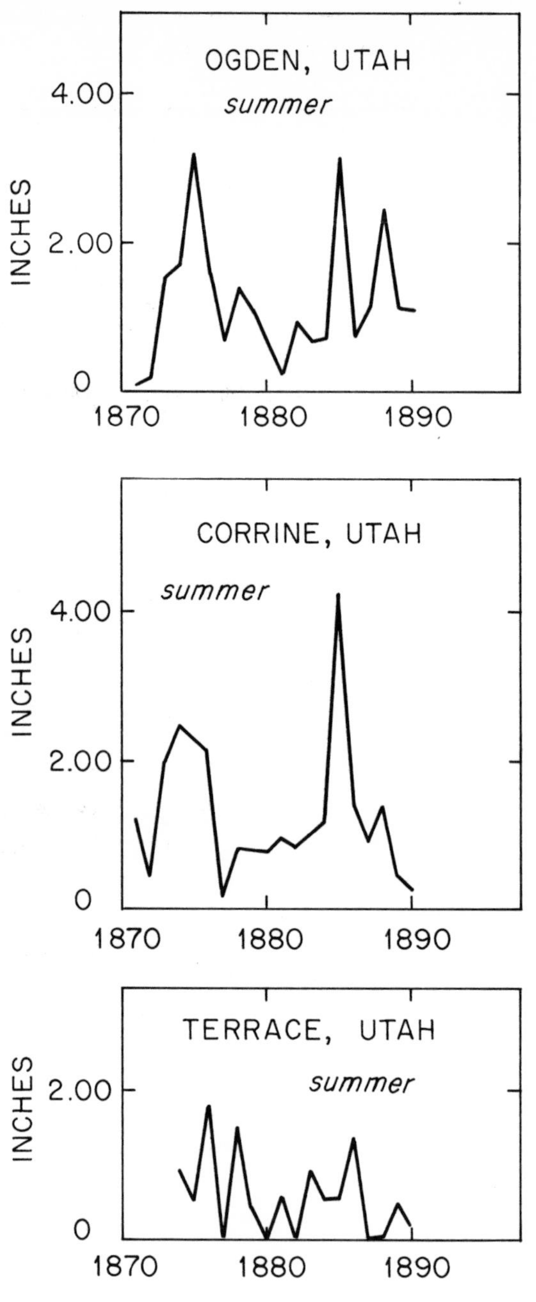

Figure 189: Utah summer precipitation

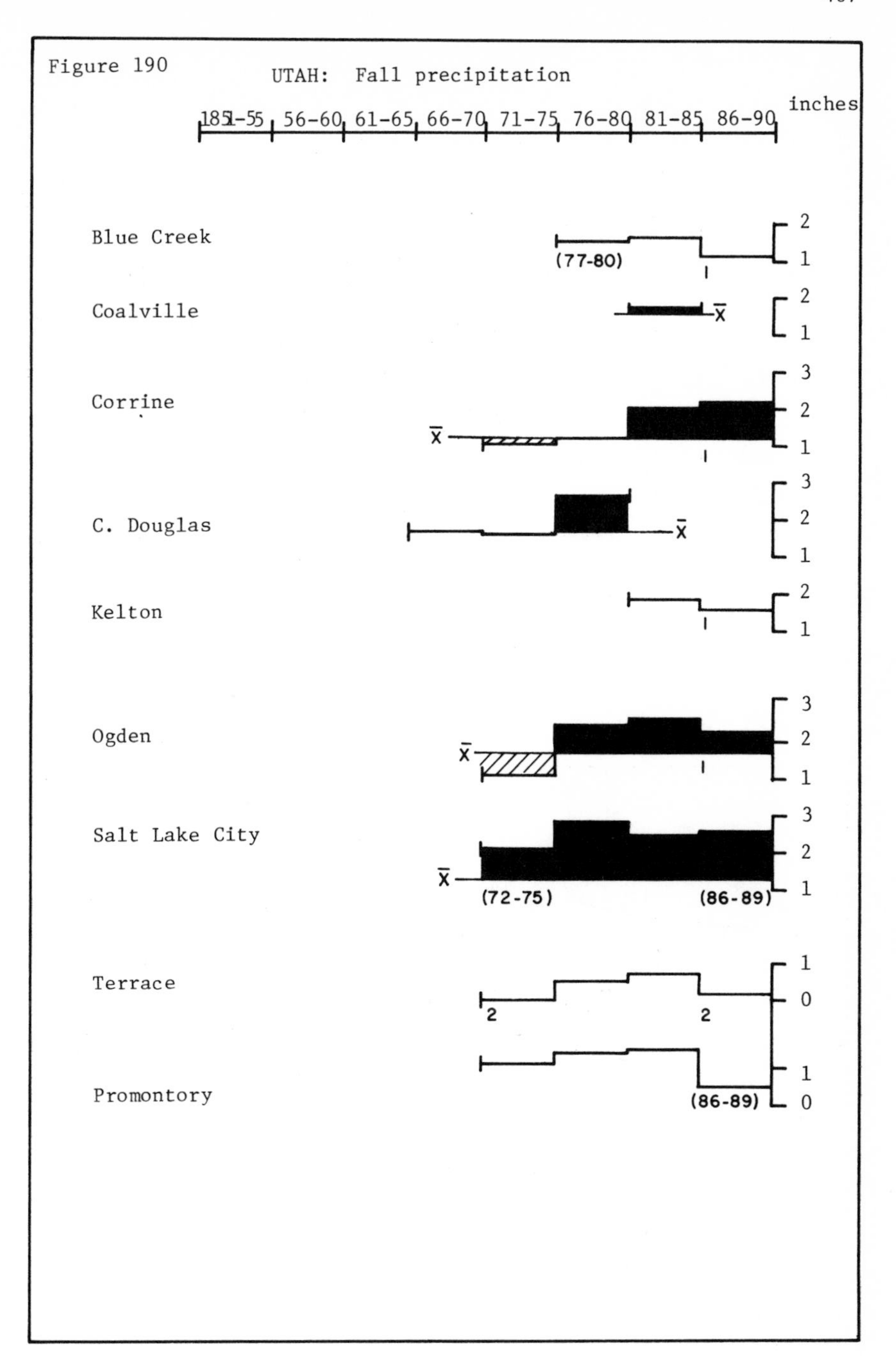
Figure 190
UTAH: Fall precipitation
1851-55
56-60
61-65
66-70
71-75
76-80
81-85
86-90
inches
Blue Creek
(77-80)
Coalville
Corrine
C. Douglas
Kelton
Ogden
Salt Lake City
(72-75)
(86-89)
Terrace
Promontory
(86-89)

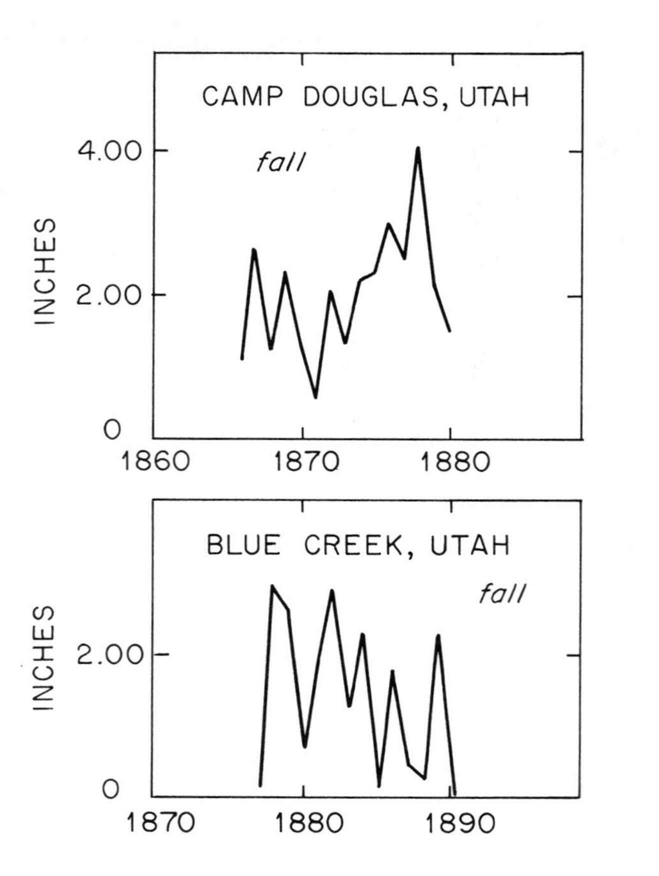

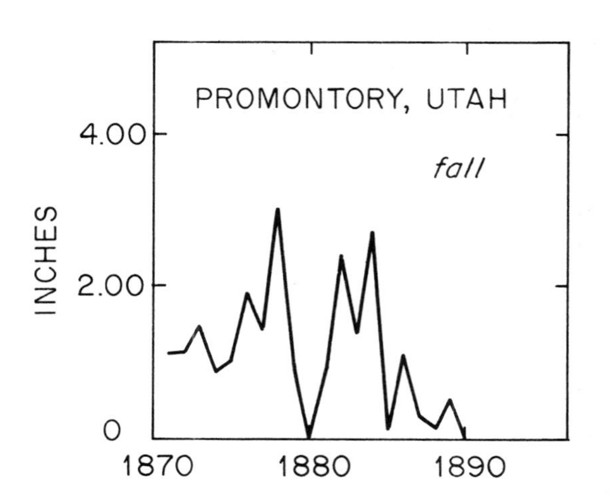

Figure 191: Utah fall precipitation

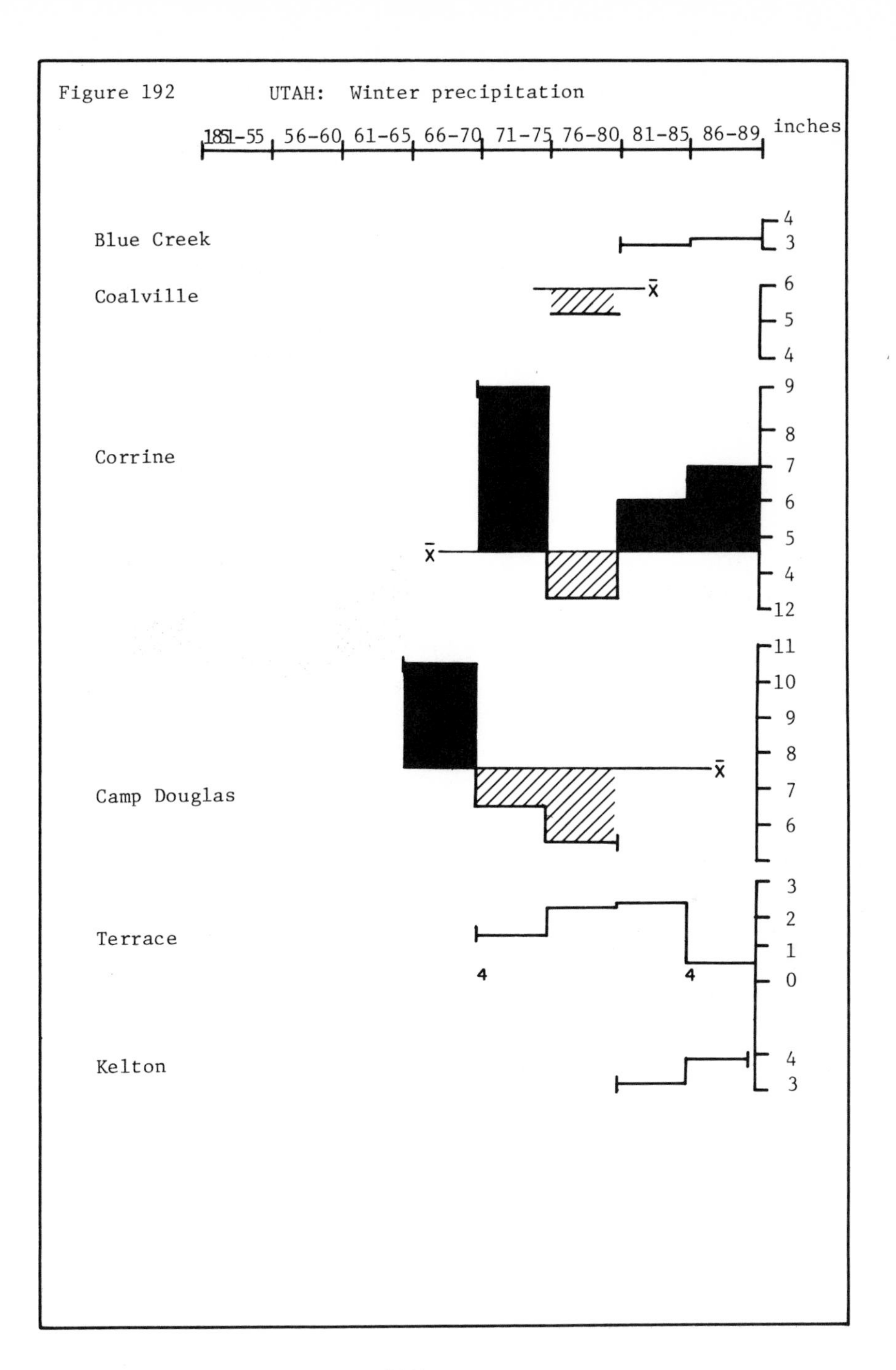
Figure 192
UTAH: Winter precipitation
1851-55
56-60
61-65
66-70
71-75
76-80
81-85
86-89
inches
Blue Creek
Coalville
Corrine
Camp Douglas
Terrace
Kelton

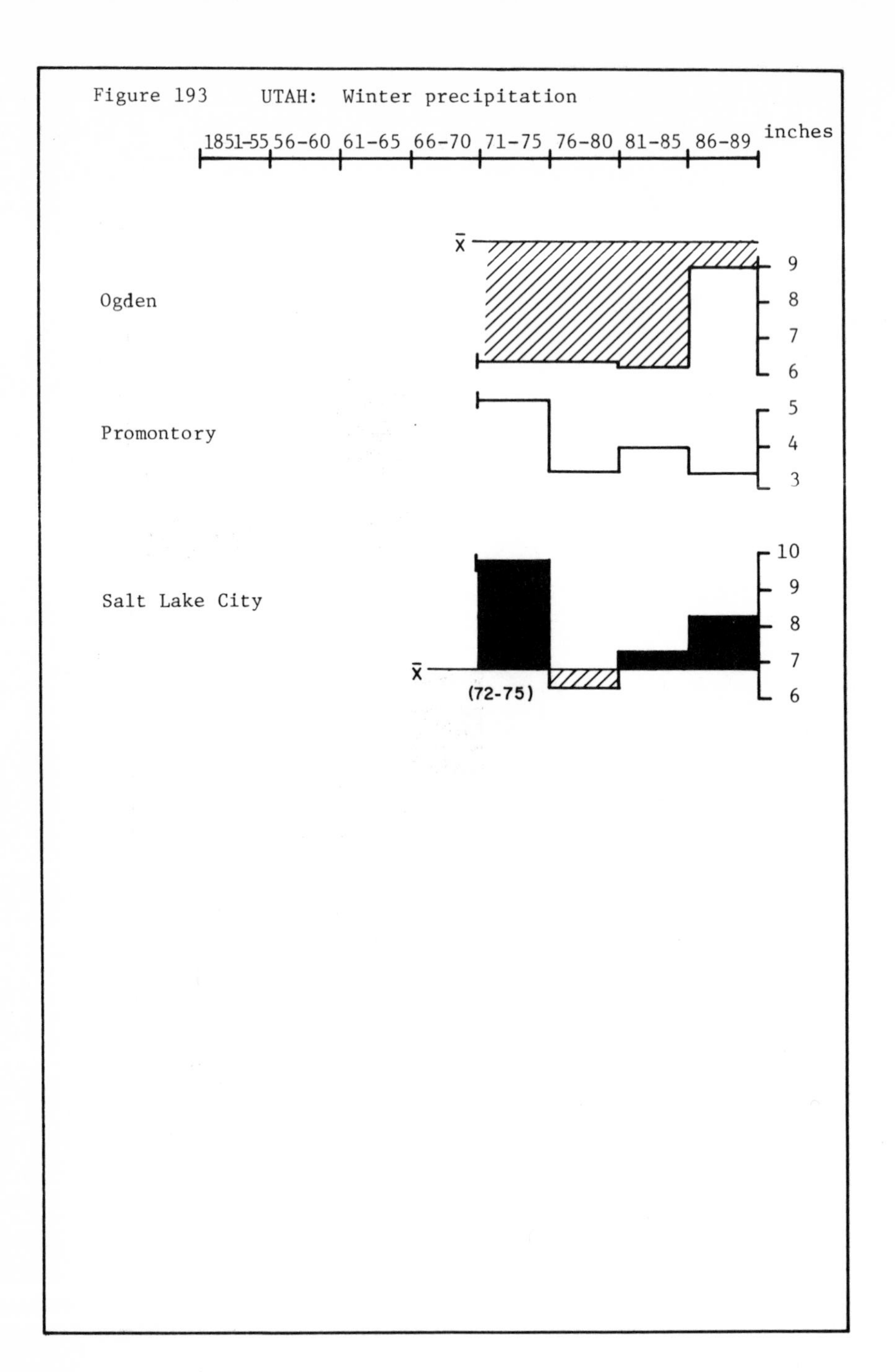

Figure 193 UTAH: Winter precipitation
1851-55
56-60
61-65
66-70
71-75
76-80
81-85
86-89
inches
Ogden
Promontory
Salt Lake City
x̄
(72-75)
9
8
7
6
5
4
3
10
9
8
7
6

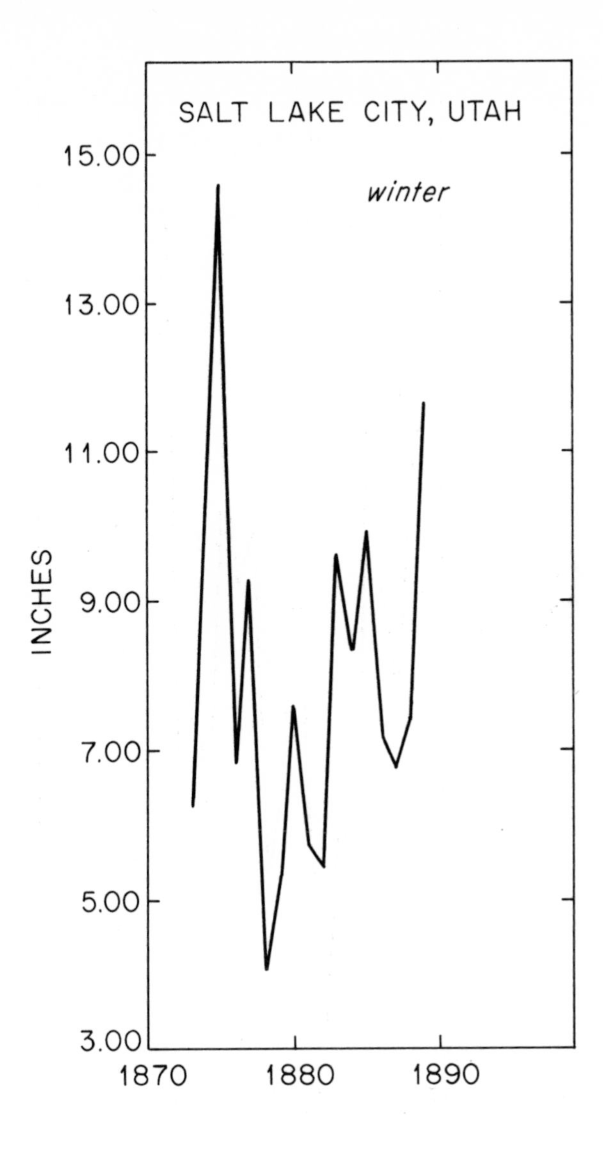

Figure 194: Utah fall precipitation

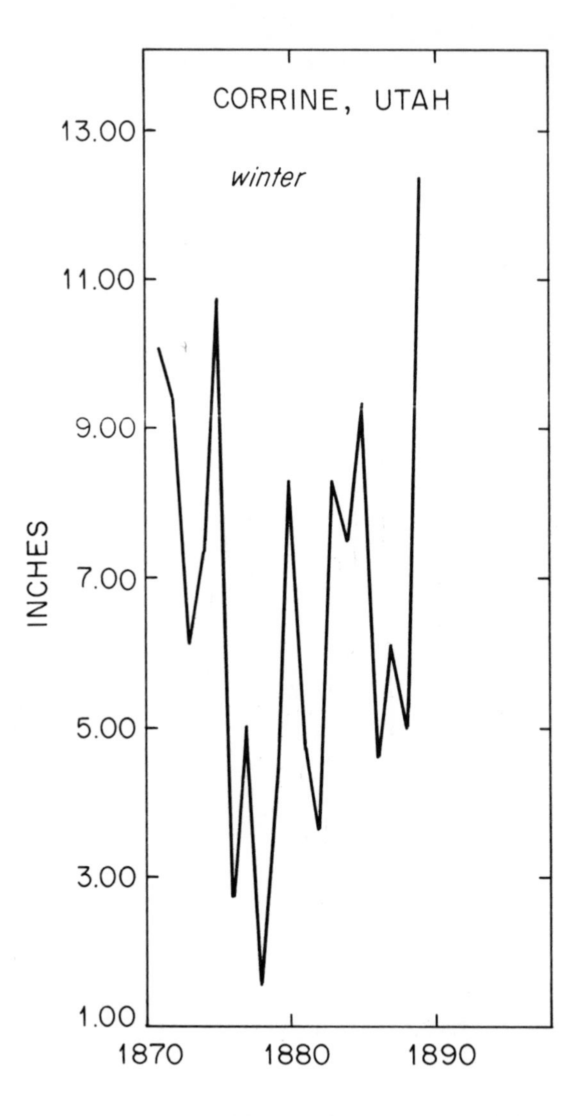

Figure 195: Utah winter precipitation

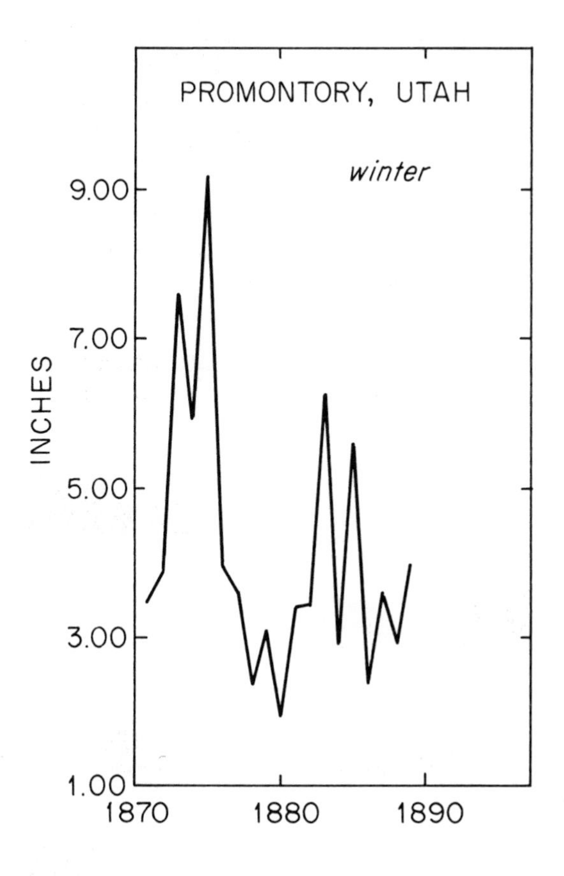

Figure 196: Utah winter precipitation

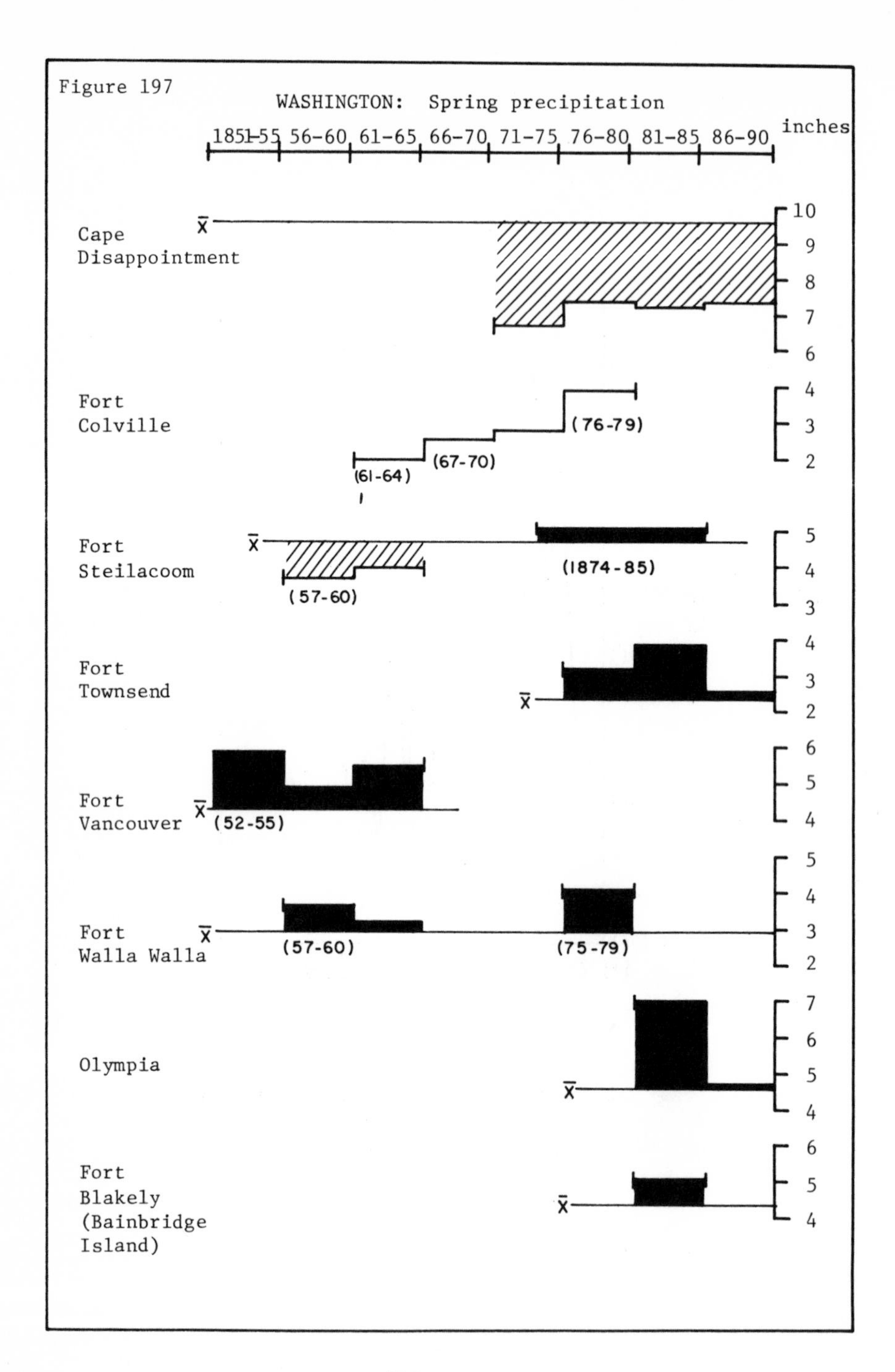
Figure 197
WASHINGTON: Spring precipitation
1851-55 56-60 61-65 66-70 71-75 76-80 81-85 86-90
inches
Cape Disappointment
Fort Colville
(61-64)
(67-70)
(76-79)
Fort Steilacoom
(57-60)
(1874-85)
Fort Townsend
Fort Vancouver
(52-55)
Fort Walla Walla
(57-60)
(75-79)
Olympia
Fort Blakely (Bainbridge Island)

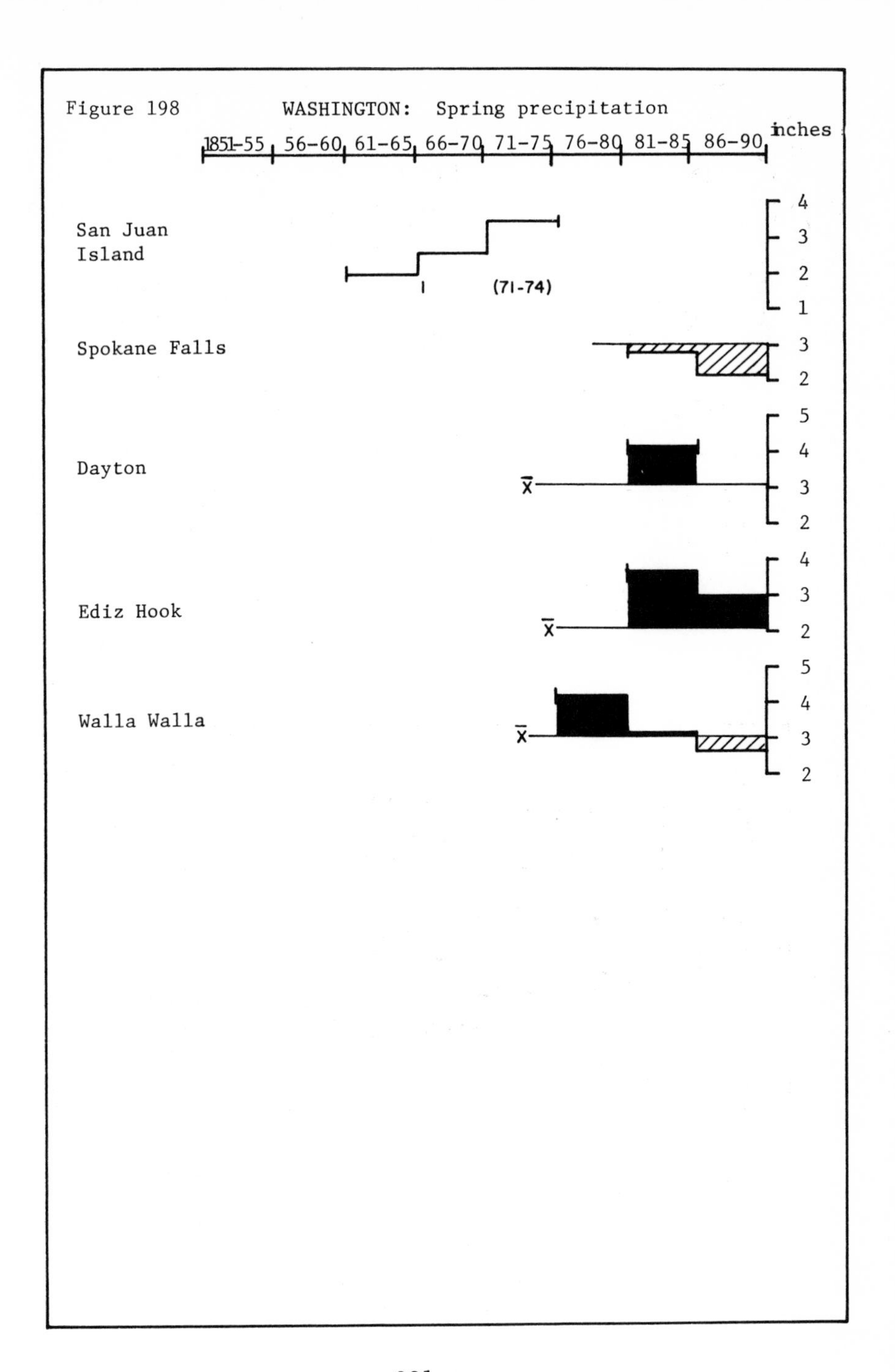
Figure 198
WASHINGTON: Spring precipitation
1851-55
56-60
61-65
66-70
71-75
76-80
81-85
86-90
inches
San Juan Island
(71-74)
4
3
2
1
Spokane Falls
3
2
Dayton
5
4
3
2
Ediz Hook
4
3
2
Walla Walla
5
4
3
2
$\bar{X}$

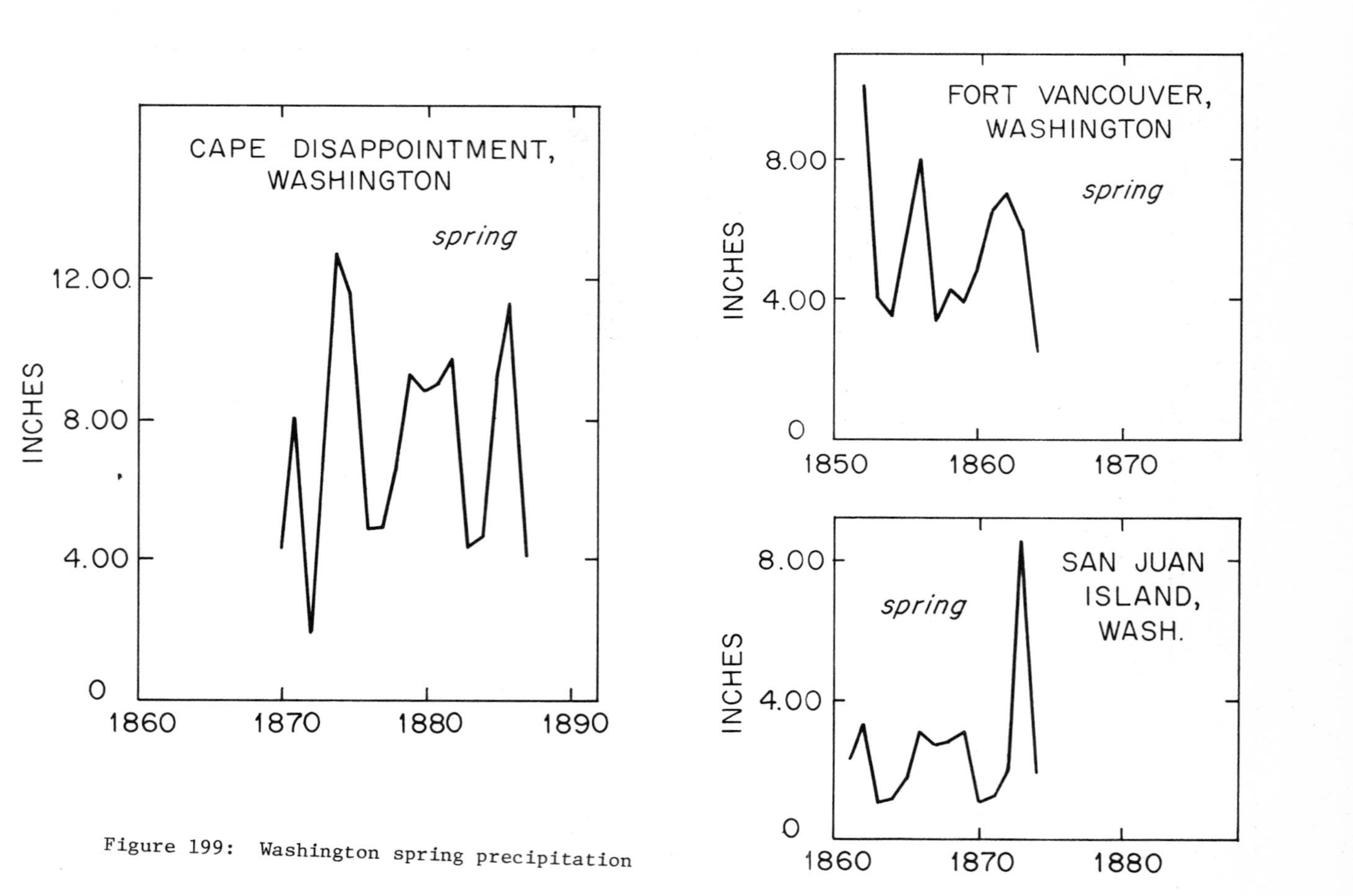

Figure 199: Washington spring precipitation

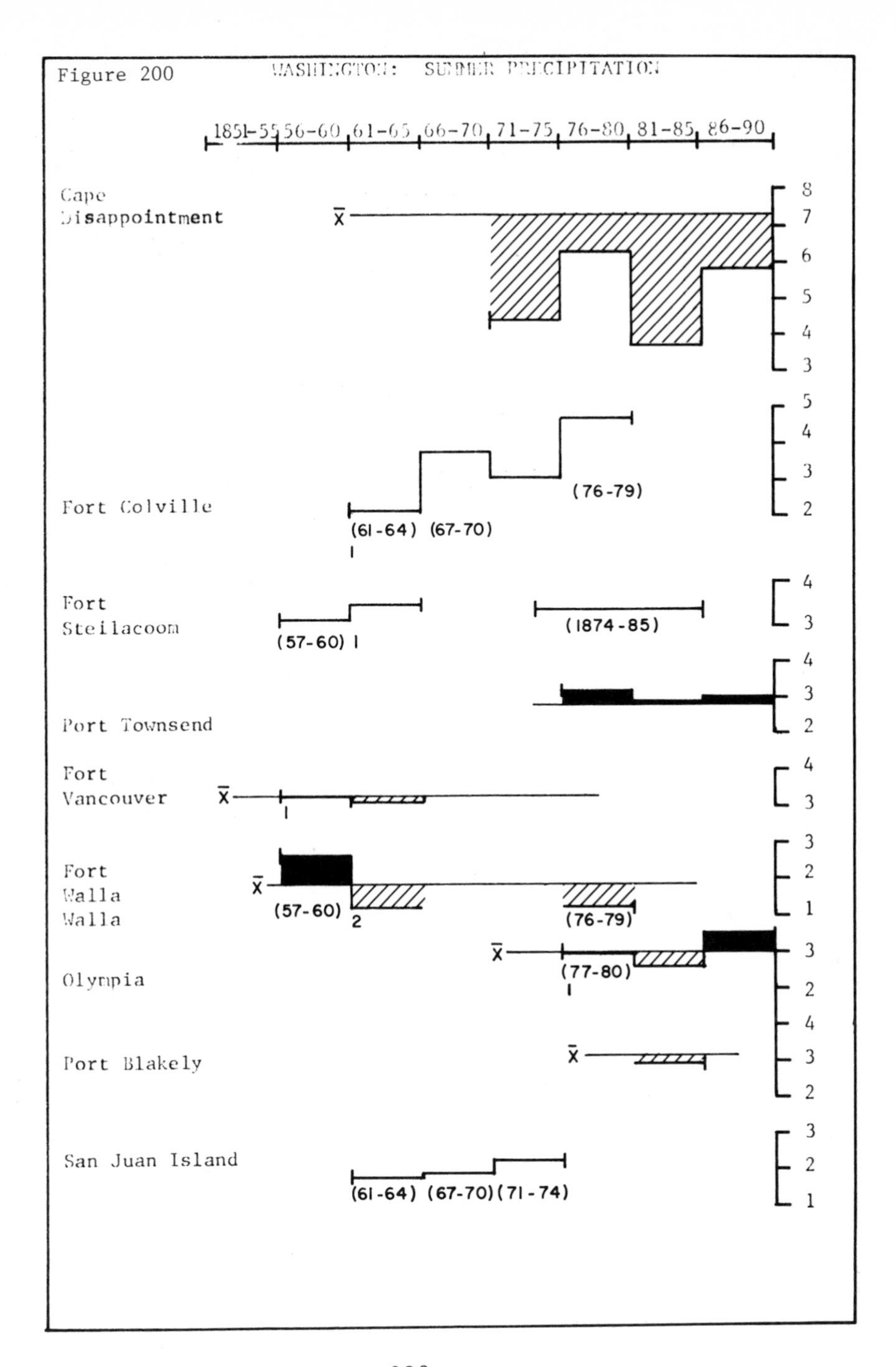
Figure 200
WASHINGTON: SUMMER PRECIPITATION
1851-55
56-60
61-65
66-70
71-75
76-80
81-85
86-90
Cape Disappointment
Fort Colville
(61-64)
(67-70)
(76-79)
Fort Steilacoom
(57-60)
(1874-85)
Port Townsend
Fort Vancouver
Fort Walla Walla
(57-60)
(76-79)
Olympia
(77-80)
Port Blakely
San Juan Island
(61-64)
(67-70)
(71-74)

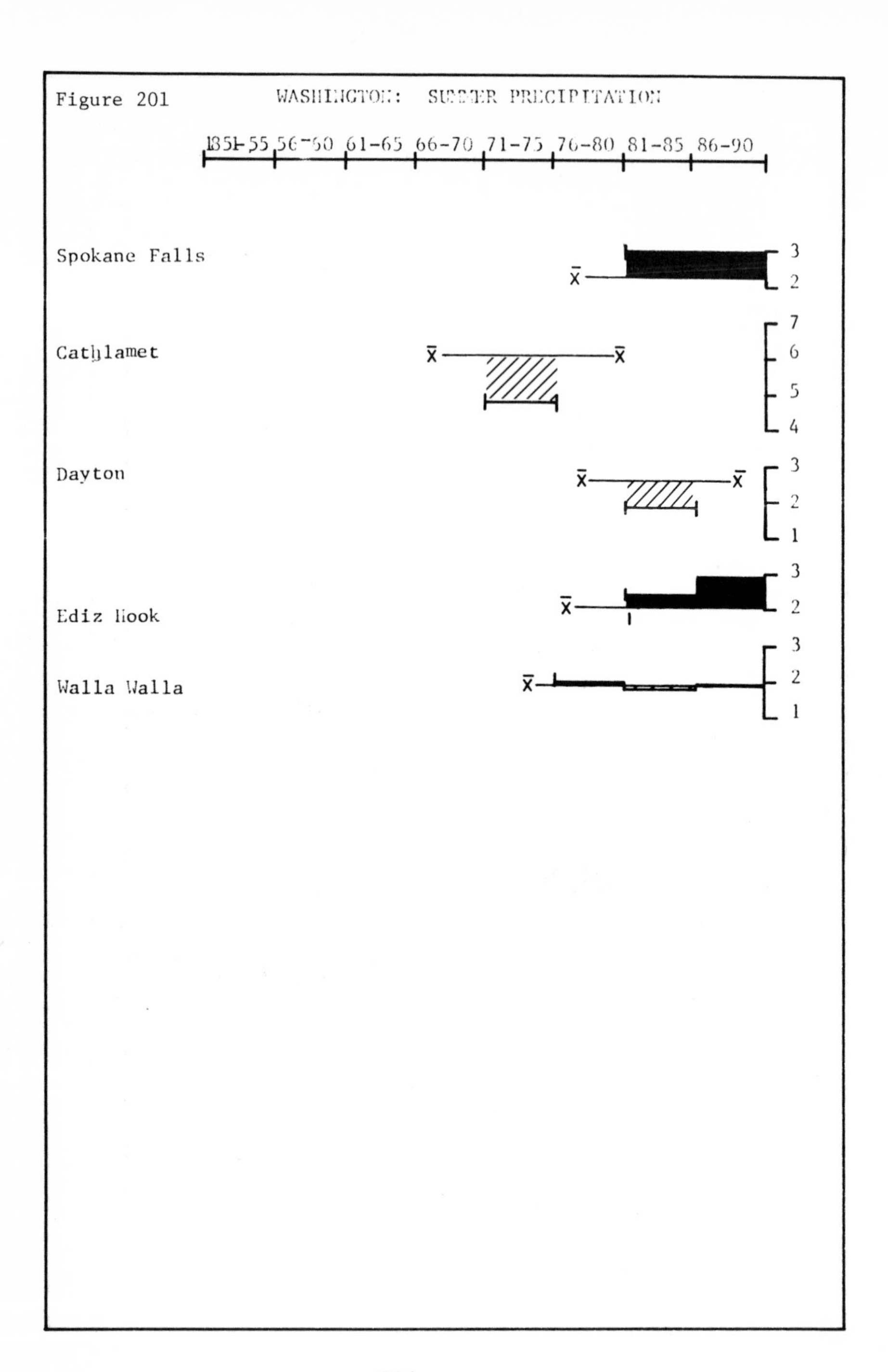
Figure 201
WASHINGTON: SUMMER PRECIPITATION
1851-55
56-60
61-65
66-70
71-75
76-80
81-85
86-90
Spokane Falls
Cathlamet
Dayton
Ediz Hook
Walla Walla
3
2
7
6
5
4
3
2
1
3
2
3
2
1

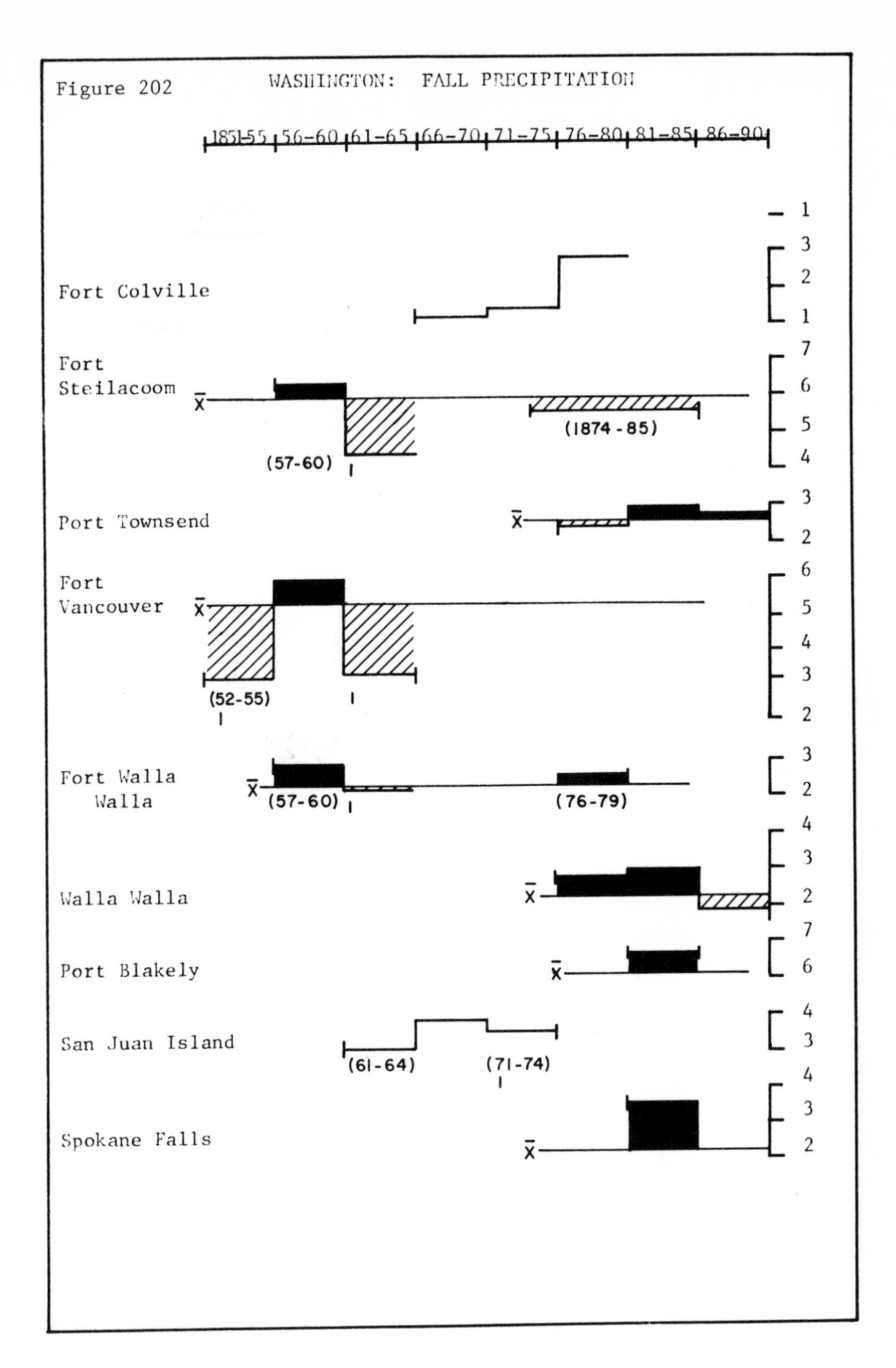
Figure 202
WASHINGTON: FALL PRECIPITATION
1851-55
56-60
61-65
66-70
71-75
76-80
81-85
86-90
Fort Colville
Fort Steilacoom
(57-60)
(1874-85)
Port Townsend
Fort Vancouver
(52-55)
Fort Walla Walla
(57-60)
(76-79)
Walla Walla
Port Blakely
San Juan Island
(61-64)
(71-74)
Spokane Falls

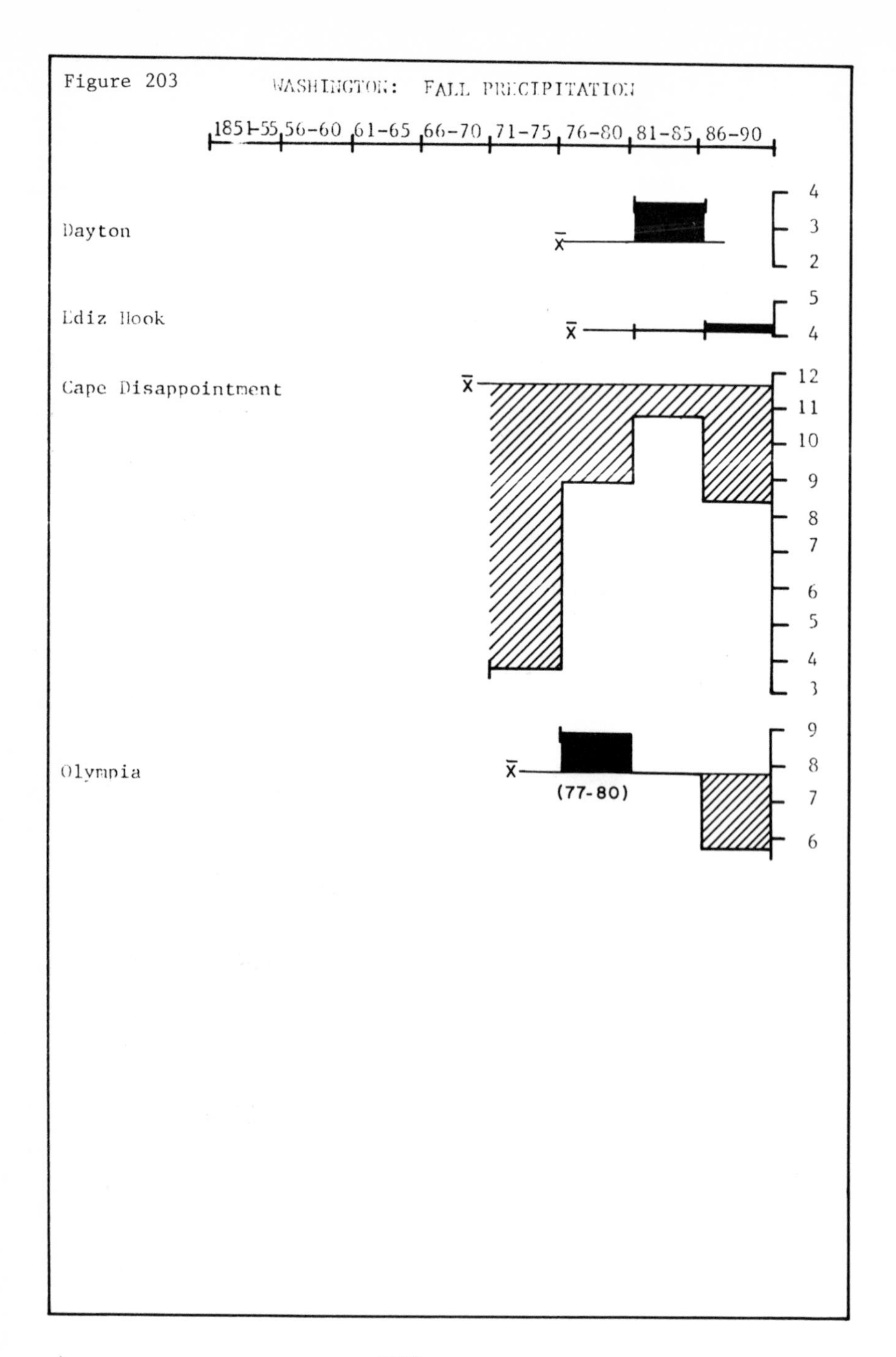

Figure 203
WASHINGTON: FALL PRECIPITATION
1851-55
56-60
61-65
66-70
71-75
76-80
81-85
86-90
Dayton
4
3
2
Ediz Hook
5
4
Cape Disappointment
12
11
10
9
8
7
6
5
4
3
Olympia
(77-80)
9
8
7
6

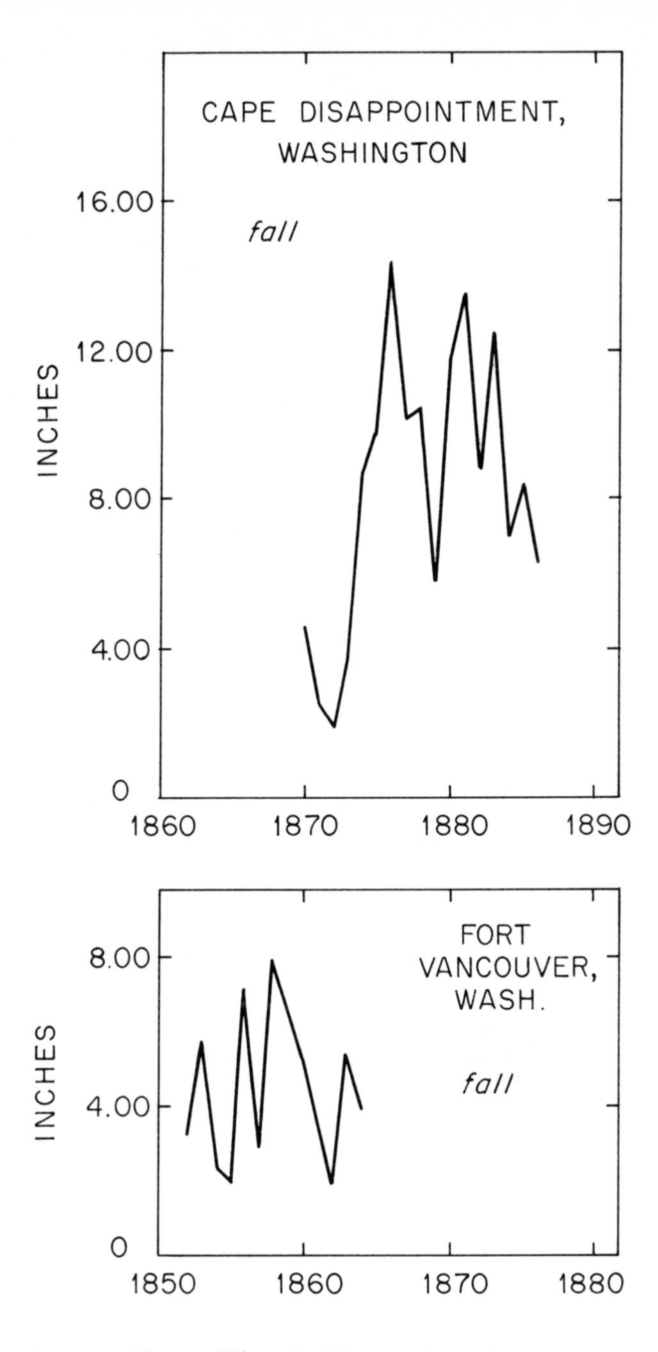

Figure 204: Washington fall precipitation

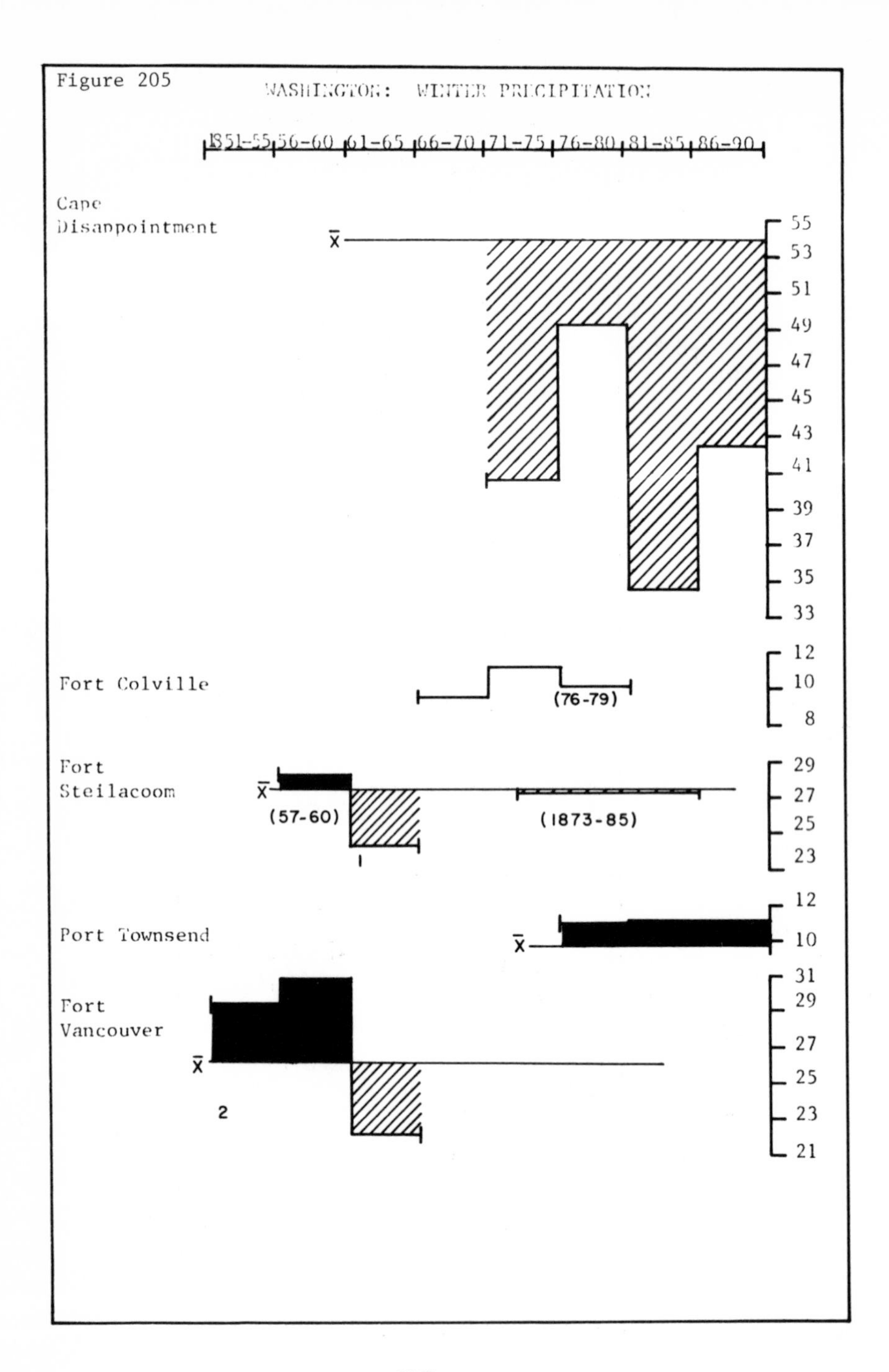
Figure 205
WASHINGTON: WINTER PRECIPITATION
1851-55
56-60
61-65
66-70
71-75
76-80
81-85
86-90
Cape Disappointment
55
53
51
49
47
45
43
41
39
37
35
33
Fort Colville
(76-79)
12
10
8
Fort Steilacoom
(57-60)
(1873-85)
29
27
25
23
Port Townsend
12
10
Fort Vancouver
2
31
29
27
25
23
21

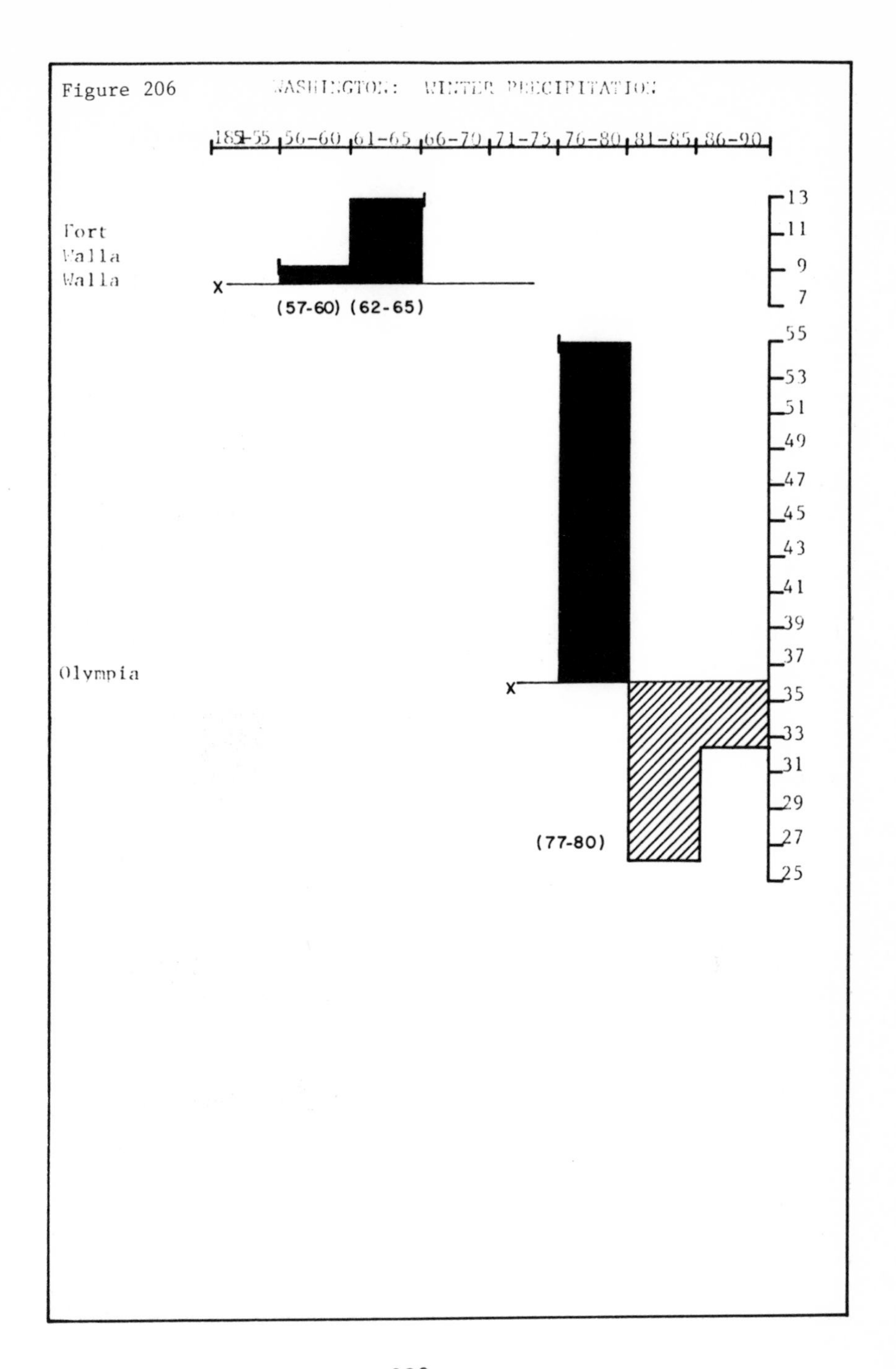
Figure 206
WASHINGTON: WINTER PRECIPITATION
1851-55
56-60
61-65
66-70
71-75
76-80
81-85
86-90
Fort
Walla
Walla
X
(57-60) (62-65)
13
11
9
7
55
53
51
49
47
45
43
41
39
37
35
33
31
29
27
25
Olympia
X
(77-80)

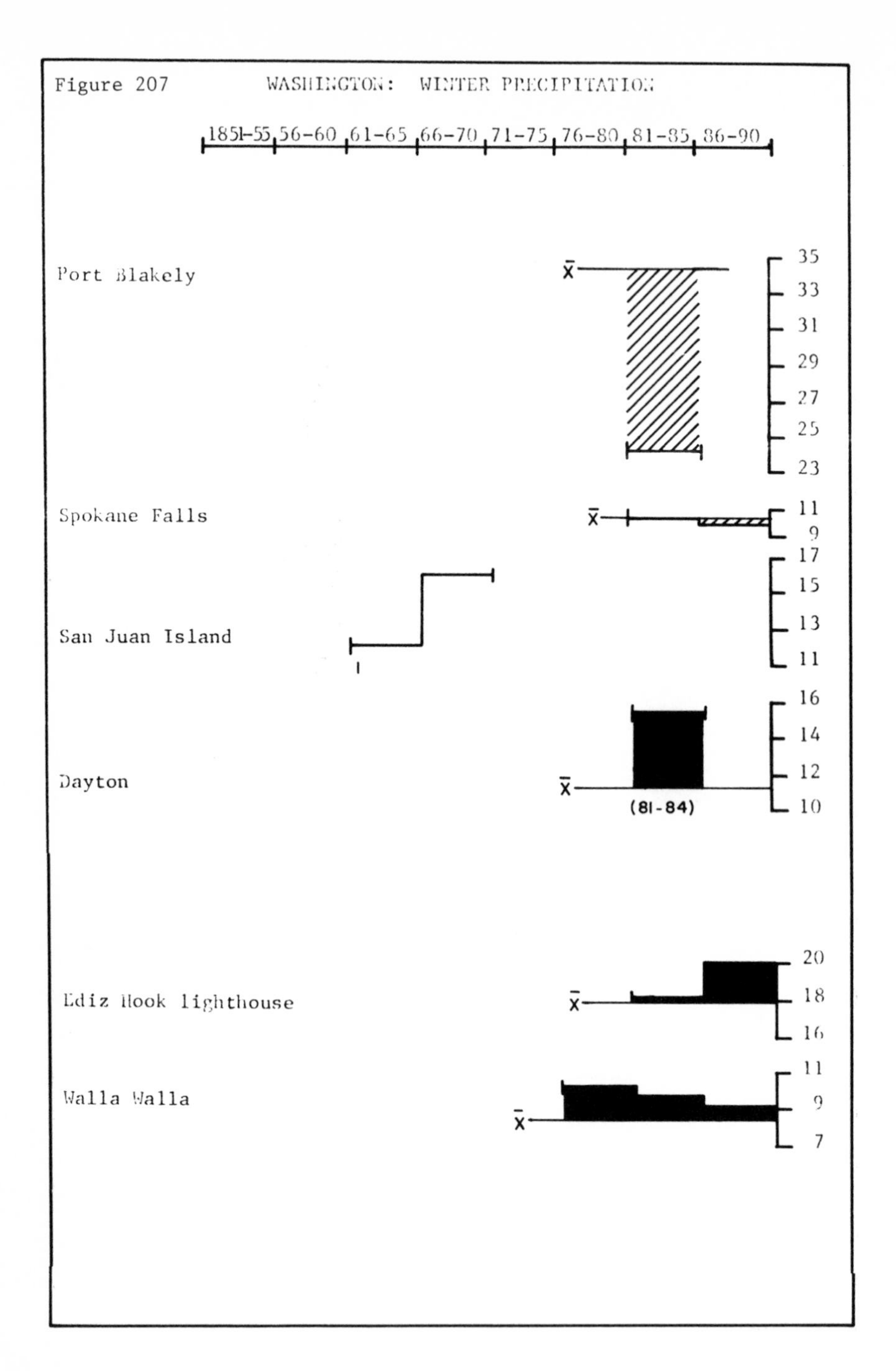
Figure 207
WASHINGTON: WINTER PRECIPITATION
1851-55
56-60
61-65
66-70
71-75
76-80
81-85
86-90
Port Blakely
35
33
31
29
27
25
23
Spokane Falls
11
9
17
15
San Juan Island
13
11
16
14
Dayton
12
(81-84)
10
20
Ediz Hook lighthouse
18
16
11
Walla Walla
9
7

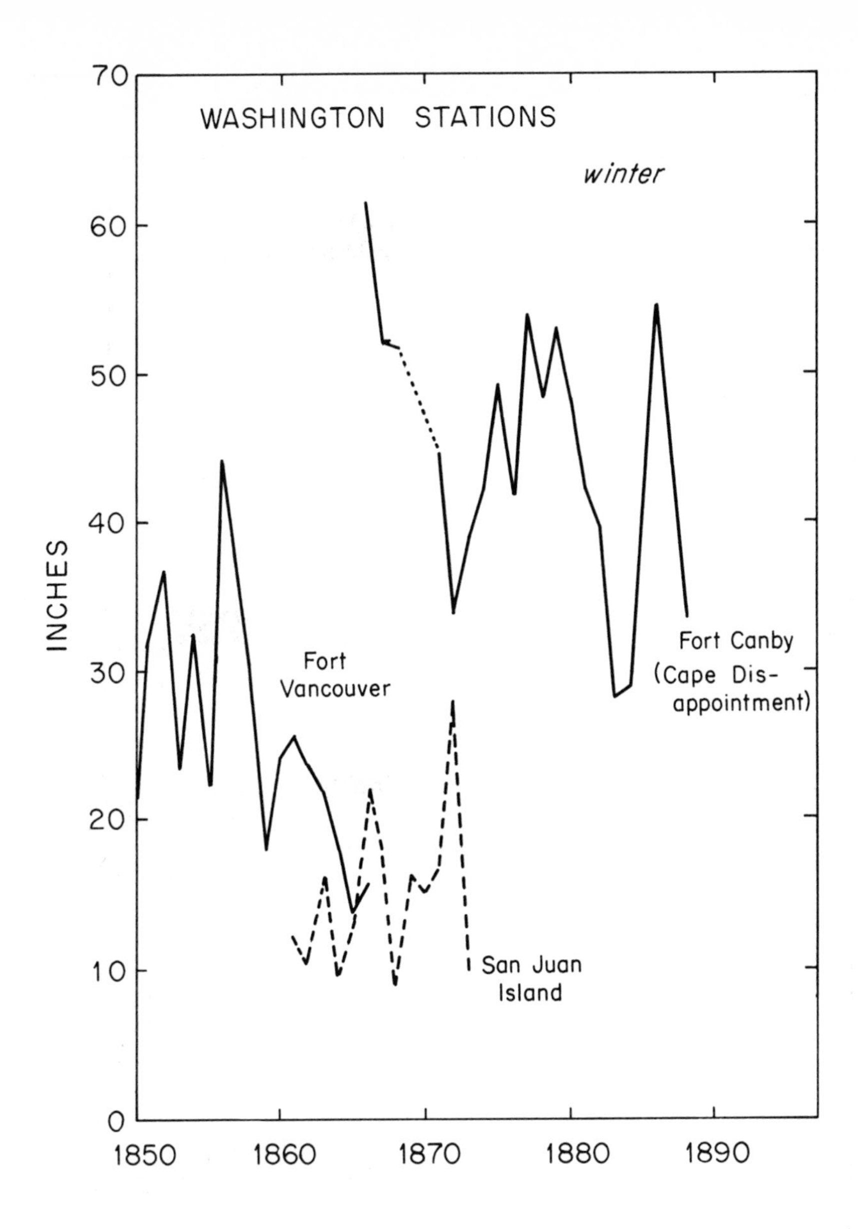

Figure 208: Washington winter precipitation

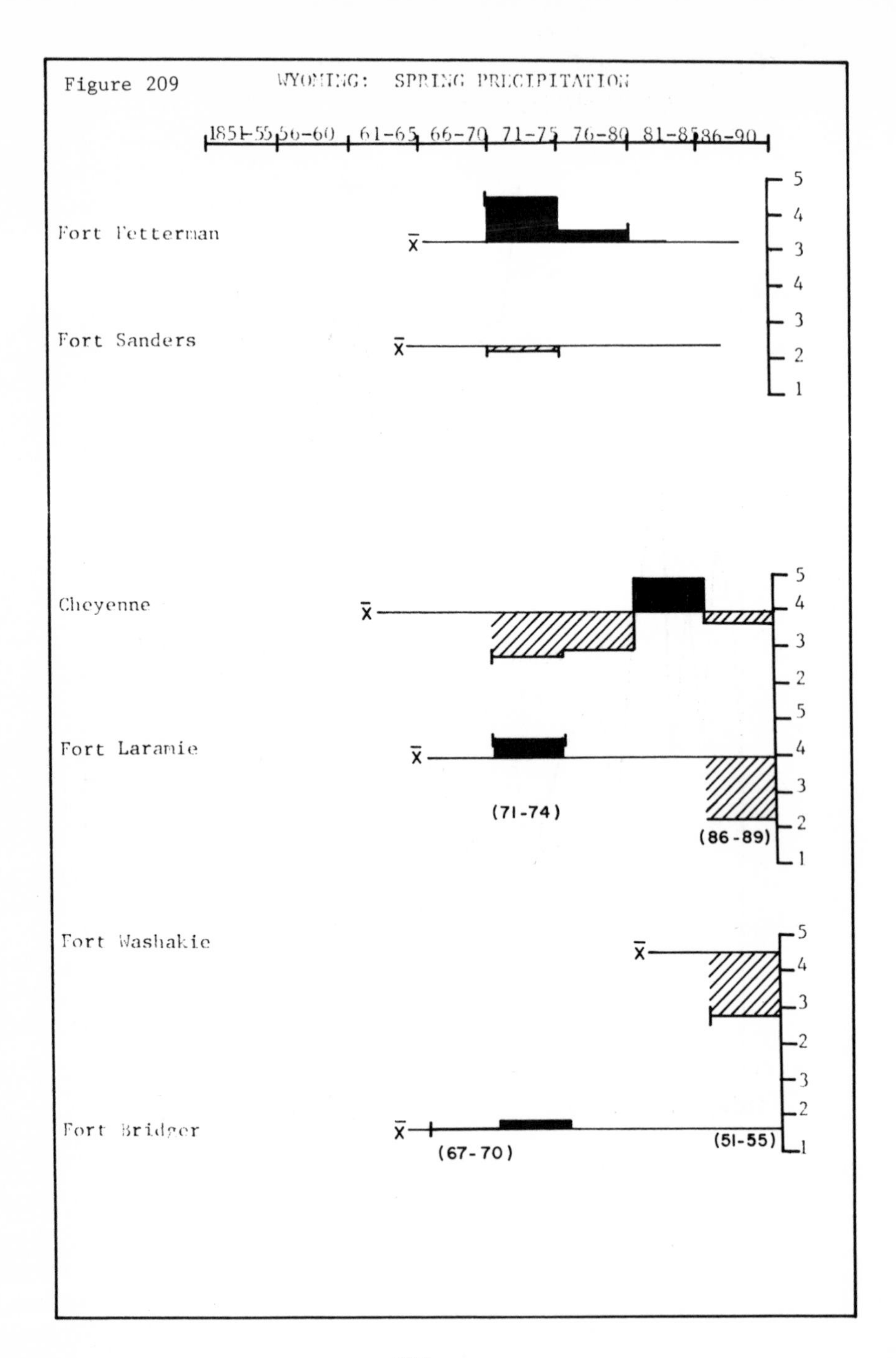
Figure 209
WYOMING: SPRING PRECIPITATION
1851-55
56-60
61-65
66-70
71-75
76-80
81-85
86-90
Fort Fetterman
Fort Sanders
Cheyenne
Fort Laramie
(71-74)
(86-89)
Fort Washakie
Fort Bridger
(67-70)
(51-55)

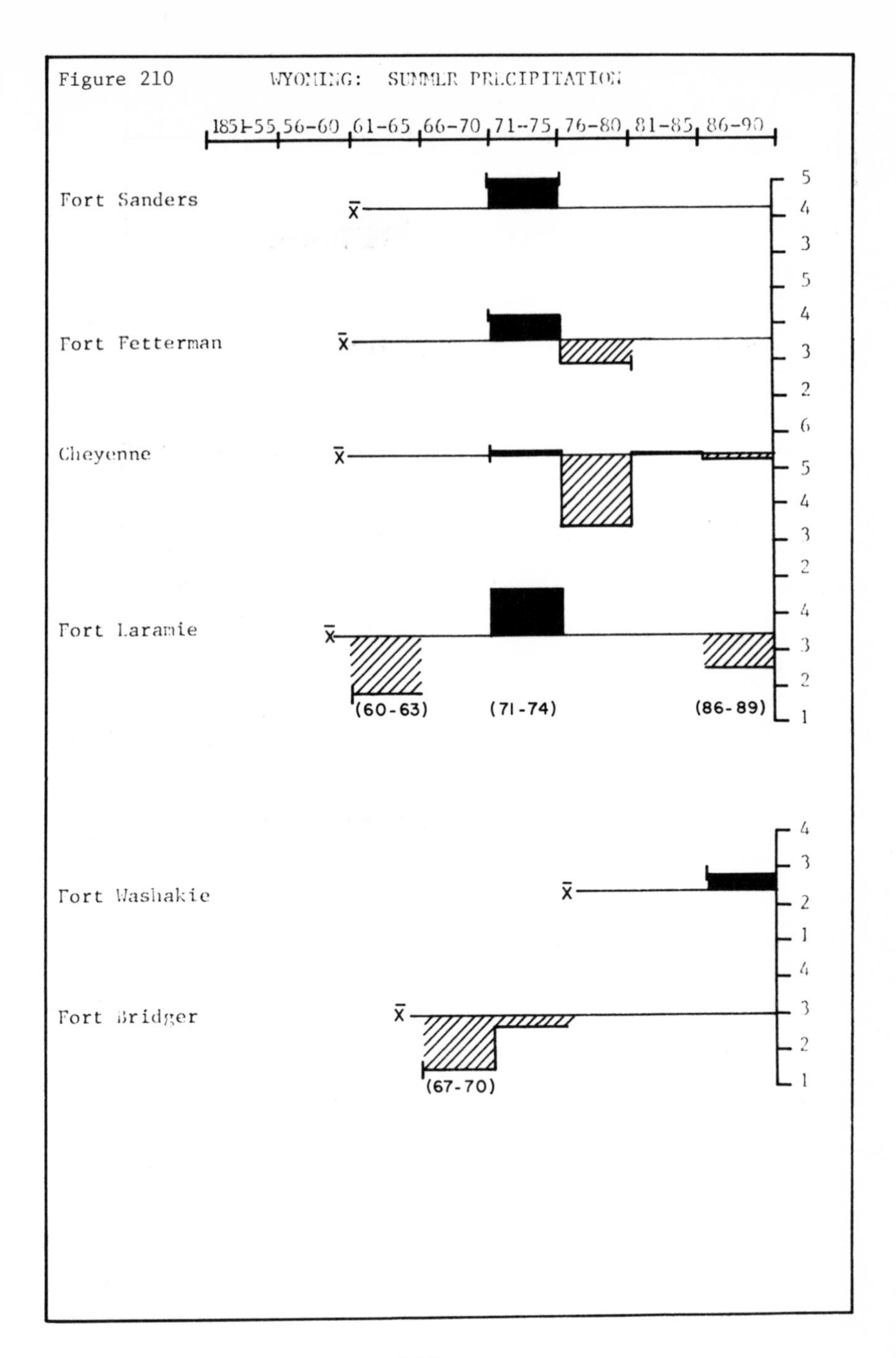

Figure 210
WYOMING: SUMMER PRECIPITATION
1851-55
56-60
61-65
66-70
71-75
76-80
81-85
86-90
Fort Sanders
Fort Fetterman
Cheyenne
Fort Laramie
Fort Washakie
Fort Bridger
(60-63)
(71-74)
(86-89)
(67-70)

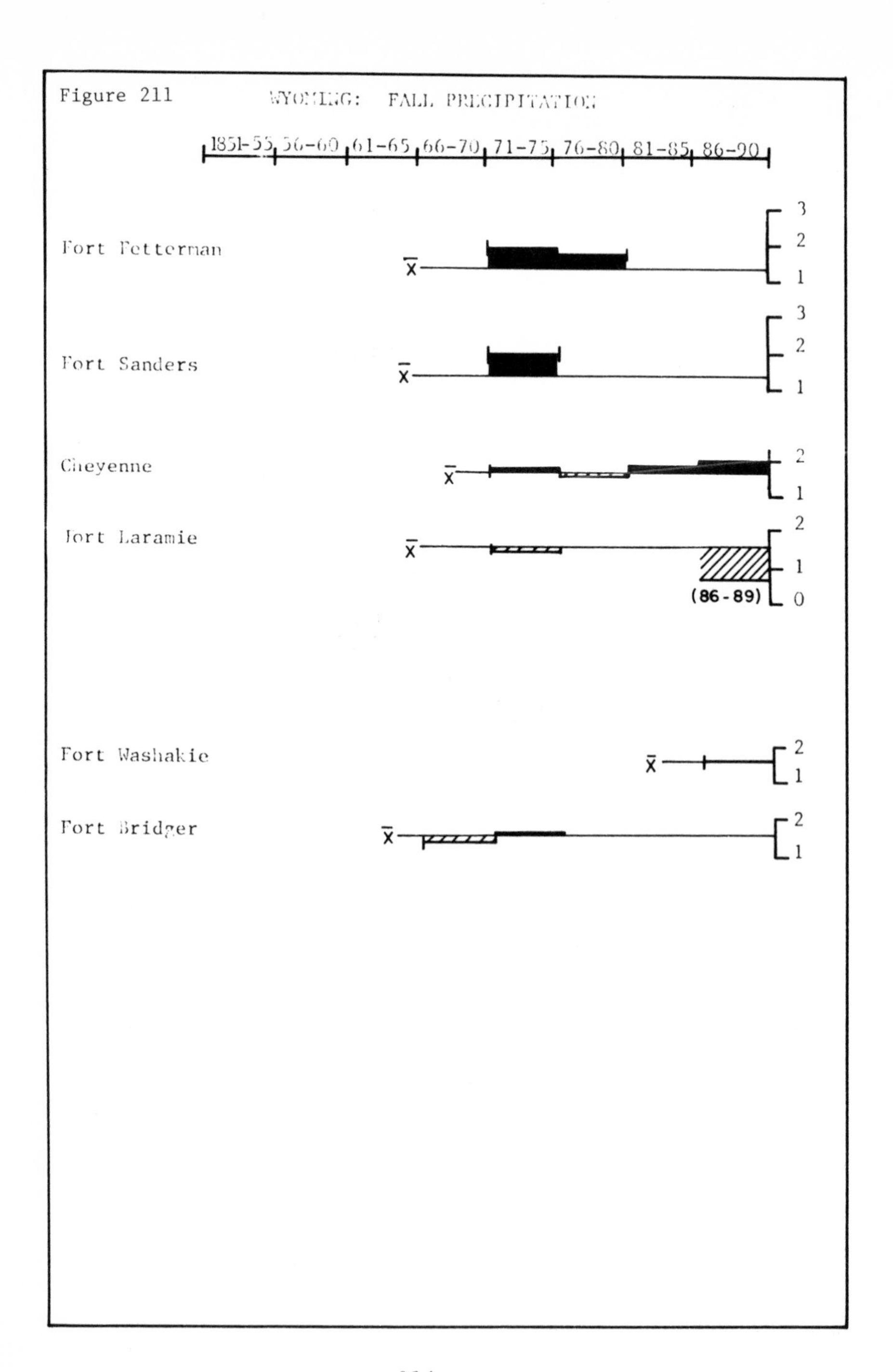
Figure 211
WYOMING: FALL PRECIPITATION
1851-55
56-60
61-65
66-70
71-75
76-80
81-85
86-90
Fort Fetterman
Fort Sanders
Cheyenne
Fort Laramie
(86-89)
Fort Washakie
Fort Bridger

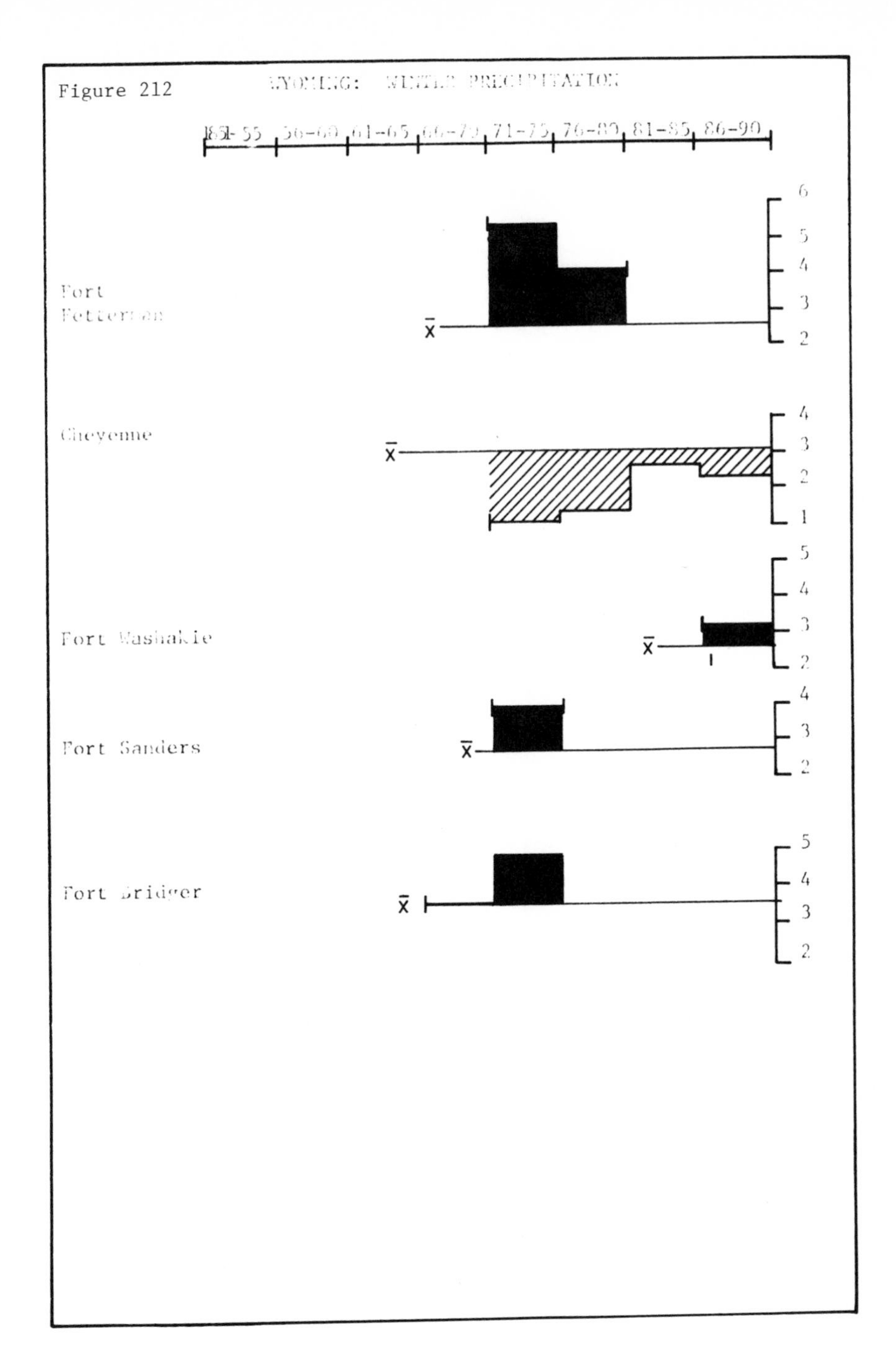

Figure 212

APPENDIX C

SYNOPTIC WEATHER TYPES OF NORTH AMERICA

(from Elliott, 1949)

TYPE A OF NORTH AMERICAN WEATHER TYPES

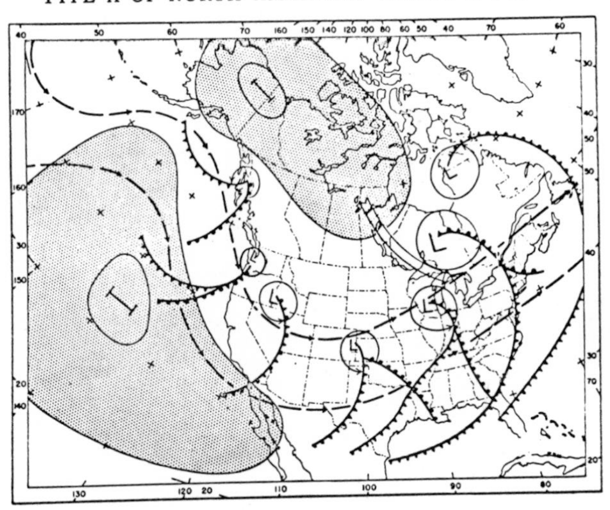

TYPE B OF NORTH AMERICAN WEATHER TYPES

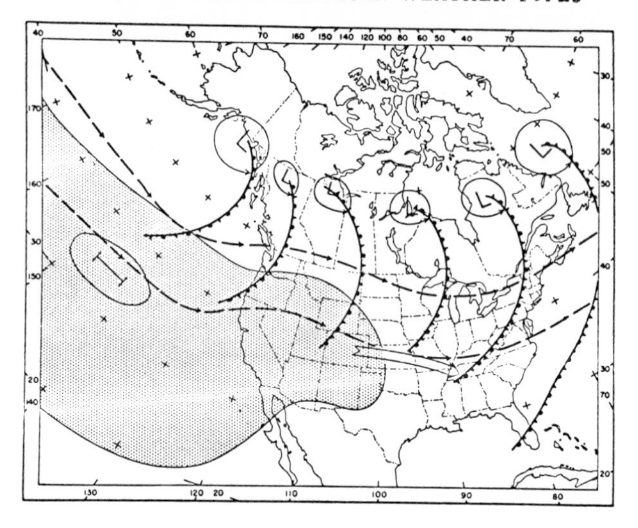

TYPE Bn-a OF NORTH AMERICAN WEATHER TYPES

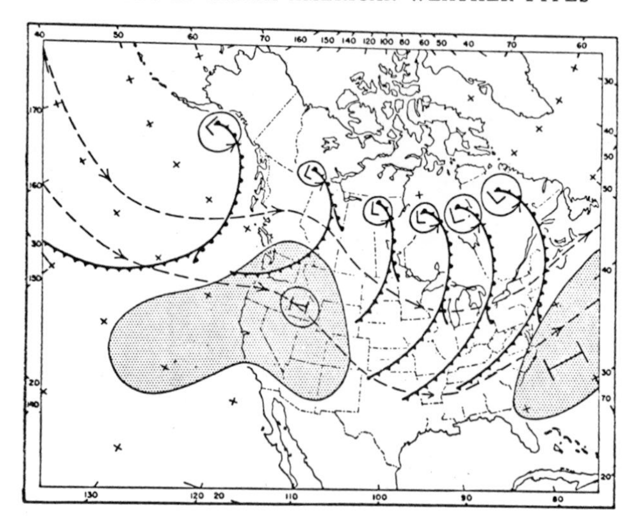

TYPE Bn-b OF NORTH AMERICAN WEATHER TYPES

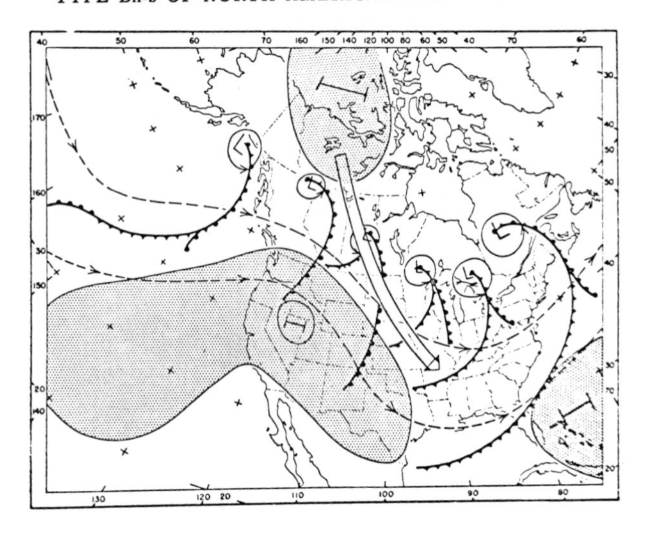

TYPE Bn-c OF NORTH AMERICAN WEATHER TYPES

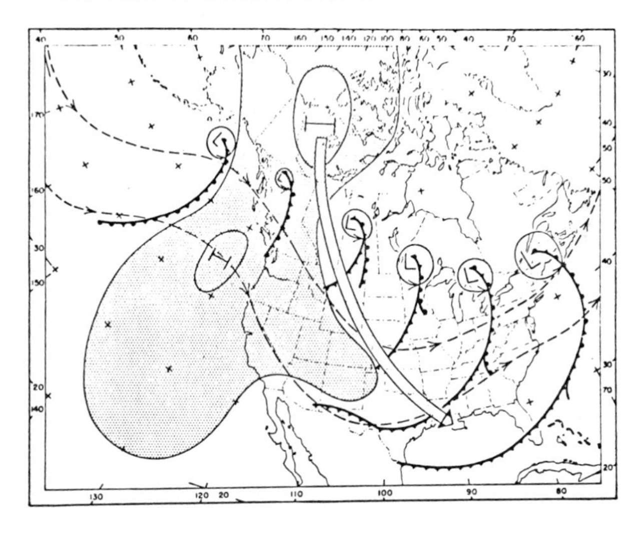

TYPE Bs OF NORTH AMERICAN WEATHER TYPES

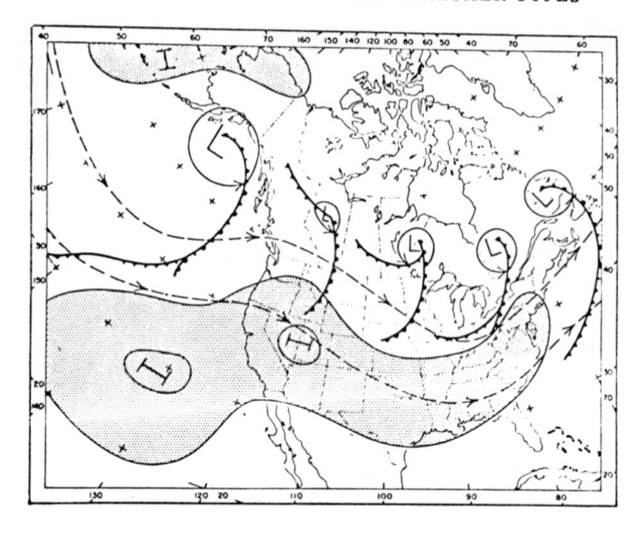

TYPE CH OF NORTH AMERICAN WEATHER TYPES

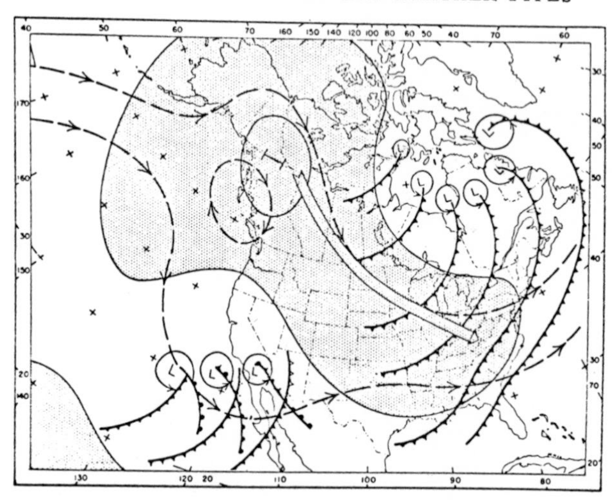

TYPE CL OF NORTH AMERICAN WEATHER TYPES

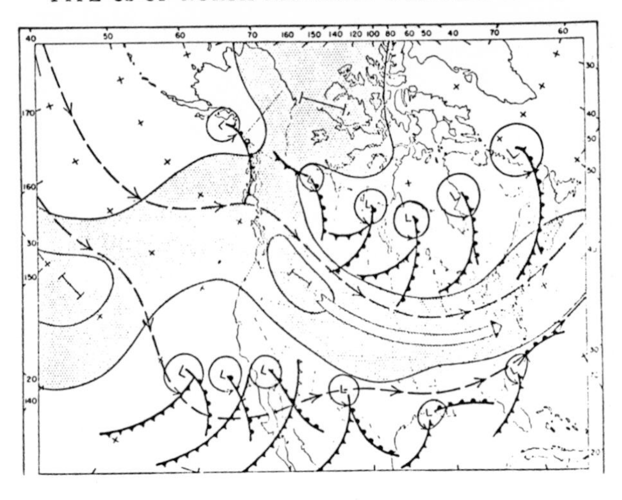

TYPE EL OF NORTH AMERICAN WEATHER TYPES

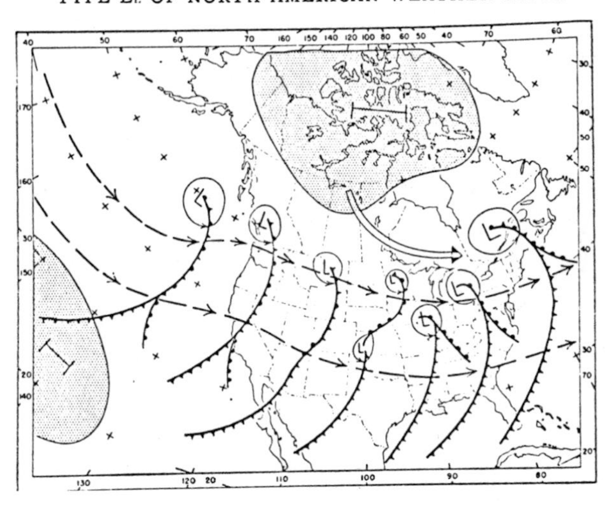

TYPE E_M OF NORTH AMERICAN WEATHER TYPES

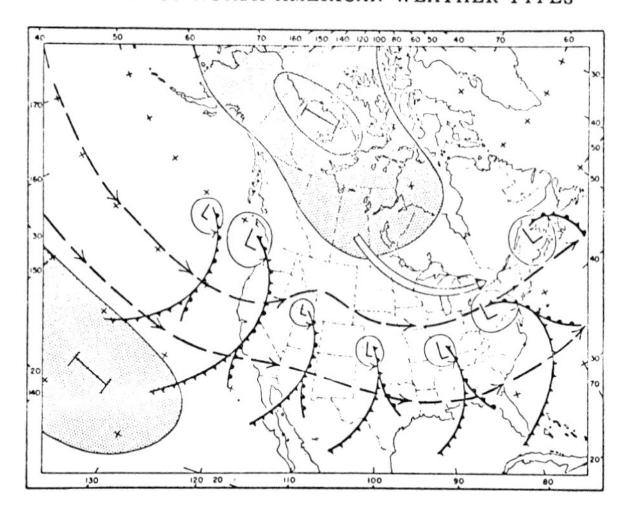

TYPE E_H OF NORTH AMERICAN WEATHER TYPES

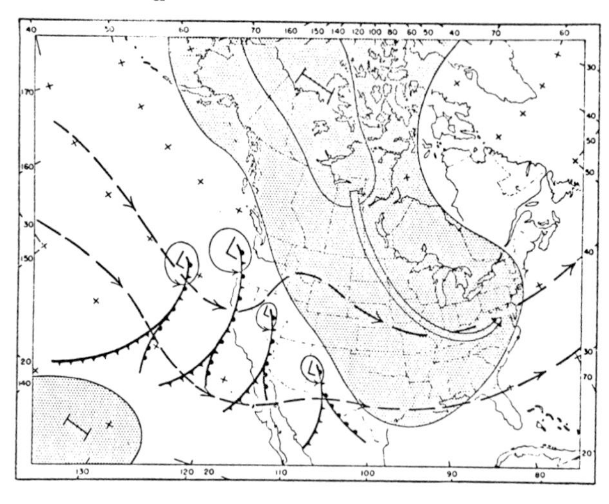

BIBLIOGRAPHY

Anderson, R. L., 1941, Distribution of the serial correlation coefficient. Annals of Mathematical Statistics, 8, 1-13.

Angell, J. K., and Korshover, J., 1962, The biennial wind and temperature oscillations of the equatorial stratosphere and their possible extension to higher latitudes. Monthly Weather Review, 90, 127-132.

Augulis, R. P., 1969, Precipitation probabilities in the western region associated with winter 500 mb map types. U.S. Weather Bureau, Western Region Technical Memorandum, WBTM WR 45-1, Salt Lake City, Utah, 91 pp.

________, 1970a, Precipitation probabilities in the western region associated with spring 500 mb map types. U.S. Weather Bureau, Western Region Technical Memorandum, WBTM WR 45-2, Salt Lake City, Utah, 75 pp.

________, 1970b, Precipitation probabilities in the western region associated with summer 500 mb map types. U.S. Weather Bureau, Western Region Technical Memorandum, WBTM WR 45-3, Salt Lake City, Utah, 73 pp.

Baker, D. G., 1960, Temperature trends in Minnesota. Bulletin of the American Meteorological Society, 41, 18-27.

Barreis, D. A., and Bryson, R. A., 1965, Climatic episodes and the dating of the Mississippian cultures. Wisconsin Archaeologist, 46, 203-220.

Barry, R. G., 1972, Climatic environment of the east slope of the Colorado Front Range. Occasional Paper No. 3, Institute of Arctic and Alpine Research, University of Colorado, 206 pp.

Barry, R. G., and Bradley, R. S., 1972, Historical climatology, in The San Juan Ecology Project. H. L. Teller, J. D. Ives and H. W. Steinhoff (eds.). Department of Watershed Sciences, Colorado State University, Fort Collins, Colorado, 294-335.

Barry, R. G., and Perry, A. M., 1973, Synoptic Climatology, Methuen, London, 555 pp.

Baur, F., 1949, Bezeihungen des Grosswetters zu kosmischen vorgangen, in Hannsuring Lehbruch der Meteorologie (S. Suflage), Teil 8: Die Erscheinungen des Grosswetters, 8, Kapitel, Band II (1951), Leipzig (Hirzel).

Benedict, J. B., 1968, Recent glacial history of an alpine area in the Colorado Front Range, U.S.A., II, dating the glacial deposits. Journal of Glaciology, 7, 77-87.

Berlage, H. P., 1957, Fluctuations of the general atmospheric circulation of more than one year; their nature and prognostic value. Mededelingen en Verhandelingen No. 69, Koninklijk Nederlandsch Meteorologisch Institut, S-Gravenhage, 152 pp.

Berson, F. A., and Kulkarni, R. N., 1968, Sunspot cycle and the quasi-biennial oscillation. Nature, 217, 1133-1134.

Bhargava, B. N., and Bansal, R. K., 1969, A quasi-biennial oscillation in precipitation at some Indian stations. Indian Journal of Meteorology and Geophysics, 20, 127-128.

Bigelow, F. M., 1902, Report of the barometry of the United States, Canada and the West Indies. Report of the Chief of the Weather Bureau 1900-01, Vol. II, Washington, D.C.

Blackman, R. B., and Tukey, J. W., 1958, The measurement of power spectra from the point of view of communications engineering. Dover Publications, New York, 190 pp.

Blasing, T. J., 1975, Methods of analyzing climatic variations in the North Pacific sector and western North America for the last few centuries. Ph.D. thesis, University of Wisconsin, Madison, Wisconsin (University Microfilm, Ann Arbor, Michigan), 177 pp.

Blodget, L., 1857, Climatology of the U.S. and of the temperate latitudes of the North American continent. J. B. Lippincott and Company, Philadelphia, 536 pp.

Bradley, R. S., and Barry, R. G., 1973a, Secular climatic fluctuations in southwestern Colorado. Monthly Weather Review, 101, 264-270.

Bradley, R. S., and Barry, R. G., 1973b, Historical climatology in The San Juan Ecology Project, H. L. Teller, J. D. Ives and H. W. Steinhoff (eds.), Department of Watershed Sciences, Colorado State University, Fort Collins, Colorado, 134-149.

Brunk, I. W., 1968, Comments on "a comparison of the climate of the eastern U.S. during the 1830's with the current normals." Monthly Weather Review, 96, 656-657

Bryson, R. A., and Hare, F. K., 1974, Climates of North America, in World Survey of Climatology, Vol. 11, H. E. Landsberg (ed.), Elsevier, New York, 420 pp.

Bryson, R. A., and Lahey, J. F., 1958, The March of the Seasons, Meteorological Department, University of Wisconsin, Madison, Wisconsin, 41 pp.

Bryson, R. A., and Lowry, W. P., 1955, Synoptic climatology of the Arizona summer precipitation singularity. Bulletin of the American Meteorological Society, 36, 329-339.

Conrad, V., and Pollack, L. W., 1950, Methods in Climatology. 2nd edition, Harvard University Press, Cambridge, Massachusetts, 459 pp.

Crandell, D. R., 1969. Surficial geology of Mount Rainier National Park, Washington. U.S. Geological Survey Bulletin 1288, 41 pp.

Darter, L. J., 1942, List of climatological records in the National Archives, Special List No. 1, National Archives, Washington, D.C., 160 pp.

Dightman, R. A., 1956, Grinnell Glacier studies; a progress report as related to climate. Unpublished manuscript, U.S. Weather Bureau, State Climatologist's Office, Helena, Montana.

Dightman, R. A., and Beatty, M. A., 1952, Recent Montana glacier and climate trends. Monthly Weather Review, 80, 77-81.

Dixon, W. J., 1973, BMD; biomedical computer programs. 3rd edition, University of California Press, Berkeley, 600 pp.

Douglass, A. E., 1928, Climatic cycles and tree growth. Vol. II. Publication No. 289, Carnegie Institute, Washington, D.C.

Dyson, J. L., 1948, Shrinkage of Sperry and Grinnell glaciers, Glacier National Park, Montana. Geographical Review, 38, 96-103.

Dzerdzeevskii, B. L., 1966, Some aspects of dynamic climatology. Tellus, 18(4), 751-760.

Elliott, R. D., 1943, Studies of persistent regularities in weather phenomena. Part 2 of Synoptic Weather Types of North America. Meteorology Department, California Institute of Technology, Pasadena, California.

________, 1949, The weather types of N. America. Weatherwise, 2, 15-18, 40-43, 64-67, 86-88, 110-113 and 136-138.

Engelen, G. B., 1972a, Two year periodicity in regional snow moisture contents of snow and ice in hydrology, in The Role of Snow and Ice in Hydrology. Proceedings of the UNESCO/WMO Conference, Banff, Alberta.

________, 1972b, Two year cycles in soil moisture recharge, snowpack and streamflow in relation with atmospheric conditions, in The Role of Snow and Ice in Hydrology. Proceedings of the UNESCO/WMO Conference, Banff, Alberta.

Fritts, H. C., 1965, Tree-ring evidence for climatic changes in western North America. Monthly Weather Review, 93, 421-443.

________, 1966, Growth rings of trees: their correlation with climate. Science, 154, 973-979.

________, 1971, Dendroclimatology and dendroecology. Quaternary Research, 1, 419-449.

________, 1974, Some quantitative methods for calibrating ring widths with variables of climate. Abstracts of the 3rd biennial meeting, American Quaternary Association, University of Wisconsin, Madison, Wisconsin, July/August 1974, 2-5.

Fritts, H. C., and Blasing, T. J., 1973, Tree ring analysis of environmental variability. Progress report on Grant GA-26581 (NSF) during 1971-73, Laboratory of Tree Ring Research, University of Arizona, Tucson, Arizona.

Fritts, H. C., Blasing, T. J., Hayden, B. P., and Kutzbach, J. E., 1971, Multivariate techniques for specifying tree-growth and climate relationships and for reconstructing anomalies in paleoclimate. Journal of Applied Meteorology, 10, 845-864.

Fritts, H. C., Mosimann, J. E., and Bottorff, C. P., 1969, A revised computer program for standardizing tree-ring series. Tree Ring Bulletin, 29, 15-20.

Fritts, H. C., Smith, D. G., and Stokes, M. A., 1965, The biological model for paleoclimatic interpretation of Mesa Verde tree ring series. American Antiquity, 31(2), Post 2, 101-121.

Gilman, D. L., Fuglister, F. J., and Mitchell, J. M., Jr., 1963, On the power spectrum of "red noise." Journal of Atmospheric Sciences, 20, 182-184.

Glock, W., and Argeter, S. R., 1966, Tree growth as a meteorological indicator. International Journal of Biometeorology, 10, 47-62.

Greely, A. W., 1889, Climate of Oregon and Washington. Senate Executive Document No. 282, 50th Congress, 1st Session.

________, 1891, Report of the climatology of the arid regions of the U.S. with reference to irrigation. House Executive Document No. 287, 51st Congress, 2nd Session.

Hardman, G., and Venstrom, C., 1941, A 100 years record of Truckee River runoff estimated from changes in the levels and volumes of Pyramid and Winnemucca lakes. Transactions of the American Geophysical Union 22, part 1, 71-90.

Havens, J. M., 1958, An annotated bibliography of meteorological observations in the United States, 1715-1818. Key to Meteorological Records Documentation No. 5-11. U.S. Department of Commerce, Weather Bureau, Washington, D.C. 14 pp.

Hess, P., and Brezowsky, H., 1952, Katalog der Grosswetterlagen Europas. Ber. dtsch. Wetterd. U.S. Zone (Bad Kissinger) No. 33, 39 pp.

Hubbs, C. L. 1957, Recent climatic history in California and adjacent areas, in Proceedings of the Conference on Research in Climatology, Scripps Institute of Oceanography, La Jolla, California, 10-22.

Huff, F. A., and Changnon, S. A., Jr., 1973, Precipitation modification by major urban areas. Bulletin of the American Meteorological Society, 54, 1220-1232.

Ives, R. L., 1954, Climatic studies in western North America. Proceedings of the Toronto Meteorological Conference, Royal Meteorological Society, London, 218-222.

Jagannathan, P., and Parthasarathy, 1973, Trends and periodicities of rainfall over India. Monthly Weather Review, 101, 371-375.

Jorgensen, D. L., Klein, W. H., and Korte, A. F., 1967, A synoptic climatology of winter precipitation from 700 mb lows for intermountain areas of the west. Journal of Applied Meteorology, G, 782-790.

Joseph, E., 1973, Time series analysis of annual temperatures. Monthly Weather Review, 101, 501-504.

Kendall, M. G., 1970, Rank correlation methods. 4th edition, Griffin, London, 202 pp.

Klein, W. H., 1965, Application of synoptic climatology and short-range numerical prediction to 5-day forecasting. Research Paper No. 46, U.S. Weather Bureau, Washington, D.C. 92 pp.

Kohler, M. A., 1949, On the use of double mass analysis for testing the consistency of meteorological records and for making required adjustments. Bulletin of the American Meteorological Society, 30, 188-189.

Korte, A. J., Jorgensen, D. K., and Klein, W. H., 1969, Charts giving station precipitation in the plateau states from 850 and 500 millibar lows during winter. ESSA Technical Memorandum WATM TDL 25 U.S. Department of Commerce, Silver Springs, Maryland, 9 pp and 2 appendices.

________, 1972, Synoptic climatological studies of precipitation in the plateau states from 850, 700 and 500 millibar lows during spring. NOAA Technical Memorandum NWS TDL-48, U.S. Department of Commerce, Silver Springs, Maryland, 130 pp.

LaMarche, V. C., Jr., 1974, Paleoclimatic inferences from long tree-ring records. Science, 183, 1043-1048.

LaMarche, V. C., Jr., and Fritts, H. C., 1971, Anomaly patterns of climate over western United States, 1700-1993 derived from principal component analyses of tree-ring data. Monthly Weather Review, 99, 138-142.

Lamb, H. H., 1963, On the nature of certain climatic epochs which differed from the modern (1900-1939) normal. Proceedings of the WMO/UNESCO Rome 1961 Symposium on Changes of Climate. Paris (UNESCO Arid Zone Research Series XX), 125-150.

________, 1966, Climate in the 1960s. Geographical Journal, 132, 183-212.

________, 1969, Climatic fluctuations in general climatology. World Survey of Climatology, Vol. 2, H. E. Landsberg (ed.), Elsevier, New York, 173-249.

________, 1972, Climate: present, past and future, Vol. 1, Methuen, London, 613 pp.

Lamb, H. H., and Johnson, A. I., 1959, Climatic variation and observed changes in the general circulation. Geografiska Annaler, 41, 94-133.

________, 1966, Secular variation of the atmospheric circulation since 1750. Geophysical Memoir, No. 110, Meteorological Office, London (H.M.S.O.).

Landsberg, H. E., 1960, Note on the recent climatic fluctuations in the United States. Journal of Geophysical Research, 65, 1519-1525.

________, 1962, Biennial pulses in the atmosphere. Beitrage zur Physik der Atmosphere, 35, 184-194.

________, 1970, Man-made climatic changes, Science, 170, 1265-1274

Landsberg, H. E., Mitchell, J. M., Jr., and Crutcher, H. L., 1959, Power spectrum analysis of climatological data for Woodstock College, Md. Monthly Weather Review, 87, 283-298.

Lawrence, E. N., 1970, Variations in weather type temperature averages. Nature, 226, 633-634.

Lawson, M. P., Reiss, A., Phillips, R., and Livingston, K., 1971, Nebraska droughts: a study of their past chronological and spatial extent with implications for the future. Occasional Paper No. 1, Department of Geography, University of Nebraska, Lincoln, 150 pp.

Leaf, C. F., 1962, Snow measurement in mountainous regions. Unpublished M.S. Thesis, Department of Civil Engineering, Colorado State University, Fort Collins, Colorado.

Leith, C. E., 1974, The natural and radiatively perturbed troposphere. Climatic Impact Assessment Program (CIAP), Monograph 4, Chapter 3, Department of Transport, Washington, D.C.

Leopold, L. B., 1951, Rainfall frequency: an aspect of climatic variation. Transactions of the American Geophysical Union, 32(3), 347-357.

LeRoy, Ladurie, E., 1971, Times of Feast, Times of Famine. Doubleday and Company, New York, 426 pp.

Mann, H. B., 1945, Non-parametric test of randomness against trend. Econometrika, 13, 245-259.

McDonald, J. E., 1956, Variability of precipitation in an arid region: a survey of characteristics for Arizona. Technical Report on the Meteorology and Climatology of Arid Regions No. 1, Institute of Atmospheric Physics, University of Arizona, Tucson, 88 pp.

Mitchell, J. M., Jr., 1961, The measurement of secular temperature change in the eastern United States. Research Paper No. 43, U.S. Weather Bureau, Washington, D.C. 70 pp.

________, 1963, On the world-wide pattern of secular temperature change, in Changes in Climate. Proceedings of the Rome Symposium, 1961, organized by UNESCO and the World Meteorological Organization, Arid Zone Research XX, UNESCO, Paris, 161-181.

________, 1965, The solar inconstant. Proceedings of the Seminar on Possible Responses of Weather Phenomena to Variable Extraterrestrial Influences. National Center for Atmospheric Research Technical Note, TN-8, Boulder, Colorado, 155-174.

Mitchell, V. L., 1969, The regionalization of climate in montane areas. Ph.D. Thesis, University of Wisconsin, Madison, Wisconsin, 147 pp.

Moss, J. H., 1951, Late glacial advances in the southern Wind River Mountains, Wyoming. American Journal of Science, 249, 865-883.

Namias, J., 1957, Characteristics of cold winters and warm summers over Scandinavia related to the general circulation. Journal of Meteorology, 14, 235-250.

________, 1959, Seasonal interactions between the north Pacific Ocean and the atmosphere during the 1960s. Monthly Weather Review, 97, 173-192.

________, 1970, Climatic anomaly over the United States during the 1960s. Science, 170, 741-743.

Parry, M., 1966, The urban "heat island" in Biometeorology, Vol. 2, S. W. Tromp and W. H. Weike (eds.), Rogamon Press, London, 616-624.

Porter, S. C., and Denton, G. H., 1967, Chronology of neoglaciation in the North American Cordillera. American Journal of Science, 265, 177-210.

Reed, R. J., Campbell, W. J., Rasmussen, L. A., and Rogers, D. G., 1961, Evidence of a downward propagating, annual wind reversal in the equatorial stratosphere. Journal of Geophysical Research, 66, 813-818.

Roden, G. I., 1966, A modern statistical analysis and documentation of historical temperature records in California, Oregon and Washington, 1821-1964. Journal of Applied Meteorology, 5, 3-24.

Rosendal, H. E., 1970, The unusual circulation pattern of early 1843. Monthly Weather Review, 98, 266-270.

Sanchez, W. A., and Kutzbach, J. E., 1974, Climate of the American tropics and sub-tropics in the 1960s and possible comparisons with climatic variations of the last millenium. Quaternary Research, 4, 128-135.

Schott, C. A., 1873, Tables and results of the precipitation in rain and snow in the U.S. and at some stations in adjacent parts of North America and Central and South America. Smithsonian Institution Contributions to Knowledge XVIII (S.I. Publication No. 222), Washington, D.C., 178 pp.

________, 1876, Tables, distribution and variation of atmospheric temperature in the U.S. and some adjacent parts of America. Smithsonian Institution Contributions to Knowledge XXI (S.I. Publication, No. 277), Washington, D.C., 360 pp.

Schott, C. A., 1881, Tables and results of the precipitation and snow in the U.S. and at some stations in adjacent parts of North America and in Central and South America. Smithsonian Institution Contributions to Knowledge XXIV (S.I. Publication No. 353), Washington, D.C., 209 pp.

Schulman, E., 1956, Dendroclimatic changes in semiarid America. University of Arizona Press, Tucson, 142 pp.

Schuurmans, C. J. E., 1969, The influence of solar flares upon the tropospheric circulation. Mededlingen en Verhandelingen No. 92, Koninklikj Nederland Meteorologisch Institut (Staatsdrukkerij), S-Gravenhage.

Scouler, M. F., 1827, On the temperature of the northwest coast of America. Edinburgh Journal of Science, 6, 251-253.

Sellers, W. D., 1960, Precipitation trends in Arizona and western New Mexico. Proceedings of the 28th Western Snow Conference, 81-94.

________, 1968, Climatology of monthly precipitation patterns in western United States, 1931-1966. Monthly Weather Review, 96, 585-595.

Shapiro, R., and Ward, F., 1962, A neglected cycle in sunspot numbers? Journal of Atmospheric Science, 27, 1021-1026.

Sigafoos, R. S., and Hendricks, E. L., 1973, Recent activity of glaciers of Mount Rainier, Washington. U.S. Geological Survey Professional Paper, 387-B, 24 pp.

Smithsonian Institution, 1893, The meteorological work of the Smithsonian Institution. Annual Report of the Regents of the Smithsonian Institution, Washington, D.C., 89-92.

Stockton, C. W., and Fritts, H. C., 1971, Conditional probability of occurrence for variations in climate based on widths of annual tree rings in Arizona. Tree Ring Bulletin, 31, 3-24.

Stockton, C. W., and Fritts, H. C., 1973, Long-term reconstruction of water-level changes for Lake Athabasca by analysis of tree rings. Water Resources Bulletin, 9, 1006-1027.

Stokes, M. A., Drew, L. G., Stockton, C. W. (eds.), 1973, Tree ring chronologies of western America, Vol. 1: selected tree ring stations. Laboratory of Tree-Ring Research, University of Arizona, Tucson, Arizona.

Sullivan, W. G., and Severson, J. O., 1966a, A study of summer showers over the Colorado mountains. U.S. Weather Bureau Central Region Technical Memorandum, No. 2, Kansas City, Missouri, 8 pp and figures.

________, 1966b, Areal shower distribution--mountain versus valley coverage. U.S. Weather Bureau Central Region Technical Memorandum, No. 3, Kansas City, Missouri, 3 pp. and figures.

Thomas, H. E., 1959, Reservoirs to match our climatic fluctuations. Bulletin of the American Meteorological Society, 40, 240-249.

Thomas, M. K., 1965, Climatic trends on the Canadian Prairies, in Water and Climate, Water Studies Institute Report No. 2, Saskatoon, Saskatchewan, 21-45.

Tyson, P. D., 1969, Spatial variation of rainfall spectra in South Africa. Annals of the American Association of Geographers, 59, 711-720.

Veryard, R. G., 1963, A review of studies on climatic fluctuations during the period of the meteorological record, in Changes in Climate. Proceedings of the Rome Symposium, 1961, organized by UNESCO and the World Meteorological Organization, Arid Zone Research XX, UNESCO, Paris, 3-16.

Von Eschen, G. F., 1958, Climatic trends in New Mexico. Weatherwise, 21(6), 191-5.

Wahl, E. W., 1968, A comparison of the climate of the eastern U.S. during the 1830s with the current normals. Monthly Weather Review, 96, 73-82.

Wahl, E. W., and Lawson, T. L., 1970, The climate of the mid-nineteenth century United States compared to the current normals. Monthly Weather Review, 98, 259-265.

Waldmeier, M., 1961. Sunspot activity in the years 1610-1960. Schulthess, Zurich

Weber, G. A., 1922, The Weather Bureau, its history, activities and organization. Institute for Government Research, Service Monographs of the U.S. Government No. 9, Johns Hopkins Press, Baltimore, 87 pp.

Willett, H. C., 1949, Long-period fluctuations of the general circulation of the atmosphere. Journal of Meteorology, 6, 34-50.

Willett, H. C., 1964, Evidence of solar climatic relationships, in _Weather and Our Food Supply._ Center for Agricultural and Economic Development, Iowa State University, Ames, Iowa, 123-151.

Williams, L. D., 1974, _Neoglacial landforms and Neoglacial chronology of the Wallowa Mountains, Northeastern Oregon._ Unpublished M.S. thesis, Department of Geology-Geography, University of Massachusetts, Amherst, Massachusetts, 97 pp.

Wing, K., 1943, Freezing and thawing dates of lakes and rivers as phenological indicators. _Monthly Weather Review,_ 71, 149-158.

World Meteorological Organization (WMO), 1966, Climatic Change. _Technical Note_ No. 79, WMO, Geneva, 79 pp.